Recent Advances in the Biochemistry of Fruits and Vegetables

Annual Proceedings of the Phytochemical Society of Europe

1. **Biosynthetic Pathways in Higher Plants**
   Edited by J. B. PRIDHAM AND T. SWAIN
2. **Comparative Phytochemistry**
   Edited by T. SWAIN
3. **Terpenoids in Plants**
   Edited by J. B. PRIDHAM
4. **Plant Cell Organelles**
   Edited by J. B. PRIDHAM
5. **Perspectives in Phytochemistry**
   Edited by J. B. HARBORNE AND T. SWAIN
6. **Phytochemical Phylogeny**
   Edited by J. B. HARBORNE
7. **Aspects of Terpenoid Chemistry and Biochemistry**
   Edited by T. W. GOODWIN
8. **Phytochemical Ecology**
   Edited by J. B. HARBORNE
9. **Biosynthesis and its Control in Plants**
   Edited by B. V. MILBORROW
10. **Plant Carbohydrate Biochemistry**
    Edited by J. B. PRIDHAM
11. **The Chemistry and Biochemistry of Plant Proteins**
    Edited by J. B. HARBORNE and C. F. VAN SUMERE
12. **Recent Advances in the Chemistry and Biochemistry of Plant Lipids**
    Edited by T. GALLIARD AND E. I. MERCER
13. **Biochemical Aspects of Plant-Parasite Relationships**
    Edited by J. FRIEND AND D. R. THRELFALL
14. **Regulation of Enzyme Synthesis and Activity in Higher Plants**
    Edited by H. SMITH
15. **Biochemical Aspects of Plant and Animal Coevolution**
    Edited by J. B. HARBORNE
16. **Recent Advances in the Biochemistry of Cereals**
    Edited by D. L. LAIDMAN AND R. G. WYN JONES
17. **Indole and Biogenetically Related Alkaloids**
    Edited by J. D. PHILLIPSON AND M. H. ZENK
18. **Nitrogen Fixation**
    Edited by W. D. P. STEWART and J. R. GALLON
19. **Recent Advances in the Biochemistry of Fruits and Vegetables**
    Edited by J. FRIEND AND M. J. C. RHODES

ANNUAL PROCEEDINGS OF THE PHYTOCHEMICAL SOCIETY
OF EUROPE NUMBER 19

Recent Advances in the Biochemistry of Fruits and Vegetables

Edited by

J. FRIEND

Department of Plant Biology, University of Hull, Hull, England

and

M. J. C. RHODES

ARC Food Research Institute, Colney Lane, Norwich, England

1981

ACADEMIC PRESS

A Subsidiary of Harcourt Brace Jovanovich, Publishers

London New York Toronto Sydney San Francisco

ACADEMIC PRESS INC. (LONDON) LTD.
24/28 Oval Road
London NW1

United States Edition published by
ACADEMIC PRESS INC.
111 Fifth Avenue
New York, New York 10003

British Library Cataloguing in Publication Data

Recent advances in the biochemistry of fruit and vegetables.—(Annual proceedings of the Phytochemical Society of Europe, ISSN 0309–0393; 19)
1. Fruit—Physiology—Congresses
2. Vegetables—Congresses
3. Botanical chemistry—Congresses
I. Friend, J. II. Rhodes, M. J. C. III. Series
634′.04 SB354

ISBN 0–12–268420–6

Filmset in Great Britain by Latimer Trend & Company Ltd, Plymouth and printed by St Edmundsbury Press, Bury St Edmunds, Suffolk

Contributors

T. Ap Rees, *Botany School, University of Cambridge, Downing Street, Cambridge CB2 3EA, England*

I. M. Bartley, *East Malling Research Station, East Malling, Maidstone, Kent ME19 6BJ, England*

E. M. Beyer, Jr, *Biochemicals Department, E.I. du Pont de Nemours & Co., Experimental Station, Wilmington, Delaware 19898, USA*

J. B. Biale, *Department of Biology, University of California, Los Angeles, California 99024, USA*

W. L. Dixon, *Botany School, University of Cambridge, Downing Street, Cambridge CB2 3EA, England*

F. Franks, *Botany School, University of Cambridge, Downing Street, Cambridge CB2 3EA, England*

D. Graham, *Plant Physiology Unit, CSIRO Division of Food Research, North Ryde, New South Wales, Australia*

D. Grierson, *Department of Physiology and Environmental Studies, University of Nottingham, School of Agriculture, Sutton Bonington, Loughborough, Leics., England*

E. Harel, *Department of Botany, The Hebrew University of Jerusalem, Jerusalem, Israel*

A. C. Hill, *ARC Food Research Institute, Colney Lane, Norwich NOR 70F, England*

G. E. Hobson, *Glasshouse Crops Research Institute, Littlehampton, West Sussex BN16 3PU, England*

M. Knee, *East Malling Research Station, East Malling, Maidstone, Kent ME19 6BJ, England*

C. Lance, *Laboratoire de Biologie Végétale IV, Université Pierre et Marie Curie, 12 rue Cuvier, 75005 Paris, France*

A. M. Mayer, *Department of Botany, The Hebrew University of Jerusalem, Jerusalem, Israel*

C. J. Pollock, *Botany School, University of Cambridge, Downing Street, Cambridge CB2 3EA, England*

M. J. C. Rhodes, *ARC Food Research Institute, Colney Lane, Norwich NOR 70F, England*

D. A. Robb, *Department of Biochemistry, University of Strathclyde, The Todd Centre, 31 Taylor Street, Glasgow G4 ONR, Scotland*

N. G. Robertson, *Department of Physiology and Environmental Studies,*

University of Nottingham, School of Agriculture, Sutton Bonington, Loughborough, Leics., England

C. F. TIMBERLAKE, *Long Ashton Research Station, Long Ashton, Bristol, England*

G. A. TUCKER, *Department of Physiology and Environmental Studies, University of Nottingham, School of Agriculture, Sutton Bonington, Loughborough, Leics., England*

L. S. C. WOOLTORTON, *ARC Food Research Institute, Colney Lane, Norwich NOR 70F, England*

S. F. YANG, *Department of Vegetable Crops, University of California, Davis, California 95616, USA*

R. E. YOUNG, *Department of Botany and Plant Sciences, University of California, Riverside, California 99024, USA*

Preface

Fruits and vegetables are valuable commodities not only for producers and consumers but also for plant biochemists. They have been used for many years as experimental material for investigations of plant respiration, cell wall composition, organic acid metabolism and many aspects of plant pigments. In the past few years plant biochemists have begun to investigate the molecular basis of the mode of action of plant hormones, the control of alternative pathways of respiration, carbohydrate and phenol metabolism, the molecular structure of plant cell walls, the mechanism of gene action and the mode of action of plant enzymes. In these endeavours fruits and vegetables have been shown to be ideal experimental material and it thus seemed appropriate that the Phytochemical Society of Europe should have chosen "Recent Advances in the Biochemistry of Fruits and Vegetables" as the topic for its Symposium in 1980.

This choice of subject enabled us to invite speakers from the United States and Australia as well as Europe to survey some of the exciting progress made in the ten years since Dr Hulme published his monumental two-volume book "The Biochemistry of Fruits and their Products".

It was a great pleasure to welcome the speakers to the meeting at Norwich. In the present volume the speakers have given us revised versions of their original presentations which we believe have captured the excitement of the new developments that each of them conveyed to the audience.

In the opening chapter J. Biale and the late Roy Young review the work of the past ten years on respiration and fruit ripening and draw together some of the themes developed later in greater detail. In the following chapters T. Ap Rees, C. Lance, S. F. Yang and E. M. Beyer Jr review low temperature sweetening, cyanide-insensitive respiration and the biosynthesis and mode of action of ethylene. The next group of chapters deals with cell walls, in which G. Hobson and M. Knee examine the alterations in cell wall-degrading enzymes and the composition of cell walls which are related to the changes of texture occurring during the ripening process; then D. Grierson explains mainly his own recent work on the molecular biological control of the changes in synthesis of cell wall-degrading enzymes which occur during ripening.

Phenolic compounds, including leuco-anthocyanins and those compounds related to chlorogenic acid which occur widely in many plant tissues, as well as the anthocyanin pigments, are important factors in the acceptability of fresh fruits and vegetables by the consumer. In many cases phenolic oxidation after damage leads to the development of unacceptable changes; under these circumstances there is a need to inhibit or prevent oxidation. The changes in polyphenol oxidases during the ripening of fruits, and the mechanism of action of the whole tyrosinase group of enzymes are surveyed by A. Mayer and D. Robb respectively. It might seem strange that the same group of enzymes are given different names in successive chapters; nevertheless, this difference in nomenclature is a reflection of the varying practices of plant physiologists and biochemists as opposed to enzymologists.

The increased biosynthesis of phenolic compounds in plants subjected to physiological stress are discussed by M. J. C. Rhodes. The role of anthocyanins as pigments in fruits and vegetables is then described by C. F. Timberlake. In the final chapter D. Graham looks back at the themes of the symposium and outlines topics for future research.

It is a great sadness that we have to record the death of Professor Roy Young, the co-author of the opening chapter, since the Symposium. Roy Young had been a collaborator of Professor Jacob Biale for many years; we had hoped that he would have been able to attend the Norwich meeting but he withdrew because of illness and died about nine months later.

We should like to thank Drs Graeme Hobson and Michael Knee for their help in the initial planning of the Symposium programme, the authors for their cooperation in the production of their chapters and also ICI Plant Protection Ltd., Marks and Spencer Ltd. and Unilever Ltd. for their generous donations to the Phytochemical Society of Europe which helped to defray some of the expenses of the meeting.

Finally we wish to thank the staff of Academic Press for their help and cooperation in the preparation of this volume.

October 1981

J.F.
M.J.C.R.

Contents

Contributors v
Preface vii

CHAPTER 1

Respiration and Ripening in Fruits—Retrospect and Prospect

Jacob B. Biale and Roy E. Young

I. Introduction 1
II. Patterns of Respiration 2
A. The Climacteric Phenomenon—Ideas and Features 2
B. The Non-climacteric Pattern—Criteria 5
C. Classification—Revised Lists 7
III. The Ethylene Factor in Respiration and Ripening 13
A. Stages in Development of Knowledge 13
B. Exogenous Ethylene Effects 14
C. Ethylene Evolution in Relation to Respiratory Rise 16
D. Endogenous Ethylene in Climacteric and Non-climacteric Fruits . 21
E. Autocatalytic Ethylene Production 23
IV. The Ripening Process—Metabolic Aspects 25
A. Ripening as an Interplay of Synthetic and Degradative Reactions . 25
B. Carbohydrate Metabolism and Enzymatic Regulation in a Climacteric Fruit 26
C. Ripening Trends in a Non-climacteric Fruit 28
D. Pectic Substances and Pectolytic Enzymes 29
E. Regulation of Ripening at the Mitochondrial Level 32
F. Phosphate Esterification during Ripening 33
G. Protein and Nucleic Acid Metabolism 34
V. Concluding Remarks 35
References 37

CHAPTER 2

Low Temperature Sweetening of Higher Plants

T. Ap Rees, W. L. Dixon, C. J. Pollock and F. Franks

I. Introduction 41
II. Occurrence of Low Temperature Sweetening 42
III. Significance of Low Temperature Sweetening 44

IV. Metabolic Pathways of Low Temperature Sweetening 46
A. Sugar Synthesis 46
B. Source of Substrates for Sugar Synthesis 47
C. Starch–Sugar Balance 48
D. Intracellular Location of Pathways of Sweetening 49
V. Possible Mechanisms of Low Temperature Sweetening 49
A. Compartmentation 50
B. Coarse Control 51
C. Fine Control 52
VI. Conclusions 58
References 60

CHAPTER 3

Cyanide-insensitive Respiration in Fruits and Vegetables

Claude Vance

I. Introduction 63
II. The Cyanide-insensitive Pathway 64
A. Distribution 64
B. Structure 65
C. Basic Problems 69
III. Participation of the Alternative Pathway. 73
A. The Alternative Pathway in Potato Tuber 73
B. Contribution of Both Pathways 75
IV. Physiological Significance of the Alternative Pathway 77
A. Thermogenesis in the Araceae 77
B. Cyanogenic Plant Species 78
C. Elimination of Reducing Power 79
D. Role in Malate Oxidation 81
E. Integration of the Alternative Pathway in Cell Metabolism . . 82
V. Conclusions 84
References 84

CHAPTER 4

Biosynthesis of Ethylene and its Regulation

S. F. Yang

I. Introduction 89
II. Pathway of Ethylene Biosynthesis 90
III. Mechanism of Auxin-induced Ethylene Production in Vegetative Tissue 95
IV. Regulation of Ethylene Production in Ripening Fruits 95
V. Regulation of Stress-induced Ethylene Production 97
VI. Inhibition of Ethylene Production 101
VII. Conclusion 102
Acknowledgement. 104
References 104

CHAPTER 5

Ethylene Action and Metabolism

E. M. Beyer Jr

I. Introduction 107
II. Early Studies of Ethylene Action 108
III. Early Studies of Ethylene Metabolism 109
IV. ^{14}C-Ethylene Purification 110
V. Tissue Preparation and Treatment 111
VI. ^{14}C-Ethylene Metabolism in Peas 112
A. ^{14}C-Ethylene Conversion to $^{14}CO_2$ 112
B. ^{14}C-Ethylene Conversion to ^{14}C-Ethylene Oxide 113
C. Changes in Metabolism during Pea Seedling Growth 115
D. Factors Influencing Ethylene Metabolism in Peas 115
E. ^{14}C-Ethylene Tissue Metabolites 116
F. ^{14}C-Ethylene Tissue Metabolism in Other Tissues 117
VII. Physiological Significance 117
A. Ethylene Inactivation 117
B. Mechanism of Action 118
References 120

CHAPTER 6

Enzymes and Texture Changes During Ripening—G. E. Hobson 121

CHAPTER 7

Composition and Metabolism of Cell Wall Polysaccharides in Ripening Fruits

M. Knee and I. M. Bartley

I. Introduction 131
II. Cell Wall Structure 132
A. General Features of Plant Primary Walls 132
B. Structure of the Middle Lamella 132
C. Special Features of Fruit Cell Walls 133
III. Wall Degradation during Fruit Ripening 137
A. Cellulose 137
B. Hemicelluloses 137
C. Loss of Galactose Residues 137
D. Increase of Soluble Pectin 139
IV. Role of Synthetic Processes in Cell Wall Changes 140
A. Synthesis versus Degradation in Ripening Fruit 140
B. Cell Expansion and Other Cell Separating Systems 143
V. Model of the Cell Separation Process 144
References 145

CHAPTER 8

The Molecular Biology of Ripening

D. Grierson, G. A. Tucker and N. G. Robertson

I. Introduction . . . 147
II. Tomato Ripening Mutants . . . 148
III. Fruit-Softening Enzymes . . . 148
IV. Purification and Properties of Polygalacturonase . . . 149
V. Conversion of PG2→PG1 . . . 153
VI. Regulation of Polygalacturonase Activity During Ripening . . . 154
VII. Messenger-RNA Synthesis During Ripening . . . 155
VIII. Translation of Polygalacturonase m-RNA *in vitro* . . . 156
IX. Conclusions and Prospects . . . 158
Acknowledgement . . . 158
References . . . 158

CHAPTER 9

Polyphenol Oxidases in Fruit—Changes During Ripening

A. M. Mayer and E. Harel

I. Introduction . . . 159
II. Catechol Oxidase in Apples . . . 166
III. Catechol Oxidase and Laccase in Peaches . . . 169
IV. Catechol Oxidase in Grapes . . . 171
V. Catechol Oxidase in Olives . . . 173
VI. Callus and Suspension Cultures as Model Systems . . . 174
VII. The Function of Catechol Oxidase in Fruits . . . 176
References . . . 177

CHAPTER 10

Molecular Properties of Plant Tyrosinases

D. A. Robb

I. Introduction . . . 179
II. Tyrosinase . . . 180
A. Historical Perspective . . . 180
B. Molecular Properties . . . 181
C. Spectral Properties: Oxytyrosinase . . . 183
D. Binuclear Copper . . . 184
E. Ligands to the Copper . . . 185
F. Inhibition by Anions . . . 186
III. Concluding Remarks . . . 188
References . . . 189

CHAPTER 11

Changes in Phenolic Metabolism in Fruit and Vegetable Tissues Under Stress

M. J. C. Rhodes, L. S. C. Wooltorton and A. C. Hill

I. Introduction 191
II. Phenolic Responses to Mechanical Damage in Brassica Roots . . 191
III. Responses in Phenylpropanoid Metabolism to Low Temperature Treatments 206
References 216

CHAPTER 12

Anthocyanins in Fruits and Vegetables

C. F. Timberlake

I. Introduction 219
II. Chemistry and Structural Transformations 220
III. Stabilization and Colour Augmentation 223
IV. Biosynthesis 227
V. Vegetable Anthocyanins 231
A. Anthocyanins of the Cruciferae 231
B. Anthocyanins of the Umbelliferae 232
C. Anthocyanins of Other Vegetables 233
VI. Fruit Anthocyanins 234
A. Anthocyanins of the Rosaceae 234
B. Anthocyanins of the Saxifragaceae 236
C. Anthocyanins of the Vitaceae 236
D. Anthocyanins of the Moraceae 239
E. Anthocyanins of the Ericaceae 240
F. Anthocyanins of Other Fruits 240
References 242

CHAPTER 13

Fruit Biochemistry: Highlights, Perspectives and Prospects—D. Graham . 247

Author Index 255
Subject Index 267

CHAPTER 1

Respiration and Ripening in Fruits—Retrospect and Prospect

JACOB B. BIALE AND ROY E. YOUNG†

Departments of Biology and of Botany and Plant Sciences, University of California, Los Angeles and Riverside, California, USA

I. Introduction 1
II. Patterns of Respiration 2
A. The Climacteric Phenomenon—Ideas and Features 2
B. The Non-climacteric Pattern—Criteria 5
C. Classification—Revised Lists 7
III. The Ethylene Factor in Respiration and Ripening 13
A. Stages in Development of Knowledge 13
B. Exogenous Ethylene Effects 14
C. Ethylene Evolution in Relation to Respiratory Rise . . . 16
D. Endogenous Ethylene in Climacteric and Non-climacteric Fruits . 21
E. Autocatalytic Ethylene Production 23
IV. The Ripening Process—Metabolic Aspects 25
A. Ripening as an Interplay of Synthetic and Degradative Reactions . 25
B. Carbohydrate Metabolism and Enzymatic Regulation in a Climacteric Fruit 26
C. Ripening Trends in a Non-climacteric Fruit 28
D. Pectic Substances and Pectolytic Enzymes 29
E. Regulation of Ripening at the Mitochondrial Level 32
F. Phosphate Esterification during Ripening 33
G. Protein and Nucleic Acid Metabolism 34
V. Concluding Remarks 35
References 37

I. Introduction

The distinction between two respiratory patterns was advanced two decades ago to account for the post-harvest drifts in respiration observed in ripening fruits of various climatic zones. The suggestion that the responses to ethylene could serve as distinguishing features between the two patterns coupled with accumulating evidence for the phytohormonal role of ethylene has generated

† Deceased 13 January 1981.

a great deal of research activity in the past twenty years. During this period the central framework for ripening phenomena, the long established climacteric pattern, has been the subject of scrutinizing definitions and of analytical work involving the disciplines of biochemistry and molecular biology. Comparisons between climacteric and non-climacteric fruits led to valuable results not only in matters pertaining to maturation and ripening, but also in the physiology of fruits during the developmental stages.

The objective of this chapter is to comment on selected topics of the physiological and biochemical aspects of the relationship between respiration and ripening. We do not propose to present a comprehensive account, but to take a retrospective view of the nature of the distinctions between the climacteric and non-climacteric fruits in the light of the research accomplishments of the last two decades. The re-examination of concepts proposed previously will form the main body of this chapter. Updating the information on the two patterns and a renewed analysis of the differences hopefully will indicate new avenues for prospective investigations. Gaps in our knowledge of the relationship between respiration and ripening doubtless will surface, more so on non-climacteric than on climacteric fruits. To find a frame of reference for the former that was so helpful in the studies with the latter is a challenge that should be pursued.

Of special relevance to this discussion are the reviews by Biale (1960), Burg (1962), Biale and Young (1962), Burg and Burg (1962a), Hansen (1966), Pratt and Goeschl (1969), McGlasson (1970), Rhodes (1970), Biale and Young (1971), Sacher (1973), Pratt (1975), Coombe (1976), McGlasson *et al.* (1978), and Hobson (1979). The two volumes on "The Biochemistry of Fruits and Their Products", edited by Hulme (1970, 1971) certainly served as helpful references during the past decade.

II. Patterns of Respiration

A. The Climacteric Phenomenon—Ideas and Features

Kidd and West (1922, 1930) observed the upsurge in carbon dioxide evolution in apples at the end of the maturation phase and termed this process as the "climacteric" rise in respiration. In their studies, as well as in studies that followed with other fruits, this pattern of respiration was associated with ripening. In the course of time new interpretations were advanced for the concept of the climacteric. Some deleted the respiratory aspects and placed full emphasis on the rise of ethylene evolution or the autocatalytic ethylene production as the all important event. Other views were broader in scope. Rhodes (1970), for example, suggested that the term climacteric should be applied to "the whole of the critical phase in the life of the fruit which is triggered by ethylene and during which many changes are occurring."

McGlasson *et al.* (1978) called attention to the confusion in the literature about this term and expressed preference for the "respiratory climacteric" as the designation for the gaseous exchange that accompanies ripening. Of special relevance to the discussion in this chapter is the question whether any upsurge in respiratory activity of fruits should be referred to as the climacteric rise. We are inclined to reserve this expression for the physical, chemical, physiological and metabolic changes that are associated with the increase in the rate of respiration and which occur during the transition phase from the developmental stages of growth and maturation to the stage of senescence, breakdown and death. Ethylene, which will be discussed in the next section, plays a crucial role in this phase, but may not be the sole factor or the exclusive phytohormone involved. The expression for the climacteric proposed here is all inclusive for the various ripening parameters, but does exclude increases in respiration in other developmental stages or those brought about by experimental conditions and manipulations. As envisaged here the climacteric is a signal for the beginning of the end in the life of the fruit at the cellular level and consequently entails general biological interest as well as the concerns of the physiologist, the horticulturist, the shipper and the consumer of fruits.

A close examination of the respiratory course of a climacteric fruit at a suitable temperature shows a declining trend to the lowest value termed as the "pre-climacteric minimum" followed by a more or less sharp rise, depending on species, to the "climacteric peak" and a subsequent "post-climacteric" decline in rate of respiration. Normally the fruit enters the ripe edible stage at or shortly after the peak. Fruit softening, colour changes, development of desirable taste and aromatic flavour and a number of other parameters of the ripening process are associated with the climacteric cycle and will be discussed under a separate heading. At this juncture we wish to concentrate on similarities and differences in the respiratory behaviour of several fruits.

In Fig. 1 the most striking upsurge in oxygen uptake is represented by the breadfruit, native to Polynesia and grown in several tropical countries as a source of starch. The perishability of this fruit is reflected by the sharp slope of the climacteric rise. It is known that the conversion of starch to sugar in this fruit is very rapid, but no data are available to show whether this transformation coincides with the upsurge in respiration as is the case in the banana. Physiological studies on the breadfruit were carried out by Biale and Barcus (1970) under special conditions on the S.S. *Alpha Helix*, a biological research vessel, in the Amazonian forest. Despite the proximity to the source of material the respiratory activity of the breadfruit was so high at 20°C that the trend toward the pre-climacteric minimum escaped notice. In this case a lower temperature would have been more appropriate to follow all events during the climacteric cycle. However, sufficient information was obtained to suggest that the breadfruit may turn out to be a good source for biochemical investigations of the kind that have been conducted on the potato and for

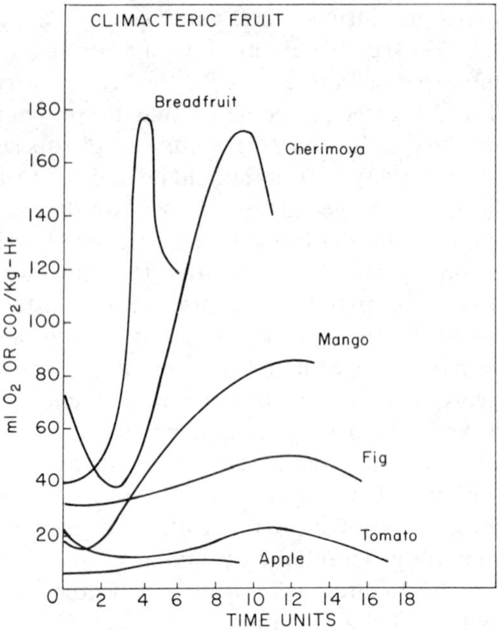

FIG. 1. Patterns of respiratory climacteric. Based on: apple—Smock (1972); breadfruit—Biale and Barcus (1970); cherimoya—Kosiyachinda and Young (1975); fig—Marei and Crane (1971); mango—Burg and Burg (1962a); tomato—Lyons and Pratt (1964). Time units: fig 1 unit = 2 days; rest 1 unit = 1 day.

physiological studies similar to those carried out on the more common climacteric fruits.

In contrast to the breadfruit, the downward trend in respiration following harvesting was observed in the cherimoya, a fruit that is thought to have originated in the highlands of Ecuador and Peru and to have spread from there northward and southward. The flesh of this fruit is creamy, custard-like, has a pleasant blend of sweetness, and is mildly acidic with many embedded black seeds that separate readily from the flesh. This fruit has a well-developed vascular system which is suitable for experiments that require injection of metabolites and inhibitors. It is promising material for metabolic and physiological studies, though its availability is restricted. Kosiyachinda and Young (1975) followed the course of the climacteric in the Chaffey variety of the cherimoya and noticed two peaks in respiration, with the second one more pronounced than the first. Biale and Barcus (1970) also reported a double peak in two other species of the same family, Annonaceae, and ascribed this phenomenon to possible variability within the tissue of an aggregate fruit consisting of numerous carpels. This explanation, however, does not account for the separation between the peaks.

The time lapse between harvesting and the pre-climacteric minimum or the climacteric peak is a function of state of maturity of the fruit and of the conditions of experimentation, such as temperature and rate of air flow. This can be demonstrated with the varieties of avocado that do not ripen while attached to the tree. When, under southern California conditions, a Fuerte avocado is picked in December and placed at 15°C it may take two weeks or more for the respiration rate to peak, while three or four months later the same response will occur in five to seven days. This phenomenon has to be kept in mind when making comparisons and may account for some of the differences between the several fruits shown in Fig. 1. It is likely that the Gardener Red Delicious apple variety included here from the study by Smock (1972) peaked early because of an advanced state of maturity. It exhibited a slow, barely detectable rise due to low respiration rate and the large scale on which the results are drawn. The inclusion of this fruit in Fig. 1 was prompted by the fact that its rate of respiration was probably the lowest reported and thus served as a striking contrast with the high values and sharp rise of the breadfruit and the cherimoya. The variety of apple for which data in Fig. 1 are plotted is probably one of the metabolically least active members of the *Malus sylvestris* species. Other varieties of apple such as Bramley's Seedling and McIntosh have a considerably higher rate of respiration though much lower than the tropical fruits of the climacteric class.

B. THE NON-CLIMACTERIC PATTERN—CRITERIA

The question of the universality of the climacteric pattern of respiration came up as we studied the post-harvest physiology of citrus fruit concurrently with the behaviour of the avocado, the banana and other sub-tropical and tropical fruits. In a study with lemons (Biale and Young, 1947) we noted that the respiratory trend of fruit kept in air at 15°C was a steady downward drift for periods as long as six months. Later, similar observations were made on oranges (Biale, 1961). We decided, therefore, to propose the existence of a second class of fruits showing a non-climacteric pattern and characterized by relatively slow changes in the various ripening parameters. Since citrus fruits served as the paramount example of this class, we assumed that a low rate of respiration is a characteristic feature of non-climacteric fruits. With more data gathered on other fruits higher activities were reported, as can be seen from Fig. 2. Some of the fruits shown had a respiratory activity comparable with several climacteric fruits.

In the case of oranges the downward drifts were observed in the Valencia as well as the Washington Navel cultivar though the actual rates were considerably higher in the latter. The gas exchange in all species of citrus depends primarily on the rind as a source of respiratory substrates. This tissue has a high content of carbohydrates, which are depleted at a slow rate during

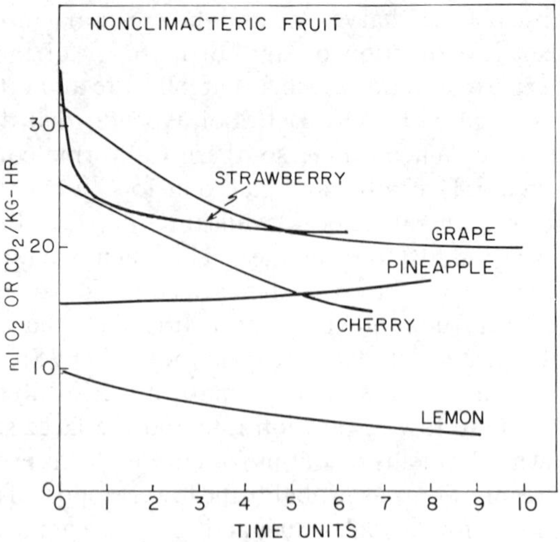

FIG. 2. Non-climacteric patterns of respiration. Based on cherry—Blanpied (1972); grape—Geisler and Radler (1963); lemon—Young *et al.* (1962); pineapple—Dull *et al.* (1967); strawberry—Uota *et al.* (1955). Time unit: cherry 1 unit = 1 day; grape 1 = 4 days; lemon 1 = 7 days; pineapple 1 = 1 day; strawberry 1 = 0·5 days.

prolonged storage at moderate as well as low temperatures. Bain (1958) found a continuous fall in respiratory activity of Valencia oranges from fruit set to full maturity. Trout *et al.* (1960) maintained that oranges placed at 4·4 to 10°C did show a typical respiratory climacteric. They realized that their irregular curves at the low temperatures might have been due to chilling injury. Of the six samples placed at the somewhat higher physiological temperature of 10°C only three showed a rise immediatedly after harvesting. These three samples were from early season picks which might have been immature. No sugar–acid ratios were reported and no mention was made of the state of maturity. In no case was there recorded a drift to pre-climacteric minimum followed by a rise. In fact a close examination of the curves suggests the occurrence of a non-climacteric pattern. The term climacteric was used also by Aharoni (1968) for the course of respiration in several samples of immature oranges and grapefruit at 15°C and at 20°C. No rise was detected in fruit that reached full size and maturity. Eaks (1970) followed the respiratory trends in several species of citrus throughout ontogeny. Small, immature oranges and grapefuit showed a rise in respiration and in ethylene production at 20°C several days after harvest. As the fruit weight increased the peaks in both CO_2 output and in ethylene evolution decreased. No respiratory rise was observed in mature fruit or in fruit approaching maturity. Lemons exhibited continuously decreasing respiratory trends at all stages of development. This evidence coupled with responses to ethylene led Eaks to conclude that all citrus fruits tested belong to the non-climacteric class. Rhodes (1970) correctly emphasized that "in considering the case of citrus fruit, we must return to our definition of the climacteric as a period of enhanced metabolic activity occurring at the transition from the growth phase of the fruit to its senescence". In the category of rises in respiration that should not be coined as climacteric belong the observations by Biale and Young (1947) on the increase in CO_2 evolution of lemons subjected to oxygen levels higher than air, and the report of Young *et al.* (1962) about the stimulation of oxygen uptake in lemons exposed to atmospheres enriched with CO_2. In none of these cases is the upsurge in respiration related to the ripening process or to a transition phase. It is caused rather by a change in experimental conditions.

Similar considerations should be applied to the grape which appears in Fig. 2 as an example of a non-climacteric berry. During the growth period of the grape there is a stage termed "veraison", when, after a stationary phase, the size of the berry increases dramatically. A rise in respiration was noted by Peynaud and Riberau-Gayon (1971) in association with the rapid growth and was termed a "rudimentary climacteric". In our opinion the use of the word climacteric in this connection is inappropriate, because veraison is an integral part of the growth stage, while the post-maturation respiratory trend of the grape is of the type shown in Fig. 2, based on the study by Geisler and Radler (1963) and on evidence by others (see Peynaud and Riberau-Gayon, 1971).

As to the pineapple there appears to be some controversy about the respiratory drift. Bose *et al.* (1962) reported the occurrence of the climacteric pattern, but Dull *et al.* (1967) took the opposite view. The latter examined the rates of CO_2 evolution during six stages of development in the presence and in the absence of ethylene. Ripe pineapple were characterized as fruit with desirable internal composition and with 50% yellow shell surface. These fruits showed a slight upward trend in respiration, but not a pattern of the climacteric type. To counter the objection of pursuing work on a complex multiple fruit, Dull *et al.* (1967) examined two other members of the pineapple family, *Bromelia fastuosa* and *Bromelia pinguin*, in which the individual fruitlets are not fused. A non-climacteric drift in CO_2 evolution was observed in both cases.

C. CLASSIFICATION—REVISED LISTS

Biale (1960) reported two lists of fruits distinguished by respiratory patterns and emphasized that placing a fruit in class B, the non-climacteric category, calls for a more tentative decision than listing one in the climacteric class A. This is simply due to the fact that positive evidence can be obtained experimentally more readily than convincing negative proof, which may require testing samples under a variety of conditions and at several stages of maturity. It is not surprising, therefore, that some changes had to be made in the original lists. Thus, the cantaloup melon studied by Lyons *et al.* (1962), the Honey Dew melon by Pratt and Goeschl (1968) and the fig by Marei and Crane (1971) are now considered climacteric fruits. Otherwise, the original lists remained unchanged. McGlasson *et al.* (1978) added several fruits to these lists and we propose now to submit a new revision by adding five climacteric and ten non-climacteric fruits based on reports of the last decade. In extending the lists it should be stressed that some of the evidence must be considered as preliminary until more data are obtained on both respiration and ethylene production.

In both classes listed in Tables I and II we observe fleshy fruits that originated from a diversity of floral structures. The botanist and the horticulturist distinguish between simple and accessory fruits. A simple fruit formed from a single pistil may be either a "berry" consisting of parenchymatous cells containing none to many seeds, or a "drupe" (also known as stone) made up of a fleshy mesocarp and a hard stony endocarp. Examples of berries are avocado, tomato and papaya among the climacteric, and grape and orange among the non-climacteric. Citrus fruits are a special kind of berry, known as "hesperidium", in which juice and pulp originate from the proliferations of the endocarp, the innermost tissue of the ovarian wall, while the thick, leathery rind stems from the exocarp and mesocarp. Among the drupes we find the climacteric peach, plum and mango and the non-climacteric cherry, blueberry and olive.

Accessory fruits are products of carpellary and extracarpellary tissue, such as receptacle, peduncle and pedicel. The climacteric melons as well as the non-climacteric cucumber are derived from tissue of the inferior ovary combined with receptacle. A "multiple" is made up of several to many flowers of an inflorescence including peduncle. Examples of this structure are the climacteric fig and breadfruit and the non-climacteric pineapple.

A few brief comments may be in order about some of the rare fruits not previously included in the lists of Biale (1960) or of McGlasson *et al.* (1978). The breadfruit, referred to earlier, belongs to the Moraceae, the family that includes also the fig and the mulberry. The fruit is a syncarp formed from the whole inflorescence; the fleshy portion is the product of the receptacle rather than ovary, mostly seedless with a large central core surrounded by abortive ovules. The mature fruit is ovoid, 10–15 cm in diameter, 13–20 cm long and 1–4·5 kg in weight. The yellowish green rind is divided into a series of projections and the edible pulp is white or yellowish. It is used as a vegetable and can be boiled, baked, roasted, fried or made into soup. It is a fair source of thiamine and of ascorbic acid, but it is primarily popular for its high carbohydrate content. In the tropics it is looked upon as bread gathered as a fruit. The *biriba* (in Portuguese; no English name known) is similar in shape and consistency of flesh to the cherimoya, but is inferior in taste and differs from other anonas in that colour change from green to yellow does take place upon ripening. The soursop belongs to the same family, has a dark green skin

TABLE I
Fruits showing the respiratory climacteric during ripening

Common name	Scientific name	Reference
Apple	*Malus sylvestris*	1
Apricot	*Prunus armeniaca*	1
Avocado	*Persea americana*	1
Banana	*Musa sapientum*	1
Biriba	*Rollinia orthopetala*	2
Breadfruit	*Artocarpus altilis*	2
Cherimoya	*Annona cherimola*	1
Chinese gooseberry	*Actinidia chinensis*	4
Feijoa	*Feijoa sellowiana*	1
Fig	*Ficus carica*	4
Guava	*Psidium guajava*	3
Mammee apple	*Mammea americana*	3
Mango	*Mangifera indica*	1
Musk-melon—cantaloup	*Cucumis melo*	4
Honey Dew	*Cucumis melo*	4
Papaw	*Asimina triloba*	1
Papaya	*Carica papaya*	1
Passion fruit	*Passiflora edulis*	1
Peach	*Prunus persica*	1
Pear	*Pyrus communis*	1
Persimmon	*Diospyros kaki*	4
Plum	*Prunus americana*	1
Sapote	*Casimiroa edulis*	1
Soursop	*Annona muricata*	2
Tomato	*Lycopersicon esculentum*	1
Watermelon	*Citrullus lanatus*	4

References: 1, Biale (1960); 2, Biale and Barcus (1970); 3, Akamine and Goo (1973, 1978, 1979); 4, McGlasson *et al.* (1978).

that does not turn yellow but is covered with numerous fleshy spines, may weigh 2–2·5 kg, has a pleasant subacid taste which makes refreshing drinks or pleasant flavouring for sherbet and ice cream. Guavas are native to the American tropics. The fruit is a berry surmounted by calyx lobes, ovoid- or pear-shaped, 4–12 cm long. It is exocarp pale green to bright yellow; mesocarp fleshy of varying thickness, white, yellow, pink, or red; and has numerous seeds embedded in pulp. The ripe, juicy, sweet types are eaten fresh; the fleshy mesocarp may be stewed, made into pies, preserves, juice and nectar; and the vitamin C content is 2–5 times higher than in orange juice.

The list of non-climacteric fruits includes two tomato mutants on which the several parameters of ripening were studied by Herner and Sink (1973) and reviewed by Tigchelaar *et al.* (1978). The growth pattern and the compositional changes associated with maturation turned out to be normal, as was seed maturation. They were found to differ from the common cultivars by

TABLE II
Fruits showing the non-climacteric pattern of respiration

Common name	Scientific name	Reference
Blueberry	*Vaccinium corymbosum*	4
Cacao	*Theobroma cacao*	2
Cashew apple	*Anacardium occidentale*	2
Cherry		
Sweet	*Prunus avium*	4
Sour	*Prunus cerasus*	4
Cucumber	*Cucumis sativus*	1
Grape	*Vitis vinifera*	1
Grapefruit	*Citrus paradisi*	1
Java plum	*Eugenia cumini*	3
Lemon	*Citrus limonia*	1
Lychee	*Litchi chinensis*	3
Mountain apple	*Eugenia malacensis*	3
Olive	*Olea europea*	5
Orange	*Citrus sinensis*	1
Pineapple	*Ananas comosus*	4
Rose apple	*Eugenia jambos*	3
Strawberry	*Fragaria*	1
Surinam cherry	*Eugenia uniflora*	3
Tamarillo	*Cyphomandra betacea*	4
nor-Tomato	*Lycopersicon esculentum*	6
rin-Tomato	*Lycopersicon esculentum*	6

References: 1, Biale (1960); 2, Biale and Barcus (1970); 3, Akamine and Goo (1973, 1979); McGlasson *et al.* (1978); 5, Maxie *et al.* (1960); 6, Tigchelaar *et al.* (1978).

the absence of the climacteric cycle, an absence of ethylene production in the *rin* and a greatly reduced level in the *nor* mutant. The capacity to produce ethylene autocatalytically was lacking totally. Total carotenoids as well as β-carotene content were reduced in both mutants; synthesis of lycopene was reduced and delayed in *nor* and only a trace observed in *rin*. Of the pectolytic enzymes involved in ripening, pectinmethylesterase activity was normal in both, while only traces of polygalacturonase were found. As a result firmness and shelf life were retained for a long time. The discovery of the blocking mechanism could advance our knowledge of the ripening process considerably and countribute greatly towards further understanding of the distinctions between climacteric and non-climacteric fruits.

In Table II the olive is listed for the first time among the non-climacteric fruits based on the study by Maxie *et al.* (1960). This fruit has served for thousands of years as the chief source of edible oil for the inhabitants of the Mediterranean basin. It is thought to have originated in the mountains of that region as a hybrid between wild types. Seeds of the olive dating back to about 4000 B.C.E. were found along the coast of the eastern Mediterranean. From there the olive moved westward, establishing secondary centres of diversification in the Aegean area and tertiary habitats in southern Italy, Tunisia and Spain. Olives were known to the Romans before the sixth century B.C.E. The trees are famous for longevity due to their high capacity for regeneration.

The mature fruit may reach an oil content of 20% on fresh weight basis in about six to eight months after full bloom and it can hang on the tree much longer. In the course of maturation the colour changes from straw and pink to red and black. Oil content appears to be a better index of maturity than colour change. Olive oil is highly unsaturated as 85% of it is the triglyceride of oleic acid. Linoleate makes up 4% and palmitate 6–9% of total oils. The olive resembles the avocado in the high content of unsaturated fatty acids. Both fruits also share the unique feature of the absence of starch in the edible mesocarp. Ripening of olives does not seem to be related to a transformation of an insoluble carbohydrate to a soluble one; neither do we have information about the changes in pectic substances. Ripening of green olives was not accelerated by ethylene, as reported by Maxie *et al.* (1960).

Much of the olive crop is grown for oil, but a large share is destined also for pickling and canning. Processing of olives includes the hydrolysis by sodium or potassium hydroxide of glucosides that contribute bitterness to the fruit. After this treatment olives are fermented in brine either anaerobically to retain the bright green colour or in the presence of air to impart the "ripe" dark colour due to the oxidation of orthodihydroxyphenolic compounds. Both pickled and canned olives are a rich source of vitamin A. Canned olives are a richer source of energy than pickled, because they are harvested normally in a riper state and because they are not subjected to fermentative losses. Since olives are not consumed fresh they have attracted much less attention on the part of fruit physiologists than the avocado. Comparative physiological and biochemical studies between these two species could yield valuable information.

In Table II are listed for the first time among the non-climacteric fruits several members of the Eugenia genus. According to Chandler (1958) the Surinam cherry, also known as the pitanga, is one of the more desirable fruits of this group. Mature fruits vary in form, size, colour and flavour when grown on seedling bushes. They may be oblong and 8-ribbed or ribless, from 1–2 in ($\simeq$ 2·5–5 cm) in diameter with one spherical or two hemispherical seeds. In ripening they change from light green through yellow and orange to dark crimson. The mild acidity combined with a blend of sweetness and aromatic oil flavour render some strains as attractive for use in jellies, sherbets or drinks.

The fruit of the cacao, also listed in Table II, is known botanically as a pod of elliptic–ovoid shape. A large specimen may be 25–30 cm long and 10 cm in large diameter. The rind is thick, hard and leathery while the pulp is pinkish and contains 20–40 seeds referred to as "beans". Upon ripening the skin changes to a dark brown colour. The commercial importance of this fruit is for the endosperm of the bean which is the source of chocolate.

The cashew is another fruit of special botanical interest and it is best known for its high quality kidney-shaped nut. The pericarp consists mostly of kernel and is attached externally to the cashew apple. The fleshy edible structure is made up of a swollen pedicel, varies from white to yellow, is juicy and slightly astringent. It is used also in preserves and can be fermented into wine. The cashew apple is one of the most metabolically active fruits as shown by its high rate of respiration (Biale and Barcus, 1970). Several tropical fruits of non-climacteric type exhibit this feature. A comparison of respiratory activities of representative fruits from the two classes is made in Table III.

TABLE III
Rates of respiration of several fruits of the two classes (ml O_2 or CO_2 kg^{-1} h^{-1})

Fruit	Variety	Temp. (°C)	Min.	Max.	Reference
		Climacteric			
Apple	Bramleys Seedling	23	10	16	Kidd and West (1930)
Avocado	Fuerte	20	35	155	Biale *et al.* (1954)
Banana	Gros Michel	20	20	60	Gane (1936)
Fig	Mission	20	22	40	Marei and Crane (1971)
Mango	Kent	20	15	82	Burg and Burg (1962a)
Peach	Elberta	20	25	35	Biale (1960)
Pear	Bartlett	20	12	33	Hansen (1942)
Tomato	Grand Rapids	18	8	20	Clendening (1942)
		Non-climacteric			
		Temp. (°C)	Steady state		
Cherry	Montmorency		15		Blanpied (1972)
Grape	Vi 5861	20	20		Geisler and Radler (1963)
Lemon	Eureka	20	5		Young *et al.* (1962)
Orange	Valencia	21	13		Haller *et al.* (1945)
Orange	W. Navel	21	17		Haller *et al.* (1945)
Pineapple	Cayenne	20	17		Dull *et al.* (1967)
Strawberry		15	21		Uota *et al.* (1955)

The failure to demonstrate the occurrence of the upsurge in respiration in detached fruit is by itself insufficient evidence to evaluate the past history. Both climacteric and non-climacteric fruit may not experience a respiratory rise on the tree. We referred earlier to the avocado which may hang for months on the tree in a mature but unripe state. Even after harvesting it may show resistance to a ripening agent, such as ethylene, for a day or two (Gazit and Blumenfeld, 1970). In many climacteric fruits detachment is not a requirement but it hastens ripening. In others, such as the Honey Dew musk-melon, the respiratory climacteric and ripening occur equally well on and off the plant provided the proper state of maturity was attained. With some fruits, especially those grown on vines and on small plants, it is experimentally feasible to follow the course of respiration in succession from the attached to the detached condition. Clendening (1942) did this for the tomato and was able to demonstrate the continuity of the climacteric trend after removal. The detachment phenomena discussed here have been explained by proposing interactions between ethylene and ripening inhibitors supplied from leaves or other parts of the plant. It is possible also to use the response of fruit to ethylene as a distinguishing marker between climacteric and non-climacteric fruits. This and other aspects of the ethylene factor are taken up later.

III. The Ethylene Factor in Respiration and Ripening

A. Stages in Development of Knowledge

The historical aspects of the role of ethylene in fruit ripening were discussed two decades ago by Biale (1960) and by Burg (1962). Review and updating of our knowledge in this field can be done by delineating several stages of development in a more or less chronological fashion. These are:

(i) The use of oil heaters in the transport of citrus fruits and the subsequent discovery that ethylene was the active agent as an incomplete combustion product responsible for accelerated colouring of fruit.

(ii) Studies on the effects of ethylene on ripening, respiration and chemical changes in fruits and vegetables. Concurrent observations of ethylene induced abnormalities in seed germination, potato sprouting, growth of seedling and leaf epinasty. The adaptation of some of these responses into a bio-assay for semi-quantitative estimation of ethylene concentration.

(iii) Acceleration of ripening and respiration of unripe fruit by volatiles given off by ripe fruit. Hastening of colouring and CO_2 evolution of citrus fruits by emanations of some common fungi, notably of *Penicillium digitatum*.

(iv) Identification by chemical procedures of ethylene in the emanations of ripe fruit; especially apple, pear, banana and avocado and of the common green mould, *Penicillium digitatum*. Development of quantitative methods, including the manometric procedure, for determining rates of ethylene evolution.

(v) Studies on the relationship between ethylene production and the onset of the climacteric rise.

(vi) Introduction of flame ionization gas chromatography. Estimation of internal ethylene levels and advancement of the hormonal idea of ripening by ethylene.

(vii) Ethylene precursors and biosynthetic pathways.
(viii) Ideas on the mechanism of ethylene action in fruit ripening.
(ix) Interaction of ethylene with other phytohormones leading to a re-examination of its role as the sole factor in respiration and ripening.

In this chapter we shall limit ourselves to a discussion mainly of those aspects of the ethylene factor which fall under (v), (vi), and (ix) with special attention directed to the features that distinguish between climacteric and non-climacteric fruits. For a recent comprehensive treatment of the subject of the role of ethylene in fruit ripening, the reader is referred to the review by McGlasson *et al.* (1978).

B. EXOGENOUS ETHYLENE EFFECTS

The ethylene factor as related to repiratory pattern is summarized in Table IV. In this sub-section the differentiation between the two classes of fruits will centre on (a) time of application, and (b) response to a range of concentration.

TABLE IV
The ethylene factor as related to respiratory pattern

	Climacteric	*Non-climacteric*
Response to application	Prior to respiratory rise only	Throughout post-harvest life
Magnitude of response	Concentration independent	Function of concentration
Endogenous level	Variable, low to high	Low
Autocatalysis	Pronounced	Absent

In climacteric fruits it was shown that treatment with low concentrations of ethylene brings about a shift in the time when onset of the rise occurs without altering the shape of the curve of the climacteric cycle. Ethylene is effective only if applied during the pre-climacteric stage prior to the burst of ethylene production by the fruit. On the other hand, in non-climacteric fruits stimulation of the rate of respiration has been observed throughout the post-harvest life (Biale and Young, 1962). There is also a difference in the reversibility of the process. In a climacteric fruit if a sufficient concentration over a long enough period is applied to cause a respiratory rise, no return to a pre-climacteric stage will occur upon removal of the gas. In non-climacteric fruits discontinuing ethylene treatment at any time during storage will result in a return of the rate of gaseous exchange to the level of untreated control. These differences, as we shall see later, may be ascribed to the differences in the ethylene generating capacities between fruits of the two classes.

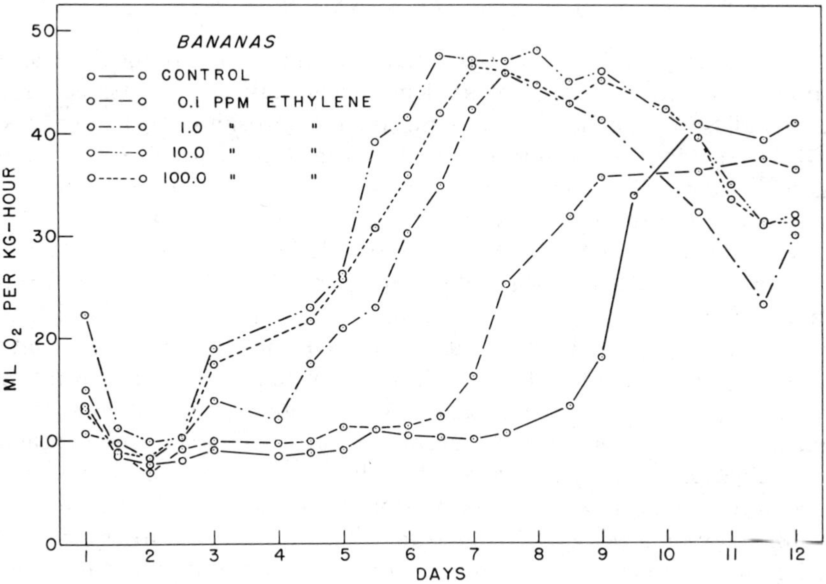

FIG. 3. Effect of ethylene on respiration of Gros Michel bananas.

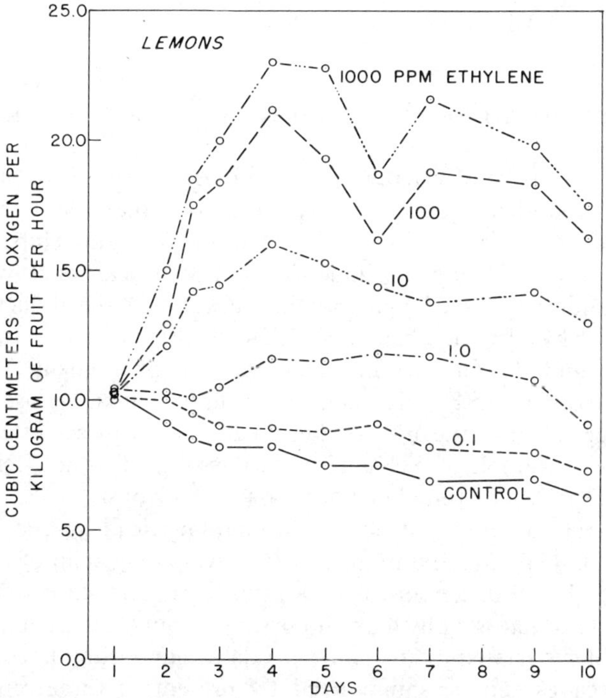

FIG. 4. Effect of ethylene on respiration of lemons.

In addition to time of application and reversibility, fruits of the two classes respond differently to a range of ethylene concentrations as can be seen from Figs 3 and 4. In the banana the onset of the rise was shifted by 0·1 p.p.m. of ethylene, the lowest concentration used, but maximum acceleration in time was achieved with 1·0 p.p.m. The somewhat higher peak values of the treated fruit are probably due to greater uniformity in a composite sample exposed to ethylene. An investigation with avocados (Biale and Young, 1971) gave similar results except that it required 10 p.p.m. of ethylene to produce the immediate response. Avocados are known to have a relatively high resistance to ethylene. Gazit and Blumenfeld (1970) have shown that if they treated avocados with 50 p.p.m. of the gas for 48 h immediately after removal from the tree, ripening was not initiated. We too found that the onset of the climacteric rise was delayed considerably when early season Fuerte avocados were subjected to ethylene right after harvest for one day only at a concentration that accelerated respiration when the exposure was for two days or more. Peacock (1972) stated that bananas exposed to 60 p.p.m. of ethylene for six hours did not commence ripening nor was there any reduction in green life. The suggestion has been made repeatedly, especially in the case of avocados which ripen only after detachment, that a ripening inhibitor from the tree must disappear or be rendered ineffective before the action of ethylene can commence. The nature of the inhibitor has not been elucidated except for the suggestion that IAA may play a role.

C. ETHYLENE EVOLUTION IN RELATION TO RESPIRATORY RISE

The marked effects of externally applied ethylene on both respiration and ripening and the identification of ethylene among the volatile products of fruits led directly to investigations of the quantitative relationship between the respiratory rate and the rate of production of the hydrocarbon. It was thought that these studies will shed light on the question of the induction of the climacteric rise by ethylene. Biale *et al.* (1954) reported on some fruits, such as the avocado and the banana, in which the two phenomena were closely associated. In others, such as the cherimoya and the feijoa, the rise in ethylene evolution lagged behind the onset of the climacteric rise by about three days at 15°C. A typical example of such lag is shown in Fig. 5 for the Wickson plum, studied about the same time but not reported previously.

The early studies were conducted by the manometric technique which had a sensitivity of 0·05 p.p.m. and required a relatively long span of time for the absorption of the gas in mercuric perchlorate. The introduction of much more sensitive gas chromatography made it possible to pursue with more precision various problems related to ethylene. The data that accumulated during the past two decades can be summarized by presenting three representative studies in which external ethylene production and internal tissue con-

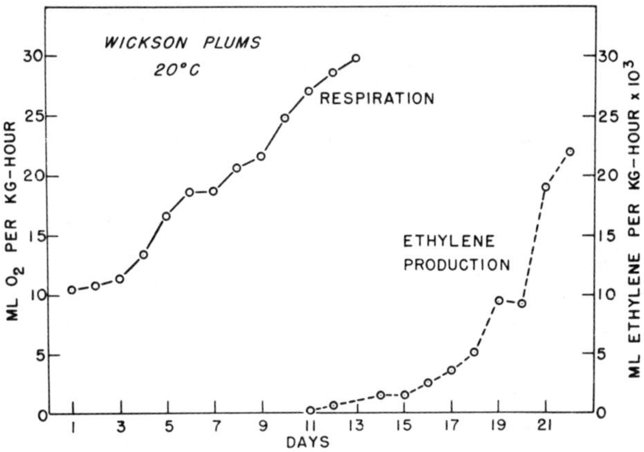

FIG. 5. Relationship between respiration and ethylene production in Wickson plums.

centrations were followed closely with the determinations of CO_2 evolution or oxygen uptake.

The role of ethylene in ripening and respiration of the Honey Dew muskmelon was studied in great detail by Pratt and Goeschl (1968) and by Pratt *et al.* (1977). These workers paid special attention to the state of maturity of the fruit by tagging the flowers at anthesis and thus knowing the chronological age precisely. It took approximately 35 days after flowering for minimum horticultural maturity to be reached and 45 days for ripening to initiate on the vine under summer conditions at Davis, California; five days later the full ripe stage was attained. Pratt and co-workers followed the rate of CO_2 evolution colourimetrically and ethylene production by flame ionization gas chromatography. In short-term experiments the internal ethylene concentration was measured by withdrawing 1·0 ml samples from the cavity of the fruit. In the long-term tests the endogenous levels were estimated from the external production by the empirical formula: p.p.m. internal $= 3{\cdot}7 \pm 1{\cdot}2 \times$ rate of production in μl kg^{-1} h^{-1}. This estimation was checked on some fruits by the direct sampling technique. Rates of respiration and of ethylene production were determined on individual fruits that ranged in age from 7 to 54 days after anthesis. Selected data on three samples are presented in Fig. 6. Fruits 28 days from flowering and younger did not show an appreciable respiratory climacteric even when stored past the time of normal ripening. These melons did not attain the minimum ethylene level required to trigger the respiratory rise. On the other hand, in 35-day fruit ethylene production increased while respiration declined to a pre-climacteric minimum. The rise of CO_2 evolution commenced when the rate of ethylene production was 0·6 μl kg^{-1} h^{-1}, which was equivalent to an internal concentration of approximately 3 p.p.m. The amplitude of the peaks in CO_2 output and in ethylene production increased in the older melons, but the times for the occurrence of the minima and the

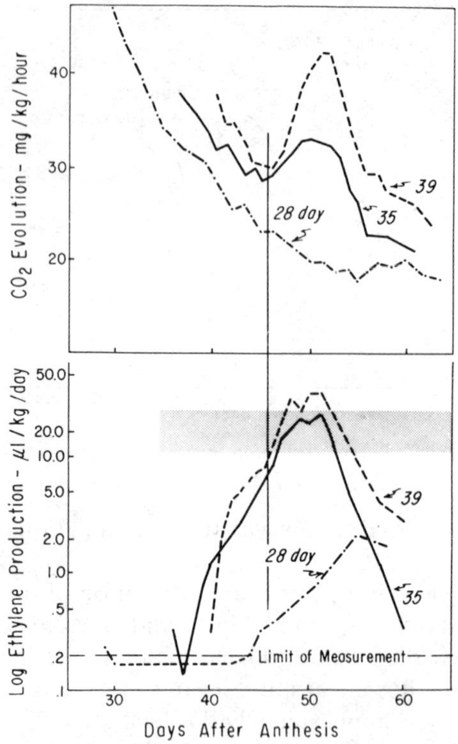

FIG. 6. Respiration in relation to ethylene production in Honey Dew musk-melon. From Pratt and Goeschl (1968). By permission.

maxima were only slightly affected by age of harvest. Tissue firmness declined sharply in 40-day fruit and older. This was the earliest observable response after the rise in ethylene production and preceded the onset of the respiratory climacteric by a week. The reduction in firmness and the upsurge in respiration appeared to be the only ripening parameters that were directly related to ethylene action. Other physiological changes, such as the decrease in growth rate and increase in soluble solids content, preceded the accelerated ethylene production. Immature melons harvested with a low sugar content of 10 to 12% did not develop the best flavour even when treated with ethylene. The conclusion with respect to Honey Dew was that while ethylene induces a respiratory climacteric in immature fruit it will not cause accumulation of sugars and flavour components typical of mature fruit.

The work with Honey Dew musk-melon presents a convincing story of an ethylene induced climacteric rise. The situation is not as definitive with the apple as shown in the studies of Reid *et al.* (1973) and of Rhodes and Reid (1975). These workers used several experimental approaches to elucidate the relationship between ethylene production and CO_2 evolution. In the first case they harvested several apples every three to five days through the ripening season. Fruit were placed at 12°C to equilibrate for 18 h and then CO_2 and ethylene production measured as well as the internal ethylene concentration. The initial values only were reported as a reflection of the activity of the fruit on the tree for that day. Mean values of fruit picked on the same day are plotted in Fig. 7, showing that the rise in respiration and the increase in ethylene coincided rather closely.

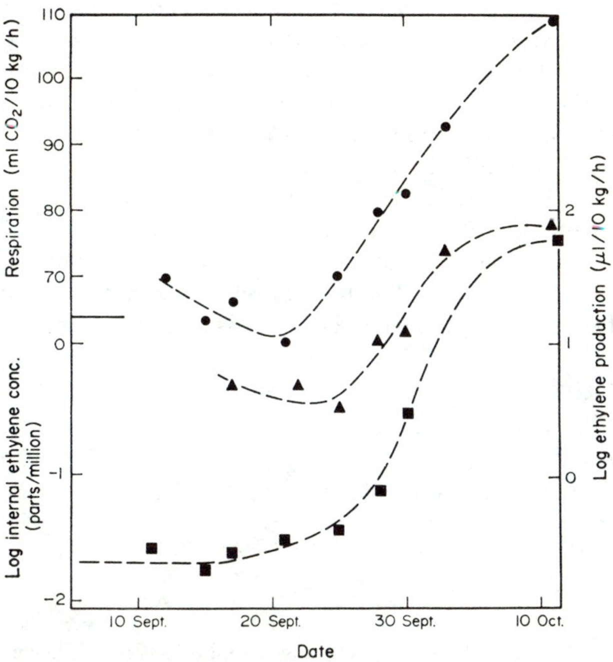

FIG. 7. Respiration in relation to ethylene production and internal ethylene concentration in Cox's Orange Pippin apples. From Rhodes and Reid (1975). By permission.

In another case a fine nylon tube was inserted into the core of several apples and each fruit placed in a separate chamber for simultaneous measurements of external production of CO_2 and ethylene and of internal concentrations of these gases. Here, again, respiratory CO_2 and ethylene levels increased about the same time. While Rhodes and Reid (1975) appreciated the technical problem connected with their experiments they did conclude that a "factor other than changes in ethylene determines the time of the onset of ripening in the apple".

Kosiyachinda and Young (1975) also came to the conclusion that ethylene does not appear to be the only controlling factor for the initiation of the climacteric in avocado and in cherimoya. They measured the two parameters with sufficient frequency of four to eight hours to detect time differences in the induction of the rise. Figure 8 illustrates this point for the avocado and similar results were obtained for the cherimoya. The onset of the rise in external ethylene production followed the onset of CO_2 rise by 0·3 to 0·9 days in the avocado and by 0·1 to 1·5 days in the cherimoya. The internal ethylene concentration (IEC) was less than 13 p.p.b. (parts of ethylene per billion parts of air) at points 1 and 2 in Fig. 8. At point 3 it averaged 0·088 p.p.m. with a range of 0·027 to 0·182 for five fruits. From the report of Biale (1960) and of Biale and Young (1971) it is obvious that higher concentrations than 1·0 p.p.m. are required to induce a rise in CO_2 output in the avocado in a short time period. In cherimoya also the IEC was less than 0·1 p.p.m. at the beginning of the climacteric rise. However, no data are available for this fruit on ethylene concentration required to stimulate respiration. The lag period found by Kosiyachinda and Young (1975) for the Chaffey cherimoya confirms the earlier observations of Biale *et al.* (1954) for the Booth variety. Similarly, the findings of these authors for the Coolidge variety of feijoa were substantiated by Reid (1975) for the Triumph. It is feasible at this juncture to make up a list of fruits for each pattern of relationships between ethylene and the rise in the respiratory climacteric. This was done and is presented in Table V.

TABLE V
Rise in ethylene production as related to respiratory rise

Precedes		Coincides		Follows	
	Ref.		*Ref.*		*Ref.*
Avocado		Apple		Avocado	
Choquette	(4)	Cox's Orange Pippin	(12)	Fuerte	(6)
Banana		Apricot		Cherimoya	
Gros Michel	(3)	Moorpark	(11)	Chaffey	(6)
Chinese Gooseberry		Guava		Feijoa	
Bruno	(10)	White, Purple	(1)	Triumph	(11)
Guava		Musk-melon		Mango	
Yellow Strawberry	(1)	Cantaloup	(7)	Haden	(3)
Musk-melon		Pear		Papaya	
Honey Dew	(9)	Anjou	(13)	Solo	(2)
Persimmon				Plum	
Atago	(5)			Wickson	(14)
				Tomato	
				Rutgers	(8)

References: 1, Akamine and Goo (1979a); 2, Akamine and Goo (1979b); 3, Burg and Burg (1962a); 4, Burg and Burg (1962b); 5, Iwata *et al.* (1969); 6, Kosiyachinda and Young (1975); 7, Lyons *et al.* (1962); 8, Mizrahi *et al.* (1975); 9, Pratt and Goeschl (1968); 10, Pratt and Reid (1974); 11, Reid (1975); 12, Rhodes and Reid (1975); 13, Wang and Mellenthin (1972); 14, Young and Biale (previously unpublished).

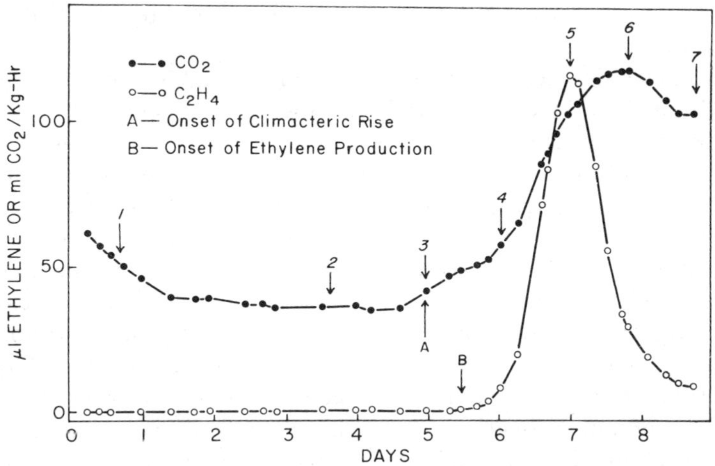

FIG. 8. Respiration and ethylene production of Fuerte avocado fruit. For explanation see Kosiyachinda and Young (1975).

In examining this table one is inclined to conclude that upsurge in ethylene synthesis cannot be considered as the critical factor in inducing the respiratory climacteric in all fruits investigated. It should be borne in mind, however, that not all studies reported in this table were carried out as meticulously as the three representative cases discussed. In view of the diversity of results it became necessary to determine whether sufficient ethylene was formed in the tissues of fruit during the pre-climacteric period to exceed the external concentration required for induction of the respiratory rise.

D. ENDOGENOUS ETHYLENE IN CLIMACTERIC AND NON-CLIMACTERIC FRUITS

There is ample evidence for the presence of ethylene in intercellular spaces, in tissues and in cavities of fruits throughout development and maturation. Akamine and Goo (1973) found high post-anthesis rates of ethylene production and CO_2 evolution in the non-climacteric lychee as well as in the climacteric avocado and mango. Accepting the recommendation of Burg and Burg (1962a) for a conversion factor of 2 p.p.m. in the internal space as being equivalent to a production rate of 1 $\mu l\ kg^{-1}\ h^{-1}$, one might expect immediately after fruit set internal ethylene concentrations of approximately 200, 100, and 6 p.p.m. for avocado, mango, and lychee respectively. The curves for ethylene production dropped sharply with fruit growth so that by the time the mango and lychee attained 4% and avocado 10% of the final weight no detectable levels of ethylene were recorded. Obviously, the ethylene present in fruits in early stages of development neither causes a respiratory rise nor does it bear any relationship to the pattern of respiration at the end of the maturation phase.

In several species measurable concentrations of endogenous ethylene were reported (cited by McGlasson, 1970) during development and maturation. The changes in internal content for several fruits in the course of the climacteric cycle are presented in Table VI. In three species listed in this table values lower than the accepted threshold level of 0·1 p.p.m. were recorded at the onset of the respiratory rise. It is highly doubtful whether 0·1 p.p.m. would induce the rise in any fruit, with the possible exception of bananas at an advanced state of maturity, in a reasonable time interval of a day or less. Data are available on avocado (Biale and Young, 1971) showing that more than 1·0 p.p.m. is required for rapid initiation of climacteric rise. Too little information has been gathered on applied ethylene to justify generalizations about the minimum induction levels. The available information on internal ethylene levels in non-climacteric fruits is also scant and covers too few fruits. Some of the values reported in Table VI are sufficient to initiate a rise in respiration given sufficient exposure time.

TABLE VI
Range in levels of internal ethylene (PPM)

Fruit	Variety	Preclim.	At onset of rise	Clim. peak	Reference
		Climacteric fruits			
Avocado	Choquette	0·04	0·75	500	2
Avocado	Fuerte	0·03	0·09	25	3
Banana	Gros Michel	0·1	1·5	40	2
Cherimoya	Chaffey	0·03	0·04	219	3
Mango	Kent, Haden	0·01	0·08	3	2
Musk-melon	Cantaloup	0·04	0·3	50	4
Musk-melon	Honey Dew	0·04	3·0	25	6
Papaya	Solo		0·1	2·8	1
Pear	Anjou	0·9	0·4	40	7
Tomato	VC-243-20	0·08	0·8	27	5
		Non-climacteric			
		Steady state			
Lemon		0·1–0·2			2
Lime		0·3–2·0			2
Orange		0·1–0·3			2
Pineapple		0·2–0·4			2

References: 1, Akamine and Goo (1979b); 2, Burg and Burg (1962a); 3, Kosiyachinda and Young (1975); 4, Lyons *et al.* (1962); 5, Lyons and Pratt (1964); 6, Pratt and Goeschl (1968); 7, Wang and Mellenthin (1972).

Peacock (1972) questioned the idea that the endogenous ethylene concentration must reach or exceed the threshold concentration to induce ripening. He succeeded in reducing the green life of bananas by ethylene levels low enough and time exposure short enough as not to induce an immediate rise in respiration. He discovered, for example, that 0·44 p.p.m. of ethylene applied for four days shortened the green life from an average of 39 to 26 days. The effectivenesss of ethylene was a function of the log of concentration, length of exposure period and of time of application after harvest. He defined the sensitivity of bananas to ethylene as the proportional loss of green life brought about by the exposure of the fruit to the gas compared with green life at harvest. As fruit approaches maturity the sensitivity decreases. To emphasize his doubts about the concept of the threshold concentration Peacock (1972) cited an experiment in which bananas were given a very high dose of ethylene—60 p.p.m.—but for six hours only, with the result that no induction of ripening or reduction of green life took place, Peacock's point was that endogenous levels of ethylene, no matter how low, exerted their ripening influence throughout the life of the bananas.

The fact that in some fruits the endogenous levels of ethylene do not change during the pre-climacteric period but ripening does eventually take place invoked the concept of changing sensitivity. Evidence that there are striking changes in sensitivity was cited from the behaviour of fruit during early stages of development in which exceptionally high levels of ethylene did not cause a rise in respiration. In these cases sensitivity may mean the action of substances such as inhibitors or phytohormones which are ethylene antagonists. This situation may prevail also in mature fruit of some species which are supplied with anti-ripening compounds from the plant prior to harvest.

E. AUTOCATALYTIC ETHYLENE PRODUCTION

Autocatalysis, in the context of fruit ripening, means the capacity of the tissue to synthesize large quantities of ethylene in response to application of low concentrations of this gas, or of any other substance, which stimulates respiration and ripening. The other substance is usually an analogue of ethylene that mimics its action such as propylene or acetylene. In using compounds of this kind it is important to know the concentrations equivalent to ethylene in producing a certain physiological response. Propylene is about 1% as active as ethylene biologically if the pea stem bio-assay is employed as a criterion (Burg and Burg, 1967). By the use of propylene as the ripening stimulant it is possible to measure ethylene production unhindered by the presence of ethylene applied externally, as the retention times of the two hydocarbons on the gas chromatograph are well separated. McMurchie *et al.* (1972) made use of this technique and exposed green bananas to 500 p.p.m. of propylene, equivalent to 5 p.p.m. of ethylene, for two and one-half days. They succeeded in inducing the typical climacteric within 10 hours after exposure with a burst of ethylene production three hours later. Lemons and oranges responded to propylene treatment with an increase in respiration but showed no increase in ethylene formation. This autocatalytic ability to synthesize ethylene in response to ripening accelerating agents does not distinguish all climacteric fruits from the non-climacteric. The tomato appears to be an exception. McGlasson *et al.* (1975) noticed that treatment of 40–80% mature tomatoes with propylene stimulated respiration but brought about no ethylene production. They concluded that "onset of ripening in normal tomato is not controlled by endogenous ethylene, although increased ethylene production is probably an integral part of the ripening process". The tomato is unique in its exceptionally high resistance to ethylene. McGlasson *et al.* (1978) discussed this phenomenon at great length. It appears that ethylene reduces the pre-climacteric life of the tomato but is not instrumental, not even in high concentrations, in inducing the autocatalytic production.

The question arises as to the essentiality of the ethylene produced autocatalytically after the initiation of ripening and respiration. Is it conceivable that this burst of ethylene during the climacteric rise is actually the "by-product of ripening"? Biale *et al.* (1954) introduced this idea as they were not able with their manometric method to measure the low levels of ethylene during the pre-climacteric stage. Some workers feel that a distinction should be made between the initiation and the development of ripening. Quazi and Freebairn (1970) maintained that autocatalytic ethylene is essential for other parameters of ripening but did not mention which ones. We know from the work of Wang and Mellenthin (1972) that softening in Anjou pears was stimulated by 0·08 $\mu l\ l^{-1}$ of ethylene, while the induction of the respiratory rise required 0·46 $\mu l\ l^{-1}$. It is also known in some fruits that colour changes involving chlorophyll destruction can be brought about by levels of ethylene of the order of 0·1 to 1·0 p.p.m. Obviously, it is necessary to separate the several parameters of ripening and to examine the requirements for their expression.

Autocatalytic ethylene production is most pronounced in mature fruit, though Sfakiotakis and Dilley (1973) observed that immature apples responsed to propylene if supplied with concentrations higher than those effective in mature fruit. It was also noted by Mapson and Robinson (1966) that the initiation of autocatalysis has a higher requirement for aerobic conditions than the continued synthesis of ethylene. This situation might be difficult to reconcile with what is now known (Adams and Yang, 1979; Yang, 1981) about the requirements for interconversion between intermediate compounds in the pathway of ethylene biosynthesis.

IV. The Ripening Process—Metabolic Aspects

A. RIPENING AS AN INTERPLAY OF SYNTHETIC AND DEGRADATIVE REACTIONS

The focus of our discussion has been directed thus far to two major features of the ripening process, namely the respiratory trends and the action of ethylene. That there is a great deal more to ripening is obvious even from a casual treatment of the subject. When the question is asked "what is ripening?", one might counter with the question "to whom"? To the consumer of fruits the all important attributes are colour, texture, taste, aroma, nutritional and aesthetic values. The fruit grower, the manager of storage plants, the shipper and the distributor have additional concerns such as keeping quality including physical and chemical changes with time, susceptibility to low temperature and resistance to fungal deterioration. The fruit physiologist must keep all of these matters in mind while exploring a wide spectrum of biochemical parameters that contribute to the complexity of the ripening process. A simplified view of two types of reactions that play a role in ripening is presented in Table VII.

TABLE VII
Ripening phenomena in fruits

Degradative	Synthetic
Destruction of chloroplast	Maintenance of mitochondrial structure
Breakdown of chlorophyll	Formation of carotenoids and of anthocyanins
Starch hydrolysis	Interconversion of sugars
Destruction of acids	Increased TCA cycle activity
Oxidation of substrate	Increased ATP generation
Inactivation by phenolic compounds	Synthesis of flavour volatiles
Solubilization of pectins	Increased amino acid incorporation
Activation of hydrolytic enzymes	Increased transcription and translation
Initiation of membrane leakage	Preservation of selective membranes
Ethylene induced cell wall softening	Formation of ethylene pathways

The attempt has been made in this table to list a degradative reaction *vis-à-vis* a biosynthetic, but this approach could not be pursued consistently in all the statements included. Chlorophyll degradation, for example, accompanies the ripening of most green fruits, but the appearance of the yellow colour may be due to unmasking of pre-existing carotenoids rather than the synthesis of these pigments. When carbohydrate metabolism is considered the hydrolysis of starch is closely related to ripening in some fruits, while in others this reaction takes place prior to the onset of ripening. The net result of starch hydrolysis is frequently a mixture of sucrose, glucose and fructose. As far as acids are concerned there is generally a decrease of total acidity with ripening, though the content of one or more acids may increase. Organic acids are metabolized, while simultaneously di- and tri-carboxylic acids play a decisive role in the operation of the Krebs cycle. As respiration is the central process in fruit metabolism there are continuous demands for fuel to keep the fruit cells in operating condition. Substrate utilization is usually coupled with ATP formation and with other synthetic reactions. Oxidative phosphorylation is the source of energy for transcription and translation that are active during certain stages of ripening, especially in climacteric fruits, and for numerous biosynthetic pathways, including ethylene formation. Selected topics cited in Table VII, especially those that received attention in our laboratories, will be taken up in this section.

B. CARBOHYDRATE METABOLISM AND ENZYMATIC REGULATION IN A CLIMACTERIC FRUIT

The compositional changes associated with ripening are particularly pronounced in bananas. Because of shipment to distant markets they are picked green in a maturity grade designated as "$\frac{3}{4}$ full" or "heavy $\frac{3}{4}$ full". Bananas attain best quality by ripening green fruit at temperatures close to 20°C and under high humidity. Ethylene may be added to accelerate the process and to achieve greater uniformity than in ordinary air. The chemical changes that occur in the course of the respiratory climacteric are of the same qualitative nature whether ethylene is used or not. The most prominent tranformation has been found to take place in the carbohydrate fraction of both pulp and peel. In the mesocarp, the edible portion, total carbohydrate may decrease during ripening by about 3% of fresh weight. The starch content which varies in the green fruit from 20–30%, depending on variety, tends to drop to 1–2% in the edible ripe banana. At the same time sugars increase from a fraction of 1% to 14–15%. In the peel the changes are not as dramatic in absolute numbers, but highly significant in relative values. Barnell (1943) reported that starch and sugar disappearance could not acount for all the dry matter lost during ripening and that hemicelluloses contributed substantially to the change. To obtain an insight into banana ripening Young *et al.* (1975) examined a number of enzymes involved in carbohydrate metabolism in relation to the events depicted in Fig. 9.

The sharp decrease in starch content took place when the respiratory rise was well on the way towards the peak, suggesting that hydrolytic enzymes are either activated or synthesized *de novo* at the onset of ripening. Young *et al.* (1975) extracted the proteins from banana pulp at seven stages of the climacteric cycle and separated them by acrylamide gel electrophoresis. They

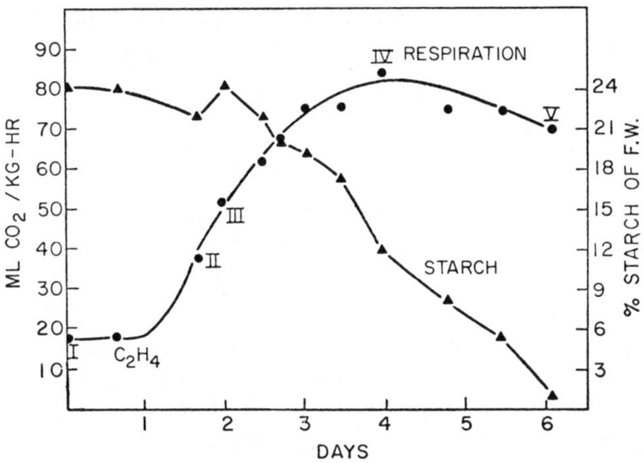

FIG. 9. Hydrolysis of starch in Valery bananas in relation to the respiratory climacteric. Reproduced from Young *et al.* (1975).

found two α-amylases, two β-amylases and three phosphorylases at all stages of ripeness. It was shown that several inhibitors prevented starch hydrolysis in extracts of pre-climacteric bananas. Assays were made also for glycolytic intermediates and enzymes. The levels of glucose-6-phosphate (G6P) and fructose-6-phosphate (F6P) were high in green fruit and changed little during ripening. Fructose-1,6-diphosphate (FDP), on the other hand, increased nine-fold in the early climacteric rise and 20-fold at the peak. A crossover plot was used to show relative changes in intermediates between pre-climacteric and climacteric peak samples. A large crossover value between these two stages was obtained for the reaction of F6P to FDP, indicating a regulatory site at phosphofructokinase (PFK). Substrate saturation studies of PFK showed a decrease in $[S]_{0\cdot5}$ from 5·38 mmol in the pre-climacteric stage to 1·72 mmol at the peak. This three-fold change was consistent with the respiratory increase. The identical V_{max} values obtained at both stages suggested that the form of the enzyme did not change during ripening. This contention was substantiated by acrylamide gel electrophoresis. Kinetic studies with substrate effects on PFK activity showing high V_{max} results led to the conclusion that the binding of the first substrate molecules inhibited the binding of subsequent molecules. This type of regulation, known as negative cooperativity, appeared to operate at different levels at both stages of the climacteric.

Solomos and Laties (1974) observed in the intact avocado fruit a pattern of changes in glycolytic intermediates which involved a rapid rise in FDP and a sharp decline in G6P, 3-PGA (phosphoglyceric acid) and PEP (phosphoenol-pyruvic acid). This pattern of changes as well as the increase in rate of respiration and of ethylene production were brought about by both cyanide and ethylene. The alternate electron pathway was implicated as operative in the avocado. However, Theologis and Laties (1978) concluded that avocado and banana slices did not utilize the alternate pathway at the climacteric peak.

C. RIPENING TRENDS IN A NON-CLIMACTERIC FRUIT

The compositional changes in the course of ripening in the pineapple are strikingly different from those in the banana. Dull (1971) reported a maximal value for starch in the pineapple of 0·1% of total dry matter. On a fresh weight basis this value would be closer to 0·01% as against 20–30% for unripe bananas. The major carbohydrate constitutents in the edible portion of the pineapple are glucose, fructose and sucrose. Sugars increase greatly during the last 40 days of maturation. Sucrose reaches a maximum value at ripeness, but reducing sugar content continues to climb into senescence. Alcohol insoluble materials dip to a minimum in the ripe fruit. Of the amino acid changes in ripening the most interesting and most spectacular phenomenon is the increase in methionine, the most widely accepted precursor for ethylene biosynthesis in higher plants. In spite of this increase the endogenous levels of ethylene in pineapple are rather low (Table VI) with no indication of any change during ripening.

The chemical changes associated with maturation of the pineapple are shown in Fig. 10. One can see in this figure four distinct stages in the development of the pineapple fruit as described by Gortner *et al.* (1967). Under Hawaiian conditions the developmental period from full bloom (anthesis) to senescence lasts approximately 110 days with the final two to three weeks characterized as the ripening phase. The most striking increases during ripening were found in reducing sugars, flesh non-protein nitrogen, carotenoids, flesh brix and shell brix (total soluble solids). Total titratable acidity reaches a peak during ripening. The increase in carotenoids is probably more apparent than real because pigments become unmasked rather than synthesized. Sharp decreases were recorded in chlorophyll content, flesh and shell pH and in protease activity. Comparing the compositional changes for the several stages of development, one should look upon ripening in the pineapple as the terminal period of maturation during which the fruit attains the most desirable quality. If the fruit is allowed to ripen on the plant it reaches a richer quality than pineapples harvested at the onset of ripening. In our opinion, ripening as a continuation of maturation is not as distinct a developmental phase in the non-climacteric pineapple as it is in the banana and in other climacteric fruits. Senescence is looked upon by Gortner *et al.* (1967) as the period following development in which biochemical ageing processes replace the "perfective" changes of ripening.

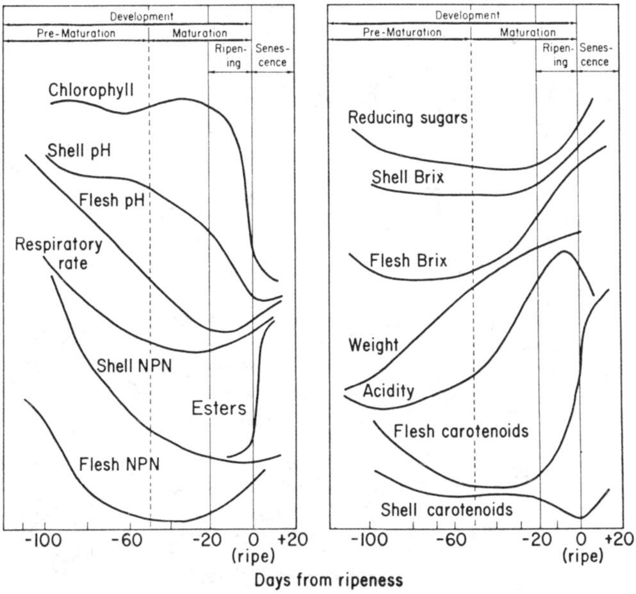

FIG. 10. Changes in physicochemical properties of pineapple fruit during the period from flowering to senescence. After Gortner *et al.* (1967). By permission.

D. PECTIC SUBSTANCES AND PECTOLYTIC ENZYMES

The decrease in fruit firmness is a general feature that accompanies ripening of both climacteric and non-climacteric fruits. Softening is brought about by changes in cell wall constituents among which pectic substances play a major role. Pilnik and Voragen (1970) refer to pectic substances as "polygalacturonides with non-uronide carbohydrates covalently bound to an unbranched chain of $1 \rightarrow 4$ linked α-galacturonic acid units". The COOH groups of the polygalacturonide chain may be partly esterified by CH_3 groups and partly or completely neutralized by one or more bases. The water insoluble component of pectic substances is called protopectin. Pectins are water soluble pectic substances with variable methyl ester content that can form gels with sugar and acid under suitable conditions. Pectic acid is composed of colloidal polygalacturonic acids which are essentially free of methyl groups. Changes from insoluble to soluble pectins accompanying ripening and the respiratory climacteric in the avocado were studied by Dolendo *et al.* (1966) and are shown in Fig. 11. Softening as indicated by pressure measurements went hand in hand with an increase in soluble pectins and a decrease in protopectin. The precise relationship of these events to the climacteric cycle could not be established since the fruit were on the rise at the start of the experiment. Pectin methyl esterase (PME) appeared to be active from the outset as indicated by changes in the degree of esterification.

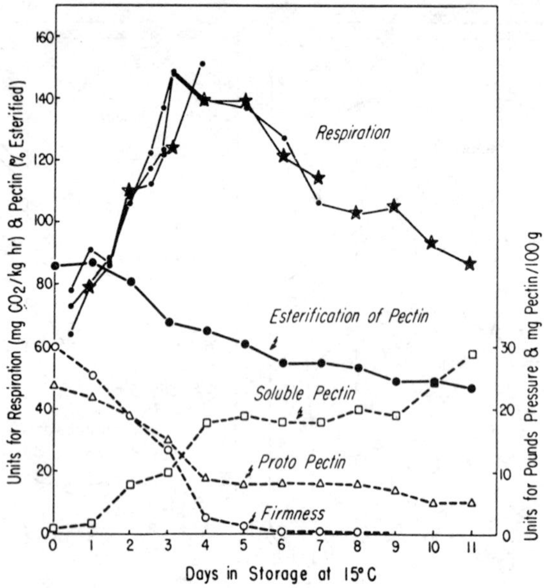

FIG. 11. Pectic changes in the avocado in relation to the climacteric pattern. Reproduced from Dolendo *et al.* (1966). By permission.

Determination of activities of key enzymes in the softening process was followed by Awad and Young (1979) in the Fuerte avocado in relation to the trends in respiration and ethylene production. These workers found great variability in the time of onset of the climacteric rise for individual fruit. They observed, however, that small plugs of tissue could be removed for analysis of enzymatic activities without affecting the pattern, the rate of respiration or the rate of ethylene production. Figure 12A shows respiration and ethylene production of one fruit from which very small samples were removed each day, while Fig. 12D presents the changes in the same parameters for an avocado that was not sampled until day 13. No significant difference was noted. Figure 12B illustrates cellulase activity of samples from Fig. 12A. Appreciable enzyme activity was present immediately after harvest. Cellulase increased with the onset of the respiratory rise and continued to increase until two days after the climacteric peak, reaching the extraordinary level of over 50,000 units per gram fresh weight. The activity of polygalacturonase (PG), the chain splitting enzyme, was not measurable until $7\frac{1}{2}$ days after picking (Fig. 12C) and then increased rapidly until the eleventh day when a value of 800 units per gram fresh weight was attained. Pectin methyl esterase (PME), also shown in Fig. 12C, was high at the time of picking, decreased at the start of the climacteric rise and came to a low level just as PG started to climb. It has been suggested that the PME might function to prepare the methylated pectin substrate for hydrolysis by PG. The decrease in PME activity could not be

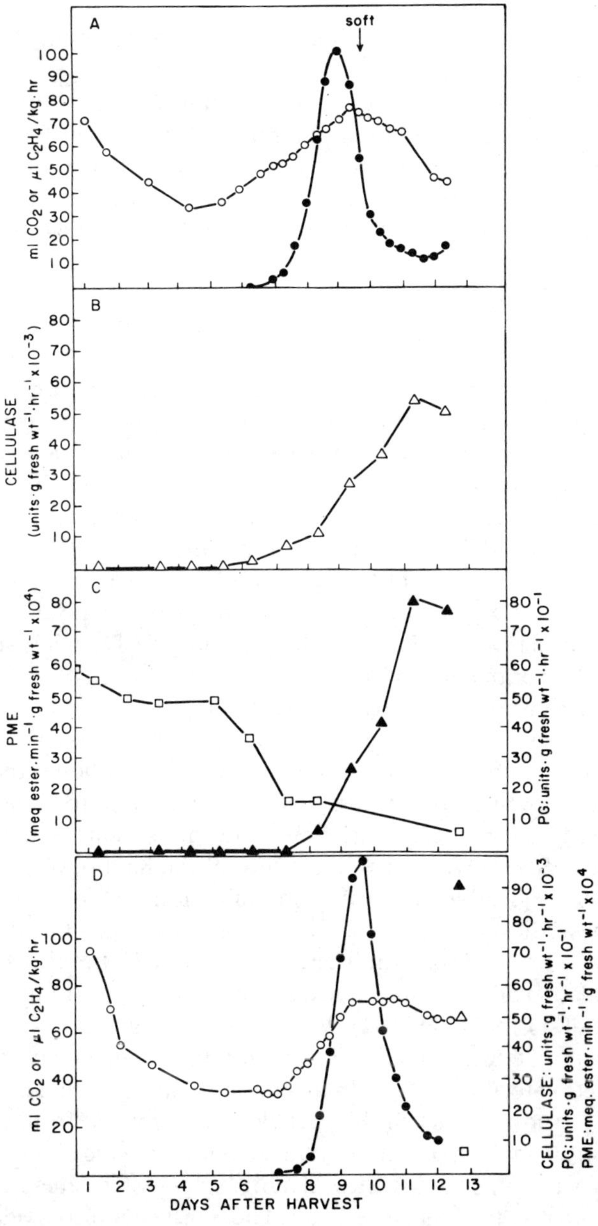

FIG. 12. Activities of cell wall enzymes of the avocado in relation to the respiratory climacteric. After Awad and Young (1979).

accounted for by the formation of an inhibitor as mixtures of ripe and unripe fruit gave the expected activity.

Activities of the three enzymes were determined on day 13 for the fruit of Fig. 12D to ascertain that the sampling procedure used in Fig. 12A did not affect the final activities of the enzymes. It appears that in the avocado it is cellulase activity rather than PG which is closely related to the climacteric rise, a behaviour that is opposite to that reported by Hobson (1964) for the tomato.

E. REGULATION OF RIPENING AT THE MITOCHONDRIAL LEVEL

Mitochondria were considered to play a significant role in the ripening process because they are the sites of respiratory enzymes and pathways and respiration is the central physiological process in detached fruits. We conducted extensive studies on the activities of mitochondria isolated from unripe and ripe avocados (for a more extensive summary and references see Biale and Young, 1971). The relevant parameters in these studies were respiratory control (RC), ADP/O ratios and thiaminepyrophosphate (TPP) as required co-factors. RC refers to the ratio of "state 3" to "state 4" oxidation. State 3 is the oxidation in the presence of ADP while state 4 is in its absence. ADP/O is an index of the degree of esterification of ADP to ATP per oxygen atom consumed in substrate oxidation.

It was observed with succinate as substrate that there was a rise in RC and in ADP/O with the climacteric rise reaching maximum values at the climacteric peak. The actual rates of oxygen uptake per unit mitochondrial protein nitrogen were not related to ripening during succinate oxidation. The response was different with other substrates of the Krebs cycle. In the case of malate as the oxidizable substrate, for mitochondria from pre-climacteric avocados the addition of ADP elicited a small response. TPP was required to obtain maximal state 3 oxidation and to achieve RC. In mitochondria from ripe fruit malate oxidation progressed at a high rate upon addition of ADP in the absence of TPP (Table VIII). When α-ketoglutarate was the oxidizable substrate, malonate was added to inactivate the succinic dehydrogenase and limit the reaction to a single step. In this case too, the requirement for TPP was much more pronounced when the particles were isolated from unripe rather than from ripe fruit. With α-ketoglutarate as substrate the ADP/O ratios were higher than with other substrates in both stages of ripeness and changed markedly from pre- to climacteric peak. The tightly coupled oxidations might be attributed to the predominance of substrate level phosphorylation with α-ketoglutarate compared to other acids in which ATP generation linked to a electron transport chain is more vulnerable to uncoupling.

The most striking results on the effects of ripening on mitochondrial processes were obtained with pyruvate as substrate (Table VIII). In the presence of all required co-factors the differences in oxidative activities were

TABLE VIII
Effect of thiaminepyrophosphate (TPP) on oxidative activity of avocado mitochondria (μl O_2 per mg mitochondrial nitrogen)

Substrate	Cofactor added	Unripe	Ripe
Malate	None	185	745
	TPP	500	1105
α-Ketoglutarate	None	110	400
	TPP	480	450
Pyruvate	None	68	740
	TPP	222	1140
μg TPP/mg mito. protein		25	67

All data except TPP content based on Lance *et al.* (1967).

large between particles of unripe and ripe fruit. The limitation in pre-climacteric fruit must be ascribed to the pyruvate carboxylase rather than the electron transport chain which was highly active with other TCA acids. Also, in the case of pyruvate, the TPP requirement was more pronounced in the pre-climacteric stage than during the climacteric peak. The difference in the levels of mitochondrial TPP suggests that either TPP is synthesized in mitochondria during ripening or that it is translocated from other sites of the cell to the mitochondria. We do not know whether TPP exerts any controlling effect in intact tissue.

F. PHOSPHATE ESTERIFICATION DURING RIPENING

The increase in oxidative phosphorylation during the respiratory climacteric in subcellular systems prompted the examination of the esterification capacity at the cellular level. Young and Biale (1967) incubated tissue discs from pre-climacteric and climacteric peak Fuerte avocado in phosphate buffer containing radioactive phosphorous. The results of this study are reported in Table IX. Ten times more ^{32}P was incorporated into climacteric peak than into pre-climacteric tissue. Of the amount incorporated 2·6% was esterified in the slices from unripe fruit and 10·8% into slices of ripe fruit. The rate of esterification into the various fractions was strikingly different for the two stages. The fact that during the pre-climacteric stage ADP showed the highest activity of the three adenylates examined, coupled with the high ADP/ATP ratio, is evidence against the theory that deficiency of phosphate acceptor is responsible for the low rates of respiration prior to the onset of the climacteric rise. The enhancement of the phosphorylative capacity with ripening sheds doubt also on the supposition that the burst of respiration during the climacteric is due to the action of an endogenous uncoupling agent. Both the mitochondrial and the tissue disc studies point to a tighter coupling of oxidative phosphorylation in preparations from ripe than from unripe fruit.

TABLE IX
Phosphate esterification in discs of avocado fruit
(P^{32} incorporation in cpm/gm F.W. × 10^{-3})

	Unripe	Ripe
Total esterified	83	4470
Sugar phosphates	37	3458
Total nucleotides	39	672
ATP	3·4	43
ADP	14·2	241
AMP	3·8	135
	μmol/gm F.W.	
ATP	89	140
ADP	207	124
ADP/ATP	2·3	0·88

Based on Young and Biale (1967).

G. PROTEIN AND NUCLEIC ACID METABOLISM

The studies on phosphate esterification by tissue slices and on oxidative phosphorylation by mitochondria clearly indicated that the machinery for energy generation is fully operative during ripening of avocado fruit. The question was raised as to what extent is the generated ATP used for biosynthesis of proteins and how does this activity vary throughout the climacteric cycle. To throw light on this problem Richmond and Biale (1966) followed the incorporation of labelled leucine and valine into proteins by tissue slices of Fuerte avocado. The results of this study are presented in Table X. Evidence that the label was incorporated into a protein fraction was obtained by proteolytic cleavage and acid hydrolysis, and by the use of puromycin, an inhibitor of protein synthesis.

Table X
Protein and nucleic acid metabolism in discs of avocado fruit in relation to ripening

	Preclim.	Early rise	Clim. peak
Leucine incorporation	110	180	10
Valine incorporation	26	120	33
P^{32} incorporation into heavy ribosomal RNA	5000	6200	130

Leucine and valine incorporation in μg/40 mg D.W./hour. P^{32} incorporation in cpm/15 discs/hour. Based on Richmond and Biale (1966, 1967).

Incorporation of label into proteins per unit of tissue weight, as well as the ratio of incorporation to uptake retained inside the permeability barrier, increased markedly from pre-climacteric to early rise and dropped drastically at the peak. Puromycin had no effect on oxygen uptake while it interfered with amino acid incorporation, suggesting that the respiratory rise was not linked to protein synthesis.

The changes in amino acid incorporation correlated with patterns of nucleic acid metabolism as shown in Table X. Tissue discs were incubated for one hour in sucrose-citrate solution, pH 6·0, containing labelled phosphate. At the end of the incubation period RNA was extracted in cold phenol and fractionated on a MAK (methylated-albumin-Kieselguhr) column. Absorbence at 260 nm and radioactivity were determined on 3 ml fractions. The pattern of ^{32}P incorporation showed marked differences in RNA synthesis along the climacteric rise. In the pre-climacteric stage and during the early rise ^{32}P appeared to be incorporated into the heavy ribosomal region. Peak activity coincided with a fraction which on comparative grounds was assumed to be DNA-like RNA, or messenger RNA. With the progress of the climacteric, when oxygen uptake reached 70–80% of the maximum, the incorporation of ^{32}P into the heavy ribosomal fraction was sharply reduced. Conceivably, senescence in the avocado fruit is brought about by failure to synthesize m-RNA.

Ribosomal RNA studies, similar to the m-RNA work on avocado, were carried out by Ku and Romani (1970) on the pear. They found the distribution of ribosomes among monomeric and polymeric forms reasonably constant until the fruit reached the climacteric peak, when a decline of polymers was observed. However, ribosomal turnover, as judged by incorporation of radioactive base, increased markedly during the climacteric rise but none could be detected at the peak. Brady and O'Connell (1976) noted increased protein synthesis as a result of treatment of bananas with ethylene rather than as a response to ripening. They obtained no evidence for the formation of a ripening specific protein. The complexity of ripening suggests that many proteins are involved and interact in the ripening reactions.

V. Concluding Remarks

As the central theme of this chapter we selected a comparative analysis between the two respiratory patterns in order to stress gaps in our knowledge while reporting the advancements in the understanding of the relationship between respiration and ripening. The last section dealing with metabolic regulation of ripening was particularly notable in deficiency of information on non-climacteric fruits. The major events occurring in climacteric fruits are depicted in Fig. 13 and may serve as a signpost for investigations to be pursued with non-climacteric fruits. By updating the lists of fruits of the two types and including certain tropical, metabolically active species, it is possible now to select material from the non-climacteric class that would compare favourably with some of the fruits of the climacteric class. However, there are other problems involved. The investigator with non-climacteric fruits is still faced with the lack of a frame of reference of the kind that the respiratory climacteric has served for decades of research. He may have to look for special parameters that are subject to a decisive change while no critical changes occur in the respiratory drift. Exploration into unique volatiles or into substances that impart aroma into ripe fruit might turn out to be useful, and it is feasible now that modern methodology for the separation of compounds present in minute quantities has become available.

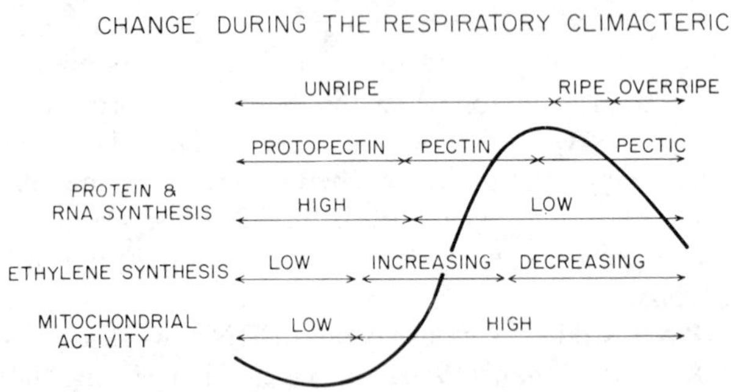

FIG. 13. Ripening events as related to the climacteric pattern.

In presenting the ethylene factor as a distinguishing feature between the two classes of fruits the question is raised as to what significance, if any, does this hydrocarbon play in the ripening of non-climacteric fruits. The very fact that it was found in small, non-changing concentrations in tissues of citrus and pineapple and that autocatalytic production is absent is in itself a strong argument for a qualitative difference in the ripening mechanisms of the two classes. However, no firm conclusion can be drawn on this point prior to further investigation into ethylene production, internal ethylene levels and exogenous ethylene effects on more fruits of the non-climacteric class.

In climacteric fruits, also, there is a noticeable gap in the link between the effects of ethylene on respiration and ripening and the metabolic events that take place during ripening. This situation is doubtless caused by the state of knowledge on the mechanism of ethylene action. In recent years it was thought that ethylene might invoke an alternate, cyanide insensitive pathway of electron transport, but this idea appears to be in doubt and subject to further clarification (Day *et al.*, 1980). The role of ethylene as an exclusive factor in the ripening of climacteric fruits has become the subject of considerable discussion. Other phytohormones, notably auxins, have been implicated in exerting an influence on the ripening process either by acting antagonistically or in concert with ethylene. Ideas on catabolism of auxins as a prerequisite for the action of ethylene in ripening were advanced by Frenkel *et al.* (1975). It is conceivable that the crucial problem of changing sensitivity of fruits to ethylene is tied up with the metabolism of auxins and other phytohormones. This subject was not touched in this chapter as it was competently covered recently in a review by McGlasson *et al.* (1978). There is certainly a need for more insight into the role of phytohormones in the ripening events of both climacteric and non-climacteric fruits.

REFERENCES

Adams, D. O. and Yang, S. F. (1979). *Proc. Natl. Acad. Sci. USA* **76**, 170–174.
Aharoni, Y. (1968). *Plant Physiol.* **43**, 99–102.
Akamine, E. K. and Goo, T. (1973). *J. Amer. Soc. Hort. Sci.* **98**, 381–383.
Akamine, E. K. and Goo, T. (1978). *J. Amer. Soc. Hort. Sci.* **103**, 308–310.
Akamine, E. K. and Goo, T. (1979a). *J. Amer. Soc. Hort. Sci.* **104**, 632–635.
Akamine, E. K. and Goo, T. (1979b). *Hort. Sci.* **14**, 138–139.
Awad, M. and Young, R. E. (1979). *Plant Physiol.* **64**, 306–308.
Bain, J. M. (1958). *Aust. J. Bot.* **6**, 1–24.
Barnell, H. R. (1943). *Ann. Botany (London)* **7**, 297–323.
Biale, J. B. (1960). *In* "Encyclopedia of Plant Physiology" (W. Ruhland, Ed.), Vol. 12 (2), pp. 536–592. Springer Verlag, Berlin.
Biale, J. B. (1961). *In* "The Orange" (W. B. Sinclair, Ed.), pp. 96–130. University of California Press, Berkeley.
Biale, J. B. and Barcus, D. E. (1970). *Tropical Sci.* **12**, 93–104.
Biale, J. B. and Young, R. E. (1947). *Amer. J. Bot.* **34**, 301–309.
Biale, J. B. and Young, R. E. (1962). *Endeavour* **21**, 164–174.
Biale, J. B. and Young, R. E. (1971). *In* "The Biochemistry of Fruits and Their Products" (A. C. Hulme, Ed.), Vol. 2, pp. 1–63. Academic Press, London and New York.
Biale, J. B., Young, R. E. and Olmstead, A. J. (1954). *Plant Physiol.* **29**, 168–174.
Blanpied, G. D. (1972). *Plant Physiol.* **49**, 627–630.
Bose, A. N., Lodh, S. B. and De, S. (1962), *Food Sci. Technol.* **2**, 117–126.
Brady, C. J. and O'Connell, P. B. H. (1976). *Aust. J. Plant Physiol.* **3**, 301–310.
Burg, S. P. (1962). *Ann. Review Plant Physiol.* **13**, 265–302.
Burg, S. P. and Burg, E. A. (1962a). *Plant Physiol.* **37**, 179–189.
Burg, S. P. and Burg, E. A. (1962b). *Nature (Lond.)* **194**, 398–399.
Burg, S. P. and Burg, E. A. (1965). *Science* **148**, 1190–1196.
Burg, S. P. and Burg, E. A. (1967). *Plant Physiol.* **42**, 144–152.
Chandler, W. H. (1958). "Evergreen Orchards". Lea and Febiger, Philadelphia.
Clendening, K. A. (1942). *Canad. J. Res.*, **C20**, 197–203.
Coombe, B. G. (1976). *Ann. Review Plant Physiol.* **27**, 207–228.
Day, D. A., Arron, G. P. and Laties, G. G. (1980). *In* "The Biochemistry of Plants" (D. D. Davies, Ed.), Vol. 2, pp. 197–241. Academic Press, London and New York.
Dolendo, A. L., Luh, B. S. and Pratt, H. K. (1966). *J. Fd. Sci.* **31**, 332–336.
Dull, G. G. (1971). *In* "The Biochemistry of Fruits and Their Products" (A. C. Hulme, Ed.), Vol. 2, pp. 303–324. Academic Press, London and New York.
Dull, G. G., Young, R. E. and Biale, J. B. (1967). *Physiologia Plant.* **20**, 1059–1065.
Eaks, I. L. (1970). *Plant Physiol.* **45**, 334–338.

Frenkel, C., Dyck, R. and Haard, N. F. (1975). *In* "Postharvest Biology and Handling of Fruits and Vegetables" (N. F. Haard and D. K. Salunkhe, Eds), pp. 19–34. Avi Pub. Co., Westport.
Gane, R. (1936). *New Phytologist* **34**, 383–402.
Gazit, S. and Blumenfeld, A. (1970). *J. Amer. Soc. Hort. Sci.* **95**, 229–231.
Geisler, G. and Radler, F. (1963). *Ber. dt. bot. Gesell* **76**, 112–119.
Gortner, W. A., Dull, G. G. and Kraus, B. H. (1967). *Hort. Sci.* **2**, 141–144.
Haller, M. H., Rose, D. H., Lutz, J. M. and Harding, P. L. (1945). *J. Agricult. Res.* **71**, 327–359.
Hansen, E. (1942). *Bot. Gaz.* **103**, 543–558.
Hansen, E. (1966). *Ann. Review Plant Physiol.* **17**, 459–480.
Herner, R. C. and Sink, K. C. (1973). *Plant Physiol.* **52**, 38–42.
Hobson, G. E. (1964). *Biochem. J.* **92**, 324–332.
Hobson, G. E. (1979). *Current Adv. in Plant Science* **37**, 1–11.
Hulme, A. C. (1970, 1971). "The Biochemistry of Fruits and Their Products", Vols 1 and 2. Academic Press, London and New York.
Iwata, T., Nakagawa, K. and Ogata, K. (1969). *J. Jap. Soc. Hort. Sci.* **38**, 84–91.
Kidd, F. and West, C. (1922). *Food Inv. Bd. Rept.* for 1921, 14–16.
Kidd, F. and West, C. (1930). *Proc. Roy. Soc. (London)* **B106**, 93–109.
Kosiyachinda, S. and Young, R. E. (1975). *Plant and Cell Physiol.* **16**, 595–602.
Ku, L. L. and Romani, R. J. (1970). *Plant Physiol.* **45**, 401–407.
Lance, C., Hobson, G. E., Young, R. E. and Biale, J. B. (1965). *Plant Physiol.* **40**, 1116–1123.
Lyons, J. M. and Pratt, H. K. (1964). *Proc. Amer. Soc. Hort. Sci.* **84**, 491–500.
Lyons, J. M., McGlasson, W. B. and Pratt, H. K. (1962). *Plant Physiol.* **37**, 31–36.
Mapson, L. W. and Robinson, J. E. (1966). *J. Food Technol.* **1**, 215–225.
Marei, N. and Crane, J. C. (1971). *Plant Physiol.* **48**, 249–254.
Maxie, E. C., Catlin, P. B. and Hartmann, H. T. (1960). *Proc. Amer. Soc. Hort. Sci.* **75**, 275–291.
McGlasson, W. B. (1970). *In* "The Biochemistry of Fruits and Their Products" (A. C. Hulme, Ed.), Vol. 1, pp. 475–519. Academic Press, London and New York.
McGlasson, W. B., Dostal, H. C. and Tigchelaar, E. C. (1975). *Plant Physiol.* **55**, 218–222.
McGlasson, W. B., Wade, N. L. and Adato, I. (1978). *In* "Phytohormones and Related Compounds: A Comprehensive Treatise" (D. S. Letham, P. B. Goodwin and T. J. V. Higgins, Eds), Vol. 2, pp. 447–493. Elsevier, Amsterdam, Oxford, New York.
McMurchie, E. J., McGlasson, W. B. and Eaks, I. L. (1972). *Nature (Lond.)* **237**, 235–236.
Mizrahi, Y., Dostal, H. C., McGlasson, W. B. and Cherry, J. H. (1975). *Plant Physiol.* **56**, 544–546.
Peacock, B. C. (1972). *Queensland J. Agric. Anim. Sci.* **29**, 137–145.
Peynaud, E. and Riberau-Gayon, P. (1971). *In* "The Biochemistry of Fruits and Their Products" (A. C. Hulme, Ed.), Vol. 2, pp. 171–205. Academic Press, London and New York.
Pilnik, W. and Voragen, A. G. J. (1970). *In* "The Biochemsitry of Fruits and Their Products" (A. C. Hulme, Ed.), Vol. 1, pp. 53–87. Academic Press, London and New York.
Pratt, H. K. (1975). *In* "Facteurs et Régulation de la Maturation des Fruits", No. 238, pp. 153–160. CNRS, Paris.
Pratt, H. K. and Goeschl, J. D. (1968). *In* "Biochemistry and Physiology of Plant Growth Substances" (F. Wightmann and G. Setterfield, Eds), pp. 1295–1302. Runge, Ottawa.
Pratt, H. K. and Goeschl, J. D. (1969). *Ann. Review Plant Physiol.* **20**, 541–584.

Pratt, H. K. and Reid, M. S. (1974). *J. Sci. Food Agric.* **25**, 747–757.
Pratt, H. K., Goeschl, J. D. and Martin, F. W. (1977). *J. Amer. Soc. Hort. Sci.* **102**, 203–210.
Quazi, M. H. and Freebairn, H. T. (1970). *Bot. Gaz.* **131**, 5–14.
Reid, M. S. (1975). *In* "Facteurs et Régulation de la Maturation des Fruits", No. 238, pp. 177–182. CNRS, Paris.
Reid, M. S., Rhodes, M. J. C. and Hulme, A. C. (1973). *J. Sci. Food Agric.* **24**, 971–979.
Rhodes, M. J. C. (1970). *In* "The Biochemistry of Fruits and Their Products" (A. C. Hulme, Ed.), Vol. 1, pp. 521–533. Academic Press, London and New York.
Rhodes, M. J. C. and Reid, M. S. (1975). *In* "Facteurs et Régulation de la Maturation des Fruits", No. 238, pp. 189–192. CNRS, Paris.
Richmond, A. and Biale, J. B. (1966). *Plant Physiol.* **41**, 1247–1253.
Richmond, A. and Biale, J. B. (1967). *Biochim. Biophys. Acta* **138**, 625–627.
Sacher, J. A. (1973). *Ann. Review Plant Physiol.* **24**, 197–224.
Sfakiotakis, E. M. and Dilley, D. R. (1973). *J. Amer. Soc. Hort. Sci.* **98**, 504–508.
Smock, R. M. (1972). *J. Amer. Soc. Hort. Sci.* **97**, 509–511.
Solomos, T. and Laties, G. G. (1974). *Plant Physiol.* **54**, 506–511.
Theologis, A. and Laties, G. G. (1978). *Plant Physiol.* **62**, 249–255.
Tigchelaar, E. C., McGlasson, W. B. and Buescher, R. W. (1978). *Hort. Sci.* **13**, 508–513.
Trout, S. A., Huelin, F. E. and Tindale, G. B. (1960). *CSIRO (Aust) Div. Food Preserv. Trans. Tech. Paper* No. **14**, 1–11.
Uota, S., Young, R. E. and Biale, J. B. (1955). Stawberry respiration (unpublished).
Wang, C. Y. and Mellenthin, W. M. (1972). *Plant Physiol.* **50**, 311–312.
Yang, S. F. (1981). This volume, Chapter 4.
Young, R. E. and Biale, J. B. (1967). *Plant Physiol.* **42**, 1357–1362.
Young, R. E., Romani, R. J. and Biale, J. B. (1962). *Plant Physiol.* **37**, 416–422.
Young, R. E., Salminen, S. and Sornsrivichai, P. (1975). *In* "Facteurs et Régulation de la Maturation des Fruits", No. 238, 271–280. CNRS, Paris.

CHAPTER 2

Low Temperature Sweetening of Higher Plants

T. AP REES, W. L. DIXON,* C. J. POLLOCK† AND F. FRANKS

Botany School, University of Cambridge, Downing Street, Cambridge, England

I. Introduction 41
II. Occurrence of Low Temperature Sweetening 42
III. Significance of Low Temperature Sweetening 44
IV. Metabolic Pathways of Low Temperature Sweetening 46
A. Sugar Synthesis 46
B. Source of Substrates for Sugar Synthesis 47
C. Starch–Sugar Balance 48
D. Intracellular Location of Pathways of Sweetening 49
V. Possible Mechanisms of Low Temperature Sweetening 49
A. Compartmentation 50
B. Coarse Control 51
C. Fine Control 52
VI. Conclusions 58
References 60

I. Introduction

It has been known for a long time that many higher plants sweeten in the autumn and early winter. Pliny the Elder (first century A.D.) commented that "turnips are believed to grow sweeter and bigger in cold weather" (libr. XVIII) and subsequently wrote "with any kind of cabbage hoarfrosts contribute a great deal to their sweetness" (libr. XIX). More specific investigations have shown that lowering the temperature below 10°C can lead to appreciable accumulation of sugars in a wide range of plants. This low temperature sweetening has attracted attention in three particular respects. First is the possibility that it contributes to frost hardiness. Second is the use of low

* Present address: Smith Kline and French Laboratories, Welwyn Garden City, Herts, England.

† Present address: Welsh Plant Breeding Station, Plas Gogerddan, Aberystwyth, Dyfed, Wales.

temperature sweetening, as a well established physiological process, for studies of the effects of cold on metabolism. Third is the economic importance of the cold-induced sweetening of one of the world's most important foods, the potato tuber. Sweetening of potatoes makes them unpalatable and has a drastic effect upon their processing quality (Burton, 1969). During the manufacture of chips and crisps reducing sugars in sweetened potatoes react with amino acids to cause considerable browning (Gray and Hughes, 1978). The economic importance of sweetening of potatoes is emphasized by the fact that the advantages of storage at low temperature, ease of maintenance of high humidity, and prevention of sprouting, senescent sweetening and the development of storage rots, are offset by the problems caused by accumulation of sugar. In this chapter we shall consider briefly the occurrence and significance of low temperature sweetening, discuss in more detail the pathways involved, and then concentrate on our understanding of the mechanisms responsible.

II. Occurrence of Low Temperature Sweetening

Low temperature sweetening is a widespread but not universal feature of higher plants. It occurs in leaves, shoots, and roots; in both Monocotyledons and Dicotyledons; in woody and in herbaceous plants, and in winter annuals (Table I). Low temperature sweetening is conceivably more widespread than is recognized at present as it could be difficult to detect in plants that normally have a high content of sugar. The sugars accumulated vary with the plant (Table I), but generally come from a small group dominated by sucrose, its components and its substitution products such as raffinose (Levitt, 1972). In certain instances sugar alcohols accumulate (Seybold, 1969). The accumulation of hexoses is probably due, in most instances, to hydrolysis of sucrose. The general lack of appropriate phosphatases makes it unlikely that hexoses arise directly from their corresponding hexose phosphates. The kinetics of the accumulation of sucrose and hexoses in potatoes (Fig. 1), and the relationship between hexose accumulation and invertase activity (Pressey and Shaw, 1966) also suggest that hexoses come from sucrose. If the above view is correct then it appears that the sugars that accumulate in the cold are those that the plant uses as storage compounds under normal conditions (ap Rees, 1977). As raffinose (Lehle and Tanner, 1973) and stachyose (Tanner and Kandler, 1966) are derived from sucrose, it is thus likely that the primary product of low temperature sweetening is sucrose.

The amounts of sugars accumulated in response to cold vary widely with the plant and the conditions, but are often relatively small (Table I). Within the same species there can be considerable variation in the extent of sweetening: this is illustrated by the behaviour of potatoes (Table II). Even within a single variety of potato the degree of sweetening varies with the conditions of

TABLE I

Occurrence of low temperature sweetening in higher plants

Plant	Part	Increase in sugars (as % initial content)	Days in cold	Major sugars accumulated	Reference
Hedera helix	Leaves	440	30	Sucrose	Steponkus and Lanphear (1968)
Pinus strobus	Leaves	130	90	Raffinose	Parker (1959)
Triticum aestivum	Shoot	650	40	Raffinose, sucrose, glucose, fructose	Green and Ratzlaff (1975)
Citrus sinensis	Shoot	900	21	—	Yelenosky and Guy (1977)
Cornus stolonifera	Twig	36	100	Raffinose, sucrose	Li *et al.* (1965)
Robinia pseudo-acacia	Bark	100	30	—	Sakai and Yoshida (1968)
Brassica oleracea	Stem	100	30	Sucrose	Carolus (1930)
Citrus paradisi	Root	75	30	Sucrose	Young (1969)
Solanum tuberosum	Tissue culture	240	3	Sucrose	Pollock and ap Rees (1975c)

TABLE II
Low temperature sweetening in potato tubers of different varieties

Variety	Sucrose	Reducing sugar	Total
	Sugar accumulation (mg/g initial fresh wt after 60 days at 2°C)		
Climax	1·2	22·4	23·6
Libertas	10·6	12·3	22·9
Record	7·1	14·3	21·4
Prudal	9·1	11·2	20·3
Industrie	5·1	11·4	16·5
Voran	4·5	10·6	15·1
Nordeling	8·3	5·1	13·4
Maritta	7·1	4·0	11·1

Modified from van Vliet and Schriemer (1960).

growth. Burton (1969) found that sugar accumulation in the variety Record varied from 1·35% to 2·29% of the fresh weight, and that the contribution of reducing sugars varied from 40% to 84% of the total sugar accumulated. The extent and timing of sweetening is illustrated in Fig. 1 by data from potatoes.

III. SIGNIFICANCE OF LOW TEMPERATURE SWEETENING

There are many instances in which a close correlation has been demonstrated between low temperature sweetening of plants and their frost hardiness. In some species, such as the *vulgare* wheats, mulberry and black locust, the correlation is very close (Levitt, 1972). In others, such as *Hedera helix* (Steponkus and Lanphear, 1968), sugar content and frost resistance both increase during hardening but there is not a close parallelism between the two increases. Further, there are also many plants in which no correlation between sweetening and hardiness is found. There are examples of sweetening that are not accompanied by frost resistance. For instance, potatoes sweeten but are not resistant: sugar cane contains large quantities of sugar but is not even resistant to chilling, let alone to freezing. Conversely, there are also examples of frost hardiness that are not accompanied by sweetening. In *Juniperus chinenesis* the development of frost resistance is actually accompanied by a fall in sugar content (Pellet and White, 1964).

The above type of observation has led to various views on the relationship between low temperature sweetening and frost hardiness (Levitt, 1972; Heber and Santarius, 1973). However, no clear picture has emerged, or indeed could be expected, from the study of the simple coincidence, or lack of it, between sweetening and hardiness. Lack of correlation between the two does not prove

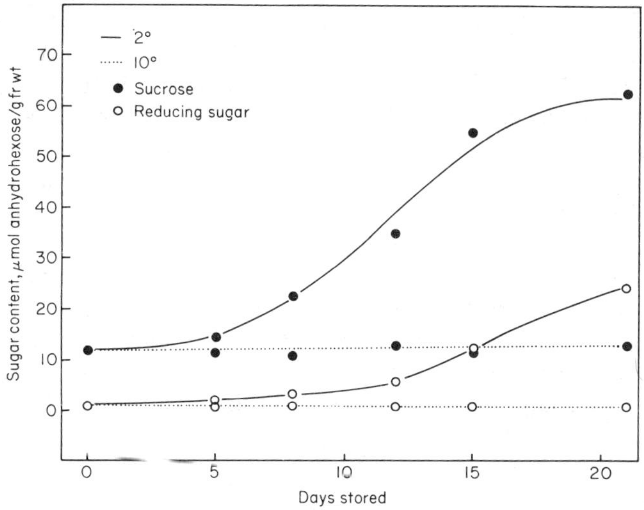

FIG. 1. Sugar content of potato tubers (Record) during storage at 2°C and at 10°C. Each point is the mean of estimates from eight different tubers (Pollock and ap Rees, 1975a).

that sweetening does not contribute to hardiness. First, measurements of the total content of sugar do not take into account compartmentation of cells and tissues. Thus significant changes of sugar in a particular compartment could be overshadowed by gross changes in the plant or organ as a whole. Secondly, not all plants need behave in the same way as the contribution of sweetening to frost resistance could vary with the plant. Finally, the observation that plants with a high sugar content are not necessarily frost resistant could mean that freezing is not the only hazard faced by plants when the temperature is lowered. The existence of chilling sensitive plants, which suffer irreparable damage at temperatures well above freezing, should emphasize this point.

In considering the significance of low temperature sweetening it is vital to appreciate that the existence of a close correlation between sweetening and hardiness does not demonstrate any causal connexion between the two. It is entirely possible that sweetening is an inevitable response of the metabolism of certain plants to low temperatures, and that such a response has nothing to do with hardiness. Two types of evidence have been used to counter the above view. First, there is evidence that in some plants feeding sugars can actually increase their resistance to cold (Tumanov and Trunova, 1963; Sakai and Yoshida, 1968): such evidence is still fairly sparse. Secondly, the sugars that accumulate during low temperature sweetening have been shown to act as cryoprotectants *in vitro*. Glucose slightly, sucrose appreciably, and raffinose markedly protect the mechanism of cyclic photophosphorylation from the effects of cold if the sugars are present when preparations of unwashed thylakoids are frozen *in vitro* (Heber and Santarius, 1971). Indications that such cryoprotection operates *in vivo* are provided by evidence that sucrose and raffinose accumulate in the chloroplasts of cabbage leaves during hardening (Santarius and Milde, 1977). However, it is not clear how such accumulation could have occurred *in vivo* as sucrose is made in the cytoplasm and does not readily cross the chloroplast envelope (Walker, 1976). Very recently Lineberger and Steponkus (1980) have provided evidence that the above *in vitro* cryoprotective effects of sugars are due to alterations in the freezing environment rather than to specific solute–membrane interactions. Direct proof that sweetening contributes to frost hardiness is still lacking. However the balance of the rather sparse evidence available favours the views that low temperature sweetening contributes to hardiness in some plants, that it is not the only mechanism of frost resistance and is not in itself necessarily a guarantee of resistance, and that its contribution to resistance varies with the plant.

IV. Metabolic Pathways of Low Temperature Sweetening

One of the major limitations to our understanding of low temperature sweetening is that we do not have a complete knowledge of the metabolic pathways involved. It is convenient to consider these pathways under the following four headings.

A. SUGAR SYNTHESIS

We have already stressed the importance of sucrose synthesis in low temperature sweetening (see section II). This synthesis could occur via either or both of the following routes.

1. Sucrose synthetase, followed by removal of sucrose to a different compartment.

   UDPG + Fructose → Sucrose + UDP

   $K = 5$ (pH 7·4, 37°C)

2. Sucrose phosphate synthetase, followed by sucrose phosphate phosphatase.

   UDPG + Fructose 6-phosphate → Sucrose phosphate + UDP

   $K = 3250$ (pH 7·5, 38°C)

   Sucrose phosphate + H_2O → Sucrose + P_i

The available evidence favours the second route overwhelmingly. It is to be preferred on thermodynamic grounds. In addition there is no direct evidence that sucrose synthetase mediates sucrose synthesis in any plant tissue, but there is direct evidence that sucrose phosphate synthetase catalyses cold-induced sweetening in potatoes. Pressey's (1970) data suggest that cold-induced sweetening shows a closer correlation with the maximum catalytic activity of sucrose phosphate synthetase than that of sucrose synthetase. Comparison of the rate of sugar accumulation in cold-stored potatoes with the maximum catalytic activities of the two enzymes shows that the activity of sucrose synthetase is too low to account for the observed rate of sugar accumulation (Table III). The activity of sucrose phosphate synthetase is well in excess of the rate of sugar accumulation. Thus we suggest that the latter enzyme is responsible for sucrose synthesis during low temperature sweetening.

TABLE III
Comparison of rate of sucrose accumulation and maximum catalytic activities of enzymes of sucrose metabolism during storage of potatoes (Record) at 2°C for 15 days

Measurement	Rate (mg sucrose day^{-1} g fresh wt^{-1})
Sucrose accumulation	1·0
Sucrose synthetase	0·8
Sucrose phosphate synthetase	15·0

Data from Pollock and ap Rees (1975a).

The conversion of sucrose to hexose during low temperature sweetening is probably mediated by an acid invertase. In cold-stored potatoes fructose and glucose accumulate in comparable amounts (Schwimmer *et al.*, 1954; Pressey and Shaw, 1966; Pollock and ap Rees, 1975a), and their accumulation closely follows changes in extractable invertase (Pressey and Shaw, 1966). As far as we know there have been no studies of the pathways of cold-induced synthesis of raffinose and stachyose but there seems to be no reason why such pathways should differ from those normally used by plants and illustrated below.

UDP-galactose + *myo*-Inositol → Galactinol + UDP
Sucrose + Galactinol → Raffinose + *myo*-Inositol
Raffinose + Galactinol → Stachyose + *myo*-Inositol

We do not know the pathways of sugar alcohol synthesis in plants as a whole, let alone during low temperature sweetening. It is likely that they are formed by reduction of hexose phosphates (Bieleski and Redgwell, 1977).

B. SOURCE OF SUBSTRATES FOR SUGAR SYNTHESIS

The arguments in the preceding sections suggest hexose 6-phosphates as the immediate source of all the sugars that accumulate in the cold. Thus the question of the source of carbon for low temperature sweetening resolves itself into the production of hexose phosphates. Starch is the classically accepted source of the sugars that accumulate in the cold and it seems clear that this is so in many instances. The most convincing demonstration of this is Isherwood's (1973) very complete analysis of potatoes. This failed to reveal any source of carbon for sugar synthesis other than starch, and showed that, when allowance is made for the small loss of CO_2 in respiration, the fall in starch is equivalent to the rise in sugars. The frequent coincidence of sweetening with a fall in starch (Levitt, 1956) strongly suggests that starch is generally the major, if not sole, source of carbon for low temperature sweetening.

The pathway of starch breakdown during low temperature sweetening is not known. The widespread distribution of α-glucan phosphorylase, α-amylase, β-amylase and maltase (ap Rees, 1974) indicates that starch breakdown could be phosphorolytic, hydrolytic, or both. It has been assumed that starch breakdown in cold-stored potatoes is phosphorolytic (Isherwood, 1973). The evidence for this view is not conclusive. No one has shown that the hydrolytic enzymes are not involved, and no one has shown that α-glucan phosphorylase from potatoes is able to initiate starch breakdown by attacking intact starch grains. Hydrolytic breakdown of starch would provide a direct route for hexose formation that did not involve synthesis of sucrose. Distinction between the possible pathways will be difficult. Perhaps the most useful approach to the question would be to compare the rate of starch breakdown *in vivo* with properly authenticated measurements of the maximum catalytic activities of the different enzymes. This might be followed by investigation of the location of the enzymes, and by a study of the products of starch breakdown *in vivo*.

A close association between sugar accumulation and a net breakdown of starch does not appear to be a necessary feature of cold-induced sweetening. For example, in shoots of ivy (Steponkus and Lanphear, 1968) and of orange (Yelenosky and Guy, 1977), sugar accumulation could not be accounted for by the net changes in starch. In these tissues hexose phosphates, the immediate products of photosynthesis, may be the source of carbon for sugar synthesis in the cold.

C. STARCH–SUGAR BALANCE

The close relationship between starch and sugar when potatoes are cooled from 10°C to 2°C and then, after an interval, warmed from 2°C to 10°C has given rise to the concept of a starch–sugar balance in which the overall change between the two compounds is seen as being reversible (Isherwood, 1973; Burton, 1978) and the steady state levels of starch and sugar as representing an equilibrium (Isherwood, 1976). This approach to the study of sweetening may turn out to be misleading. Starch and sucrose are not likely to be joined by a reversible pathway. All the available evidence suggests that sucrose is formed from starch by an irreversible pathway, and that starch is formed from sucrose by a separate but likewise irreversible route. Thus one may question whether it is wise to speak of a balance or an equilibrium between starch and sucrose as there is no direct way in which accumulation of one can influence the synthesis of the other. The observation that net breakdown of starch is not always associated with sweetening argues against the view that sweetening is due to an effect of temperature on a balance between the two compounds. Indeed, the very close relationship between starch and sucrose in stored potatoes may be due to the fact that starch is really the only possible source of carbon for sugar synthesis in the cold. It is the only major storage polysaccharide, there is no evidence of breakdown of structural polysaccharides, and the tubers are not gluconeogenic (Clegg and Whittingham, 1970).

D. Intracellular Location of Pathways of Sweetening

In order to understand the mechanism of low temperature sweetening we need to know the intracellular location of the pathways involved as temperature may affect movement of metabolites between compartments. Starch is present in plastids, thus at least the initial stages of its breakdown must also occur in plastids. In pea leaves starch is broken down to triose phosphates and 3-phosphoglycerate in chloroplasts and these compounds are then exported to the cytoplasm (Stitt and ap Rees, 1980). We do not know what happens when amyloplast starch is broken down. There have been no convincing reports of the isolation of intact amyloplasts and we do not know which of the products of starch breakdown are exported from the amyloplasts. It is almost certain that cold-induced synthesis of sucrose from starch involves movement of the products of starch breakdown from the amyloplast to the soluble phase of the cytoplasm. All the available evidence strongly suggests that sucrose synthesis occurs in the cytoplasm (Walker, 1976; Nishimura and Beevers, 1979).

V. Possible Mechanisms of Low Temperature Sweetening

This section deals almost entirely with potato tubers as there is very little information on the mechanism of sweetening from any other species. In general terms low temperature sweetening must be due to a redirection of metabolism caused by a differential effect of lowered temperature on the processes that control the formation of sucrose from starch. We consider these control mechanisms under the headings of compartmentation, coarse control, and fine control.

A. COMPARTMENTATION

It is likely that regulation of starch breakdown and of sucrose synthesis is to some extent achieved by compartmentation. It is thus possible that low temperature sweetening is, at least in part, due to effects of cooling on such compartmentation. Lowering the temperature causes lipids to undergo a transition from a liquid crystal state to that of a crystalline gel. The temperature at which this transition occurs is determined by the degree and type of unsaturation of the hydrocarbon chains of any particular lipid, and by the type of lipid (Chapman, 1975). It has been suggested that, for higher plants, cooling below 10°C can cause membrane lipids to undergo a change in phase and that this can affect the activities of membrane-bound enzymes and the movement of substrates and ions through membranes in ways that lead to a redirection of metabolism (Raison, 1973). This has led to the view that the critical temperature, below which plants fail to grow and eventually die, correlates with a temperature-induced change in both the structure and function of cellular membranes (Raison and Chapman, 1976). Whatever the general value of this hypothesis, there is, as yet, no evidence that it is important in low temperature sweetening. First there are no reports that phase changes occur in membranes during sweetening. Secondly, as far as we are aware, there has been no specific proof of the above hypothesis. No one has demonstrated that cooling a plant causes a phase change in a specific membrane, and then shown that this phase change results in alteration of the activity of an enzyme and that such an alteration is responsible for a change in flux in a metabolic pathway.

Isherwood (1976) has suggested a working hypothesis to explain low temperature sweetening in potatoes: this hypothesis is based upon a postulated effect of temperature on movement of metabolites across the amyloplast envelope. He has argued that

> the steady state levels of starch and sugars at a variety of temperatures represent equilibrium states in which there is a balance between influx and efflux of certain intermediates in the pathways linking these carbohydrates to and from the amyloplasts. This balance can be influenced by raising or lowering the temperature.

He envisaged the involvement of an electron transport system in the movement of metabolites across the amyloplast envelope, and suggested that such transport represented a balance between influx activated by electron transport and passive efflux.

The above hypothesis, or part of it, could be correct. However there is at present no direct experimental support for it. We do not know what comes out of the amyloplast, how such export occurs or how it is affected by temperature. We are also completely ignorant of whether any effect of temperature on transport across the amyloplast envelope is capable of regulating sucrose synthesis. Until we have this information there appears to be very little basis for the hypothesis. In addition there is no firm evidence that an electron transfer system is involved in metabolite transfer in any plastid. In chloroplasts, the only instance where we have any clear picture of such transport, it is clear that an electron transfer system of the type envisaged by Isherwood is not responsible for transfer of carbon to the cytoplasm (Fliege *et al.*, 1978).

B. COARSE CONTROL

By coarse control we mean regulation due to relatively slow changes in the amounts of active enzyme in a tissue. These changes may be caused by synthesis or breakdown of enzymes, or by release of enzymes from bound or inactive forms, or by production of proteins that inactivate enzymes. Such variation in the maximum catalytic activities of regulatory enzymes could lead to a redirection of metabolism. Thus, on transfer of plants to the cold, coarse control could vary the relative rates of starch breakdown, sucrose synthesis, and consumption of hexose phosphates in respiration, and this variation could cause sucrose to accumulate. Sucrose has been reported to accumulate in potatoes within six hours of transfer to 2°C (Isherwood, 1973). At this temperature appreciable changes in maximum catalytic activities of enzymes are perhaps unlikely in so short a time as six hours. Thus the rapid onset of sweetening suggests that coarse control may not be involved in the initial stages of sweetening. However, as there is often an appreciable lag in the development of the maximum rate of sugar accumulation in the cold (Fig. 1), coarse control might contribute to the establishment of this maximum rate.

Study of the role of coarse control in sweetening in potatoes is made particularly difficult by the technical problems associated with the measurement of the maximum catalytic activities of enzymes in tubers. Such measurements must be interpreted cautiously and must be accompanied by evidence of their reliability. The amount of data that meets these requirements is small. Kennedy and Isherwood (1975) have provided sound evidence that the maximum catalytic activity of α-glucan phosphorylase does not alter significantly during the onset of cold-induced sweetening in potatoes. Data for some of the enzymes involved in sucrose metabolism and in respiration are given in Table IV. In no instance did we detect any significant change in activity during the development of low temperature sweetening. Pressey and Shaw (1966) have presented evidence that the maximum catalytic activity of acid invertase increases when potatoes are transferred to 2°C, and that this increase parallels the accumulation of reducing sugars. We have already argued (see section II) that the formation of reducing sugars is probably a secondary aspect of sweetening. Thus we suggest that at present there is no sound evidence that coarse control is directly responsible for sweetening. Further, such evidence that is available argues against, but does not eliminate, a primary role for coarse control in this process.

TABLE IV
Effect of cold storage on maximum catalytic activities of enzymes in potato tubers (Record)

	Activity (nmol min $^{-1}$ g fresh wt $^{-1}$)			
Enzyme	Initial	8 days at 2°C	15 days at 2°C	15 days at 10°C
Sucrose phosphate synthetase	115 ± 5	130 ± 28	125 ± 17	125 ± 18
Sucrose synthetase	7·5 ± 0·5	8·9 ± 1·2	7·5 ± 0·4	8·7 ± 0·9
Aldolase	224 ± 12	202 ± 12	212 ± 16	258 ± 20
Glyceraldehydephosphate dehydrogenase	453 ± 40	512 ± 27	605 ± 39	597 ± 27
Glucose-6-phosphate dehydrogenase	370 ± 14	376 ± 41	459 ± 57	453 ± 37

Values are means ± S.E. of activities from 3–4 different extracts (Pollock and ap Rees, 1975a).

C. FINE CONTROL

Fine control is that caused by changes in the activities of enzymes already present in the cell. It operates in seconds or minutes as opposed to hours and is generally due to rapid alteration in the concentrations of co-factors and effectors. Such mechanisms could respond to lowered temperature in a variety of ways. Enzymatic reactions as a whole are very susceptible to changes in temperature and their temperature coefficients generally lie between 2 and 3. However these coefficients do vary according to the enzyme, the conditions under which the organism grows, the substrate concentration, and the temperature (Hochachka and Somero, 1973). Thus lowering the temperature might affect fine control of metabolism in two general ways. First, the lowered rate of metabolism as a whole could affect control mechanisms. Secondly, because of differences in detail between enzymes, lowered temperature could have a differential effect on control mechanisms that could lead to massive and possibly fatal disturbance of metabolism. It is conceivable that such effects are not only involved in sweetening but may also form the basis of much of the chilling injury suffered by plants.

Studies of low temperature sweetening in potatoes have shown that the temperature coefficients of the enzymes of carbohydrate metabolism vary according to the enzyme and the temperature (Pollock and ap Rees, 1975a). From Table V it may be seen that although the Q_{10} of sucrose phosphate synthetase did not alter significantly when the temperature was lowered, those of three key glycolytic enzymes were all found to be significantly higher at 2–10°C than at 10–25°C. Thus, at least *in vitro*, enzymes responsible for the respiration of hexose phosphates are more sensitive to cold than the enzyme responsible for sucrose synthesis. Further, the temperature at which the Q_{10} of the glycolytic enzymes alters corresponds to that which leads to sweetening. These observations suggest that preferential cold-lability of certain glycolytic

TABLE V
Temperature coefficients of enzymes of sugar metabolism from potato tubers (Record)

Enzyme	Temperature coefficient Q_{10}(10–25°C)	Q_{10}(2–10°C)	Fisher's P value 10–25° *vs.* 2–10°
Sucrose phosphate synthetase	1·92 ± 0·01	2·25 ± 0·24	N.S.
Sucrose synthetase	1·86 ± 0·03	1·96 ± 0·04	N.S.
Glucose-6-phosphate dehydrogenase	1·97 ± 0·01	2·49 ± 0·23	N.S.
Phosphofructokinase	1·47 ± 0·05	3·33 ± 0·12	<0·001
Aldolase	2·35 ± 0·08	2·65 ± 0·52	N.S.
Glyceraldehydephosphate dehydrogenase	2·48 ± 0·06	4·96 ± 0·38	<0·002
Pyruvate kinase	1·45 ± 0·05	2·38 ± 0·12	<0·001

Values are means ± S.E. of measurements made with extracts of five different unstored tubers (Pollock and ap Rees, 1975a).

enzymes could result in a disproportionate decrease in glycolysis, relative to other pathways of hexose phosphate metabolism, when the temperature was lowered to 2–10°C. Such a relative restriction of glycolysis could cause a diversion of hexose 6-phosphates to sucrose and thus contribute to sweetening.

The above hypothesis assumes that glycolytic flux in potatoes is regulated by one or more of the enzymes shown to be cold-labile. Comparative biology suggests that entry into glycolysis in potatoes is controlled by phosphofructokinase, and that pyruvate kinase controls exit from glycolysis and the oxidative pentrose phosphate pathway. However, the data available for potatoes conflict (Dixon and ap Rees, 1980a), and our understanding of glycolytic control in plants as a whole is too meagre to allow argument by analogy. Accordingly we identified the reactions that regulate glycolysis in potatoes (Dixon and ap Rees, 1980a). Sound evidence that a reaction is regulatory is provided by the demonstration that the reaction is substantially displaced from equilibrium *in vivo*, and that the substrate changes in the opposite direction to the flux when the latter is varied (Newsholme and Start, 1973).

The non-equilibrium reactions of glycolysis in potato tubers have been identified from a comparison of mass-action ratios with the apparent equilibrium constants of the enzymes. Evidence of the reliability of the measurements of substrates is provided by estimates of the recovery of known amounts of the compounds from freeze-clamped samples of potato. The amounts of the intermediates added were comparable to those found in the samples of tissue. The recoveries of all the substrates were adequate (Table VI). The data in Table VII strongly indicate that the steps catalysed by glucosephosphate isomerase, phosphoglycerate mutase and enolase are so close to equilibrium *in vivo* that they are unlikely to exert appreciable direct

TABLE VI
Effects of anoxia on amounts of glycolytic intermediates in potato tubers (Record), and recovery of intermediates added to tuber tissue

Intermediate	Content (nmol/g fresh wt) Aerobic tubers	Anoxic tubers	Fisher's *P* value Aerobic *vs.* anoxic	Recovery of added compound (%)
Glucose 6-phosphate	96·6 ± 3·4	46·8 ± 1·4	<0·001	96 ± 4
Fructose 6-phosphate	24·7 ± 0·8	13·1 ± 0·7	<0·001	89 ± 7
Fructose-1,6-bisphosphate	4·5 ± 0·2	6·1 ± 0·2	<0·001	122 ± 2
3-Phosphoglycerate	68·5 ± 1·8	57·6 ± 2·1	<0·01	93 ± 4
2-Phosphoglycerate	8·4 ± 0·5	8·2 ± 0·5	N.S.	95 ± 1
Phosphoenolpyruvate	23·0 ± 0·6	14·5 ± 0·6	<0·001	93 ± 3
Pyruvate	12·6 ± 0·5	19·6 ± 1·0	<0·001	93 ± 2

Intermediates were measured in aerobic tubers kept at 20°C, and in tubers that had been anoxic for 90 min at 20°C. The recovery experiments were carried out with aerobic tubers. All values are means ± S.E. of measurements made on six different tubers. Details in Dixon and ap Rees (1980a).

control over glycolysis. We have not yet devised a satisfactory technique for measuring ATP and ADP in the soluble phase of the cytoplasm of potatoes, and thus can not calculate the mass-action ratios for phosphofructokinase and pyruvate kinase. However, it is possible to use the lower values of the apparent equilibrium constants in Table VII, and the estimates of the ratios Fru-1,6-P_2:Fru-6-*P*; and pyruvate:*P*-enolpyruvate to calculate the ATP:ADP ratios that would have to exist in the cytoplasm if the phosphofructokinase and pyruvate kinase steps were at equilibrium *in vivo*. In aerobic tubers at 20°C this ATP:ADP ratio would have to be 2×10^{-4} for phosphofructokinase, and 3636 for pyruvate kinase. Both values are so far from that likely to exist in the cytoplasm, a value of 8·7 has been reported for

TABLE VII
Comparison of apparent equilibrium constants of glycolytic reactions with ratios of substrates in aerobic tubers of potatoes (Record)

Reaction	Apparent equilibrium constant[a]	Substrates	Ratios[b]
Glucosephosphate isomerase	0·36–0·47	Fru-6-*P*:Glc-6-*P*	0·26
Phosphofructokinase	900–1200	Fru-1,6-P_2:Fru-6-*P*	0·18
Phosphoglycerate mutase	0·1–0·2	2-P-Glyceric acid:3-P-Glyceric acid	0·12
Enolase	2·8–4·6	P-Enolpyruvate:2-P-Glyceric acid	2·74
Pyruvate kinase	2000–20,000	Pyruvate:P-Enolpyruvate	0·55

[a] Values from ap Rees *et al.* (1977).
[b] Ratios of mean values for aerobic tubers shown in Table VI.

the cytoplasm of hepatocytes (Akerboom *et al.*, 1978), that we may conclude that both enzymes catalyse non-equilibrium reactions *in vivo*.

The above arguments establish that phosphofructokinase, and to a lesser extent pyruvate kinase, could make major contributions to the regulation of glycolysis in potatoes. Evidence that they do so is provided by the changes in their substrates that occur when glycolysis is stimulated by anoxia (Table VI). Barker and el Saifi (1952a, b) have provided sound evidence that anoxia stimulates glycolysis in potatoes. The data in Table VI establish that this stimulation is accompanied by significant decreases in the contents of fructose 6-phosphate and phosphoenolpyruvate, As phosphofructokinase is the first non-equilibrium step unique to glycolysis we suggest that entry of hexose phosphate into glycolysis in potatoes is primarily controlled by this enzyme. As glycolysis is almost certainly by far the major pathway of carbohydrate oxidation in potatoes (ap Rees, 1980), it may be argued that phosphofructokinase plays a dominant role in regulating the respiration of hexose phosphates by the tubers. As pyruvate kinase is preceded by the irreversible step catalysed by phosphofructokinase, the former enzyme cannot control glycolysis directly as it can not have any direct effect on the entry of glucose 6-phosphate into glycolysis. However pyruvate kinase could regulate flow out of glycolysis and the pentose phosphate pathway and thus influence the concentration of glycolytic intermediates. As many of these can affect the activity of phosphofructokinase (Turner and Turner, 1975) variation in the activity of pyruvate kinase could also have an indirect effect on the entry of substrate into glycolysis.

The evidence that phosphofructokinase and pyruvate kinase make crucial contributions to the control of glycolysis in potatoes supports the view that the cold-lability of these enzymes could result in glycolysis being preferentially sensitive to cold. This possibility was investigated by determining the effect of cooling from 25°C to 2°C on the metabolism of [U-^{14}C]glucose by callus tissue and tubers of potato (Pollock and ap Rees, 1975b). Two aspects of the results (Table VIII) are emphasized. First, in the callus tissue cooling led to a marked decrease in the percentage of metabolized [^{14}C]-glucose that was respired to $^{14}CO_2$. Secondly, in both callus and tuber cooling caused a substantial increase in the relative labelling of the acidic components of the water-soluble substances. Analysis of this fraction revealed that most of this increase was due to the accumulation of label in hexose phosphates (Pollock and ap Rees, 1975b). These experiments strongly suggest that cooling potatoes from 25°C to 2°C does lead to a differential reduction in the movement of hexose phosphates into respiration.

If restriction of glycolysis is a cause of sweetening then the ^{14}C that accumulated in hexose phosphates in the cold in the above experiments should eventually move into free sugars. We have tested this hypothesis in a pulse and chase experiment (Dixon and ap Rees, 1980b). We emphasize the following aspects of the results that we obtained (Table IX). First, the more complete

TABLE VIII

Effects of cold on metabolism of [U-^{14}C]glucose by tissue cultures and tubers of potato (Record)

Fraction	Percentage of metabolized ^{14}C recovered per fraction Callus 25°C	2°C	Tuber 25°C	2°C
CO_2	22	4	—	—
Acidic components of water-soluble material	30	65	13	35
Basic components of water-soluble material	11	7	24	12
Fructose	5	5	6	2
Sucrose	20	13	32	12
Material insoluble in water	12	5	19	25
Incubation time (h)	3	3	6	6

Tubers were labelled by removing a minute core of tissue with a 50-μl microcapillary pipette and then placing 50 μl 0·5 mM [^{14}C]glucose (325 μCi/μmol) in the resulting hole. Values are means from 3–4 separate experiments. Details in Pollock and ap Rees (1975b).

TABLE IX

Effect of cold on metabolism of [U-^{14}C]glucose in 2-h pulse and in 18-h chase by potato tubers (Record)

Fraction	Percentage metabolized ^{14}C recovered per fraction 2°C Pulse	Chase	25°C Pulse	Chase
Sucrose	11	57	28	35
Hexose	3	11	3	5
Hexose phosphates	39	11	10	4
Organic acids	3	4	5	8
Amino acids	10	13	24	8
CO_2	0·2	1	1	11
Material insoluble in water	32	8	28	38
Protein	8	1	6	12
Starch	13	3	13	24
Non-starch polymeric glucose	0·2	1·1	4	2
Polymeric galactose	—	0·1	5	2

Tubers were labelled as described in Table VIII, after incubation for 2 h the [^{14}C]glucose was replaced with glucose and incubation was continued for a further 18 h. Values are means from 3–7 different experiments. Details in Dixon and ap Rees (1980b).

analyses than those in Table VIII show even more clearly that cooling restricts entry of labelled sugars into the respiratory pathways and causes the label to accumulate as hexose phosphates. The amount of ^{14}C that did enter the respiratory pathways may be estimated by summing that recovered in CO_2, organic acids, amino acids and protein. At the end of the pulse at 25°C, 10% of the metabolized label was in hexose phosphates and 35% had entered the respiratory pathways. At 2°C, 39% was in hexose phosphates and only 22% had entered respiration. Our second point is that during the chase at 2°C there was a very sharp decline in the label in hexose phosphates, from 39% to 11% of the ^{14}C metabolized. Detailed analysis of the distribution of label at the end of the chase at 2°C reveals the fate of the label that moved out of the hexose phosphates. Extremely little of this label can have entered the respiratory pathways as the total percentage of label recovered in the products of respiration did not increase during the chase. Nor did the label from hexose phosphates accumulate as insoluble compounds during the chase, as the proportion of metabolized ^{14}C found in this fraction fell significantly during the chase. The only substances that showed an increase in labelling during the chase, that could account for the fall in the labelling of hexose phosphates, were the free sugars. These analyses provide sound evidence that cooling tubers to temperatures that cause sweetening restricts entry of hexose phosphates into the respiratory pathways and causes these compounds to be diverted to sucrose and hexose.

The measurement of Q_{10}, the studies of the control of glycolysis, and the effects of cold on the metabolism of [^{14}C]glucose pinpoint phosphofructokinase as a major cold-sensitive step in the carbohydrate metabolism of potato tubers of the variety Record. Accordingly we have started to study the basis of the cold-lability of this enzyme. It is an oligomeric enzyme (Uyeda, 1979). Lowering the temperature could weaken hydrophobic interactions between subunits in the complex. This could lead to its dissociation and a consequent loss of catalytic activity. Cold-lability has been demonstrated in respect of phosphofructokinase from rabbit muscle (Bock and Frieden, 1974), chicken liver (Kono and Uyeda, 1973) and *Bacillus licheniformis* (Marschke and Bernlohr, 1973). In the above instances light scattering and sucrose density gradient centrifugation have provided evidence that cooling causes dissociation of the enzyme. We have used differential scanning calorimetry to study the enzyme from potatoes.

Differential scanning calorimetry monitors the heat capacity (Cp) of a system as a function of temperature. Since

$$\Delta H = \int \Delta Cp\, dT$$

the enthalpy of any thermal transition that might take place can also be obtained, provided that standard states are properly defined (Pfeil and Privalov, 1976).

The above approach required pure phosphofructokinase from potatoes.

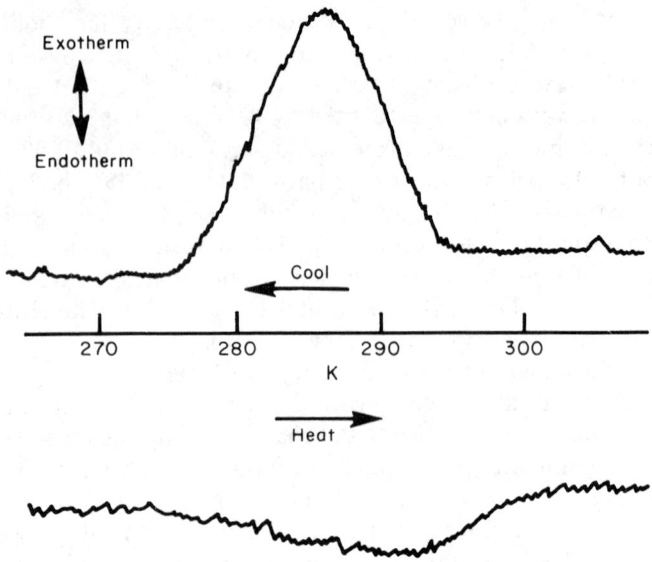

FIG. 2. Power-time differential scanning calorimetry traces for the low temperature dissociation and reassembly of phosphofructokinase from potato tubers (Record). The enzyme was cooled and heated at 10 K min^{-1} with the instrumental sensitivity at 1 mcal s^{-1} full scale deflexion. The peak height corresponds to a heat evolution of 0·15 mcal s^{-1} (Dixon, Franks and ap Rees, 1981).

This was obtained by affinity chromatography with Blue Sepharose CL-6B. The fraction used for the calorimetry revealed only one band when examined by SDS/polyacrylamide gel electrophoresis (Dixon *et al.*, 1981). Samples of pure enzyme were cooled from 293 K to 265 K at various rates within the range 5–10 K min^{-1}. After keeping the temperature at 265 K for 5 min, the samples were warmed to 293 K at the same rates. Power-time curves were recorded for cooling and warming and a typical curve is given in Fig. 2. These results demonstrate that phosphofructokinase owes its cold-lability to the denaturation of the multi-subunit enzyme complex. The fact that the dissociation is exothermic and that the corresponding entropy change is therefore negative strongly suggests that hydrophobic interactions, which contribute to the stability of the native state at ordinary temperatures, are weakened to the extent that the enzyme complex dissociates spontaneously.

VI. CONCLUSIONS

Low temperature sweetening is a widespread but not universal characteristic of higher plants. Its significance, if any, is not yet apparent. In certain instances it may contribute to frost resistance. The extent of this contribution probably varies with the plant, and it is unlikely that sweetening is ever the sole cause of frost resistance. Certainly the mere accumulation of sugar is not a guarantee of frost resistance.

There is evidence that the principal event in low temperature sweetening is the cold-induced synthesis of sucrose. The carbon for this synthesis is generally, but not always, provided by a net breakdown of starch. There is reasonable evidence that the synthesis of sucrose is catalysed by sucrose phosphate synthetase in the soluble phase of the cytoplasm. However we do not know the pathway of starch breakdown.

Our understanding of the mechanism of low temperature sweetening is still rudimentary. The fact that we do not know what leaves the amyloplasts to support the synthesis of sucrose in the cytoplasm means that we cannot assess the extent to which sweetening is due to effects of low temperature on intracellular compartmentation. As yet there is no evidence that coarse control of metabolism plays any major role in low temperature sweetening. There is some indication of how low temperature can affect fine control so as to cause sweetening in potatoes. Key regulatory enzymes of glycolysis, phosphofructokinase in particular, are more sensitive to cold than is sucrose phosphate synthetase, the enzyme responsible for sucrose synthesis. For phosphofructokinase this cold-lability appears to be due to weakening of hydrophobic interactions needed to maintain the oligomeric enzyme in an active form. This differential sensitivity of glycolytic enzymes to cold is reflected *in vivo* as there is appreciable evidence that lowering the temperature to values that lead to sweetening restricts entry of hexose phosphates into glycolysis and diverts them to sucrose. It thus seems likely that in potatoes the cold-lability of phosphofructokinase, and possibly pyruvate kinase, is at least a contributory cause of low temperature sweetening.

The wide variety of ways in which cooling could affect plant metabolism, and the considerable time it takes plants to sweeten fully in the cold, make it unlikely that cold-lability of glycolytic enzymes is the sole explanation of low temperature sweetening. It is probable that a variety of factors contribute, and it is likely that the relative contributions of each vary with the plant and with the stage of sweetening. Further investigations of the role of phosphofructokinase in sweetening will need to answer two questions in particular. The first is whether variation in the extent to which plants sweeten at low temperatures is due to variation in the cold-lability of phosphofructokinase. This may be so, but it is quite possible that cold-lability of this enzyme will be found to be widespread, and that differences in the degree of sweetening may be due to differences in the extent to which restriction of the activity of phosphofructokinase diverts hexose phosphates to sucrose. The second of the two questions is whether the contribution of the cold-lability of phosphofructokinase to sweetening varies during the time course of sugar accumulation. A particular problem is raised by the observations (Isherwood, 1976) that, although CO_2 production falls sharply when potato tubers are transferred to 2°C, the rate begins to increase after a period of 5–10 days at this temperature. It is not known whether this increase in CO_2 production reflects an increase in the amount of hexose phosphate that enters the respiratory pathways. The significance of the increase can not be assessed until we know the source of the CO_2, the pathway of its formation, and the metabolic change that accounts for the alteration in the rate of its production. The increase in CO_2 production may merely reflect a change in the proportion of respiratory substrate that is converted to CO_2. The data in Table IX show that only a small proportion of the [^{14}C]glucose that entered the respiratory pathways was released as CO_2 even after a long chase. This proportion could increase as sweetening proceeds, and thus result in increased CO_2 production. Whatever the answers to the above questions, our experiments illustrate one way in which cold can have a differential effect on plant metabolism, and the results may have wider significance in relation to sensitivity to both chilling and freezing.

REFERENCES

Akerboom, T. P. M., Bookelman, H., Zuurendonk, P. F., van der Meer, R. and Tager, J. M. (1978). *Europ. J. Biochem.* **84**, 413–420.
ap Rees, T. (1974). *MTP Int. Rev. Sci. Biochem.* **11**, 89–127.
ap Rees, T. (1977). *Symp. Soc. Exp. Biol.* **31**, 7–32.
ap Rees, T. (1980). *In* "The Biochemistry of Plants" (P. Stumpf and E. E. Conn, Eds), Vol. II, pp. 1–29. Academic Press, London and New York.
ap Rees, T., Fuller, W. A. and Wright, B. W. (1977). *Biochim. Biophys. Acta* **461**, 274–282.
Barker, J. and el Saifi, A. F. (1952a). *Proc. R. Soc. (London) Ser. B.* **140**, 362–385.
Barker, J. and el Saifi, A. F. (1952b). *Proc. R. Soc. (London) Ser. B.* **140**, 385–403.
Bieleski, R. L. and Redgwell, R. J. (1977). *Aust. J. Plant Physiol.* **4**, 1–10.
Bock, P. E. and Friedan, C. (1974). *Biochemistry* **13**, 4191–4196.
Burton, W. G. (1969). *Eur. Potato J.* **12**, 81–95.
Burton, W. G. (1978). *In* "The Potato Crop" (P. M. Harris, Ed.), pp. 545–606. Chapman and Hall, London.
Carolus, R. L. (1930). *Am. Soc. Hort. Sci. Proc.* **27**, 502–508.
Chapman, D. (1975). *Q. Rev. Biophys.* **8**, 185–235.
Clegg, C. J. and Whittingham, C. P. (1970). *Phytochemistry* **9**, 279–287.
Dixon, W. L. and ap Rees, T. (1980a). *Phytochemistry* **19**, 1297–1301.
Dixon, W. L. and ap Rees, T. (1980b). *Phytochemistry* **19**, 1653–1656.
Dixon, W. L., Franks, F. and ap Rees, T. (1981). *Phytochemistry* **20**, 969–972.
Fliege, R., Flügge, U-I., Werdan, K. and Heldt, H. W. (1978). *Biochim. Biophys. Acta* **502**, 232–247.
Gray, D. and Hughes, J. C. (1978). *In* "The Potato Crop" (P. M. Harris, Ed.), pp. 504–544. Chapman and Hall, London.
Green, D. G. and Ratzlaff, C. D. (1975). *Can. J. Bot.* **53**, 2198–2201.
Heber, U. and Santarius, K. A. (1971). *Umschau* **71**, 930–936.
Heber, U. and Santarius, K. A. (1973). *In* "Temperature and Life" (H. Precht, J. Christopherson, H. Hensel and W. Larcher, Eds), pp. 232–263. Springer-Verlag, Berlin.

Hochachka, P. W. and Somero, G. N. (1973). "Strategies of Biochemical Adaptation". Saunders, Philadelphia, London and Toronto.

Isherwood, F. A. (1973). *Phytochemistry* **12**, 2579–2591.

Isherwood, F. A. (1976). *Phytochemistry* **15**, 33–41.

Kennedy, M. G. H. and Isherwood, F. A. (1975). *Phytochemistry* **14**, 667–670.

Kono, N. and Uyeda, K. (1973). *J. Biol. Chem.* **248**, 8603–8609.

Lehle, L. and Tanner, W. (1973). *Eur. J. Biochem.* **38**, 103–110.

Levitt, J. (1956). "The Hardiness of Plants". Academic Press, New York and London.

Levitt, J. (1972). "Responses of Plants to Environmental Stresses". Academic Press, New York and London.

Li, P. H., Weiser, C. J. and van Huystee, R. (1965). *Proc. Am. Soc. Hort. Sci.* **86**, 723–730.

Lineberger, R. D. and Steponkus, P. L. (1980). *Pl. Physiol.* **65**, 298–304.

Marschke, C. K. and Bernlohr, R. W. (1973). *Archs Biochem. Biophys.* **156**, 1–16.

Newsholme, E. A. and Start, C. (1973). "Regulation in Metabolism". Wiley, London.

Nishimura, M. and Beevers, H. (1979). *Plant Physiol.* **64**, 31–37.

Parker, J. (1959). *Botan. Gaz.* **121**, 46–50.

Pellet, N. E. and White, D. B. (1969). *J. Am. Soc. Hortic. Sci.* **94**, 460–462.

Pfeil, W. and Privalov, P. L. (1976). *Biophys. Chem.* **4**, 23–32.

Pliny (first century A.D.). "Naturalis Historiae", Libr. XVIII, §130; Libr. XIX, §512.

Pollock, C. J. and ap Rees, T. (1975a). *Phytochemistry* **14**, 613–617.

Pollock, C. J. and ap Rees, T. (1975b). *Phytochemistry* **14**, 1903–1906.

Pollock, C. J. and ap Rees, T. (1975c). *Planta* **122**, 105–107.

Pressey, R. (1970). *Am. Potato J.* **47**, 245–251.

Pressey, R. and Shaw, R. (1966). *Plant Physiol.* **41**, 1657–1661.

Raison, J. K. (1973). *Symp. Soc. Exp. Biol.* **27**, 485–512.

Raison, J. K. and Chapman, E. A. (1976). *Aust. J. Plant Physiol.* **3**, 291–299.

Sakai, A. and Yoshida, S. (1968). *Cryobiology* **5**, 160–174.

Santarius, K. A. and Milde, H. (1977). *Planta* **136**, 163–166.

Schwimmer, S., Bevenue, A., Weston, W. J. and Potter, A. L. (1954). *J. Agric. Food Chem.* **2**, 1284–1290.

Seybold, S. (1969). *Flora* **A160,** 561–575.

Steponkus, P. L. and Lanphear, F. O. (1968). *Physiol. Plant.* **21**, 777–791.

Stitt, M. and ap Rees, T. (1980). *Biochim. Biophys. Acta* **627**, 131–143.

Tanner, W. and Kandler, O. (1966). *Plant Physiol.* **41**, 1540–1542.

Tumanov, I. I. and Trunova, I. (1963). *Physiol. Rast.* **10**, 176–188.

Turner, J. F. and Turner, D. H. (1975). *Annu. Rev. Plant Physiol.* **26**, 159–186.

Uyeda, K. (1979). *Adv. Enzymol.* **48**, 193–244.

van Vliet, W. F. and Schriemer, W. H. (1960). *Eur. Potato J.* **3**, 263–271.

Walker, D. A. (1976). *In* "Encyclopedia of Plant Physiology" (C. R. Stocking and U. Heber, Eds), Vol. III, pp. 85–136. Springer-Verlag, Berlin.

Yelenosky, G. and Guy, C. L. (1977). *Botan. Gaz.* **138**, 13–17.

Young, R. (1969). *J. Am. Soc. Hortic. Sci.* **94**, 252–254.

CHAPTER 3

Cyanide-insensitive Respiration in Fruits and Vegetables

CLAUDE LANCE

Laboratoire de Biologie Végétale IV, Université Pierre et Marie Curie, 12 rue Cuvier, 75005 Paris, France

I. Introduction 63
II. The Cyanide-insensitive Pathway 64
A. The Alternative Pathway in Potato Tuber 64
B. Contribution of Both Pathways 65
C. Basic Problems 69
III. Participation of the Alternative Pathway 73
A. The Alternative Pathway in Potato Tuber 73
B. Contribution of Both Pathways 75
IV. Physiological Significance of the Alternative Pathway 77
A. Thermogenesis in the Araceae 77
B. Cyanogenic Plant Species 78
C. Elimination of Reducing Power 79
D. Role in Malate Oxidation 81
E. Integration of the Alternative Pathway in Cell Metabolism . 82
V. Conclusions 84
References 84

I. Introduction

Respiration in plants presents two features which make it different from respiration in animals. First, plants can sustain relatively long periods of anoxia, from several hours to several weeks or months, whereas a few minutes of deprivation of oxygen are often fatal to animals. Secondly, exposures to such violent poisons as carbon monoxide or cyanide, that would kill an animal within minutes, have no dramatic consequences for plant respiration, at least for a limited period of time. Sometimes they even have a stimulatory effect on the rate of gas exchange.

Therefore, one of the striking features of plant respiration appears to be its resistance to respiratory inhibitors. This is exemplified by the well known so-called cyanide-resistant respiration as opposed to the normal, cyanide-sensitive respiration. This phenomenon is widely distributed in the plant kingdom. The best examples are to be found in the spadices of the Araceae, in which the cyanide-resistant respiration can display extremely high rates during the course of a respiratory crisis associated with pollination. Beside this typical situation, cyanide-insensitive respiration is also quite common in various fruits and vegetables, particularly in storage organs.

The cellular site of cyanide-insensitive respiration is the mitochondrion. Plant mitochondria can catalyse two sequences of electron transport reactions; one is that of the normal respiratory chain and the second is a cyanide-insensitive electron transport pathway. The contribution of the latter pathway to overall electron transport in plant tissues can vary considerably. General information on various aspects of cyanide insensitivity in plant respiration can be found in review articles by Hackett (1959), Bonner (1965), Ikuma (1972), Meeuse (1975), Henry and Nyns (1975), Palmer (1976) and Solomos (1977).

Rather than considering the extent and magnitude of the cyanide-insensitive respiration in fruits and vegetables, the aim of this chapter will be to attempt to find a possible significance to this particular pathway, using specific and selected examples. It will therefore consider successively the structure, the participation and the significance of the cyanide-insensitive pathway in these plant materials.

II. The Cyanide-insensitive Pathway

A. DISTRIBUTION

The resistance of plant respiration to cyanide was discovered in the late thirties as a result of the work of van Herk and Badenhuizen (1934) on the respiration of the tropical Aracea *Sauromatum guttatum* Schott. In 1950, James and Beevers extended this observation to the European *Arum maculatum* L. and *Arum italicum* Mill. Later, James and Elliott (1955) observed that the site of cyanide-resistance was localized in mitochondria isolated from the cyanide-insensitive spadix tissue of *A. maculatum*. Similar observations were made by Hackett (1957) on the American Aracea *Symplocarpus foetidus* L. These various works made the spadices of the Araceae the most suitable plant materials for the study of the mechanisms of the cyanide-insensitive respiration. A large part of our present knowledge on this subject is based on extensive studies carried out on such plant materials.

Cyanide-insensitive respiration was also observed in various higher plant organs. Among them, leaf tissues appeared to be significantly resistant to carbon monoxide and cyanide inhibition (Ducet and Rosenberg, 1962). The same observation was made for various storage organs, such as tubers of white potato, sweet potato, cassava, etc., or for fruits (avocado, banana, etc.). These studies were performed on intact organs and tissue slices or on mitochondria isolated therefrom.

The pattern of cyanide resistance can be sometimes rather complicated. When measured on whole tubers, oxygen uptake may be cyanide-resistant or even stimulated by cyanide. However, these same tubers can yield fresh slices or mitochondria that are strongly cyanide-sensitive. This is the situation in white potato tubers (Hackett *et al.*, 1960; Solomos and Laties, 1975). The level of cyanide resistance can be changed by experimental treatments. Exposure of potato tubers to ethylene, cyanide or to various volatile organic compounds can increase the rate of respiration and the level of cyanide resistance (Solomos and Laties, 1975; Rychter *et al.*, 1979; Janes *et al.*, 1979). Similarly, resistance to cyanide can be induced by the ageing of tissue slices in an aerated liquid medium. The most studied examples are the slices of white potato tuber (Hackett *et al.*, 1960; Nakano and Asahi, 1970; Dizengremel, 1975). However, in sweet potato, the respiration is initially cyanide-insensitive and ageing increases the rate, but not the cyanide resistance, of respiration (Solomos, 1977; Laties, 1978; Theologis and Laties, 1978c).

Compared to the situation in the spadix of the Araceae, in which the cyanide-insensitive pathway is always present, only varying with the development of the inflorescence (Chauveau, 1976a), the variations in cyanide resistance caused by a range of treatments permit experimental study of the control mechanisms that regulate the operation of the cyanide-insensitive pathway (Dizengremel and Lance, 1976). In this respect, the best experimental materials are undoubtedly some strains of *Neurospora* mutants in which the genetic control of cyanide resistance can be used as a tool for the study of the composition of the cyanide-insensitive pathway (Edwards, 1978; Slayman, 1978).

B. STRUCTURE

1. Development of Concepts

Soon after the localization of cyanide resistance was established to be within the mitochondria, a number of questions were raised. The cyanide-insensitive pathway became largely known as the alternative (or alternate) pathway, suggesting that two ways to reach oxygen were possible for the electrons produced by the oxidation of Krebs cycle substrates. The major questions were: are the two pathways independent or interconnected by some common intermediate? In that case, what is this intermediate? What are the electron transport components of the alternative pathway? And, most important of all, what is the nature of its terminal oxidase?

(a) *Branch point*. The alternative pathway has never been considered as being totally independent of the normal respiratory chain. At the least, both pathways must share in common the various dehydrogenases which provide electrons. Furthermore, the possible location of the branch point could be rather easily localized as a result of some general observations.

In plant mitochondria, all electron transport that is resistant to cyanide and carbon monoxide (inhibitors of cytochrome oxidase) is also resistant to antimycin, an inhibitor acting between cytochromes *b* and *c*. From this, it was concluded that the branch point was on the substrate side of cytochrome *b*. Similarly, from the observation that ADP/O ratios, which express the efficiency of oxidative phosphorylation, could be lowered to 0 for succinate or to 1 for malate in the presence of cyanide, it was also concluded that the branch point ought to be located after the first phosphorylation site (site I). This leaves a part of the respiratory chain between various flavoproteins and cytochromes *b* as the possible location of the branch point. In that incompletely defined region of the respiratory chain, a number of electron carriers such as non-haem iron compounds, ubiquinone, and also flavoproteins and cytochromes *b* are present.

Following the work of Storey (1976), most of these possibilities were eliminated to leave ubiquinone as the only electron carrier common to both pathways. This conclusion was based on the observation that ubiquinone could become reduced in the presence of inhibitors of both the cytochrome and the alternative pathways. Since then, there has been a general agreement on this point and ubiquinone has become a compound whose role in electron transport is of utmost importance (Rich, 1978; Huq and Palmer, 1978; Rustin *et al.*, 1980).

(b) *Terminal oxidase*. There have been numerous theories about the nature of the cyanide-insensitive oxidase reacting with molecular oxygen. Cytochrome oxidase, the terminal oxidase of the cytochrome pathway, is strongly inhibited by cyanide as well as by azide and carbon monoxide which have no effect on the alternative pathway oxidase. The expectation is that normal levels of cytochrome oxidase would be inhibited by cyanide.

Based on the fact that flavoproteins are generally present in high concentrations in cyanide-insensitive tissues (van Herk, 1937; James and Beevers, 1950; Lance and Bonner, 1968) an autoxidizable flavoprotein was first proposed as a potential terminal oxidase. An excess cytochrome oxidase hypothesis was also put forward (Ducet and Rosenberg, 1962), as well as a cytochrome b_7 hypothesis, based on the observation that a *b*-type cytochrome remains oxidized in the presence of cyanide (Bendall and Hill, 1956). These hypotheses have been critically reviewed by Bendall and Bonner (1971).

Observing that cyanide-insensitive electron transport could be inhibited by metal complexing agents, Bendall and Bonner (1971) suggested that non-haem iron proteins (iron–sulphur proteins) could be implicated in the alternative pathway. Support for this hypothesis came from the discovery that hydroxamic acid derivatives, such as SHAM (salicylhydroxamic acid) could specifically inhibit the alternative pathway (Schonbaum *et al.*, 1971). These compounds are potent chelating agents of transition metals.

The iron–sulphur protein theory has been further supported during the past few years by research in several laboratories (Haddock and Garland, 1971; Cammack and Palmer, 1973; Edwards *et al.*, 1976; Moore *et al.*, 1976; Henry *et al.*, 1977; Chauveau *et al.*, 1978). However, the participation of iron–sulphur proteins in electron transport in cyanide-insensitive mitochondria was seriously questioned after the failure of all attempts to observe changes in EPR signals as a function of the redox states of the mitochondria (Rich and Bonner, 1978a, b, c).

Alternative explanations have also been proposed, switching the interest towards ubiquinone, such as a protonmotive ubiquinone cycle (Rich and Moore, 1976) or an autoxidizable species or pool of quinone (Rich and Bonner, 1978a, b). A partial reduction of oxygen by the oxidase, yielding H_2O_2 instead of H_2O, was also proposed as a possible mechanism. In that case superoxide dismutase could be the oxidase or participate in the terminal oxidase reaction (Rich *et al.*, 1976). Lipoxygenase has also been suggested as a possible candidate for this function (Parrish and Leopold, 1978). At this time, the nature of the oxidase is still unknown, but the actual working hypothesis is that a quinol oxidase is responsible for the oxidation of reduced ubiquinone. Therefore, the alternative pathway would be extremely short, consisting of but one carrier, the quinol oxidase, after ubiquinone. Two attempts to isolate this protein have been reported recently (Huq and Palmer, 1978; Rich, 1978). The involvement of a protein would be consistent with the fact that in *Neurospora* the development of the alternative pathway is under the control of two nuclear genes (Edwards, 1978; Lambowitz and Zannoni, 1978).

2. *Present Picture*

(a) *Mitochondrial oxidations.* Figure 1 gives an example of the operation of the alternative pathway in plant mitochondria. Mitochondria isolated from fresh potato slices actively oxidize succinate (Fig. 1A). They respond to an addition of ADP by an increase in their rate of oxidation, which ceases when all added ADP has been converted into ATP. They are strongly sensitive to cyanide, indicating that only the cytochrome pathway is operative. By contrast, when mitochondria are isolated from tissue slices which have been aged for 24 h (Fig. 1B), cyanide is much less effective. It only inhibits 60% of the oxidation rate and the residual cyanide-resistant electron transport is then fully inhibited by SHAM. The experiment could have been conducted the other way round: SHAM being added before cyanide; in that case, one would have observed a SHAM-insensitive electron transport that would have been inhibited by cyanide. Therefore, a single inhibitor inhibits only a part of the

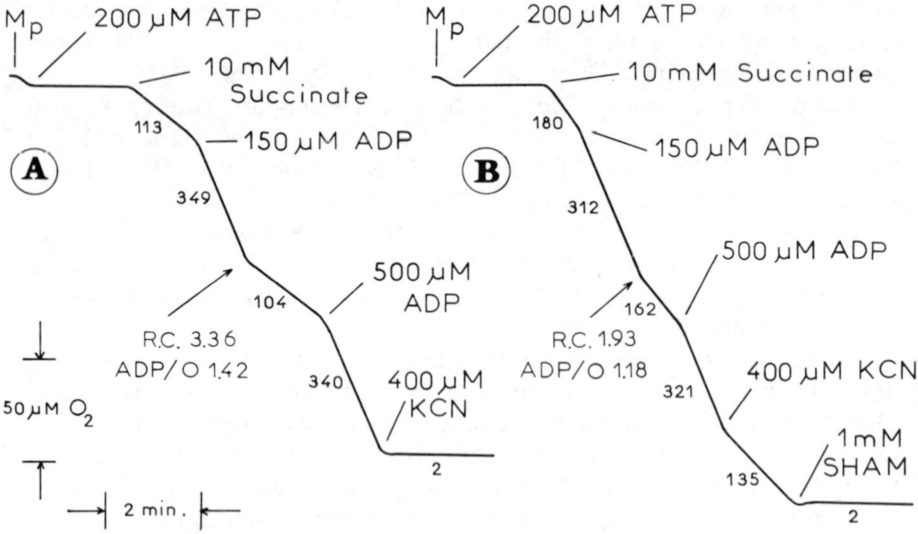

FIG. 1. Oxidation of succinate by mitochondria isolated from (A) fresh, or (B) aged white potato slices. Mp: purified mitochondria. From Dizengremel and Lance (1976).

whole electron transfer and both inhibitors have to be present simultaneously in order to bring all electron transport to a halt. This indicates that two pathways participate in the electron transport in mitochondria from aged tissue slices.

Figure 1 also shows that more oxygen is used up during the fast rate of oxidation (known as state 3 rate) associated with the phosphorylation of ADP by mitochondria isolated from aged tissues than by those isolated from fresh tissues. This can be directly traced to a decrease in the efficiency of oxidative phosphorylation as a consequence of the participation of the alternative pathway in the electron transport of aged slice mitochondria, as shown by the values of the ADP/O ratios (Fig. 1).

(b) *Electron transport pathways.* The behaviour of the white potato mitochondria in Fig. 1 can be easily interpreted according to the scheme in Fig. 2, which is a simplified picture of the electron transport system operating in plant mitochondria. As it can be seen, it is a branched pathway, with ubiquinone playing a central role in the distribution of electrons between the cytochrome and the alternative pathways. As a matter of fact, the cytochrome branch is more complex than indicated on the scheme, particularly at the cytochrome *b* level (Chance *et al.*, 1968; Palmer, 1976; Solomos, 1977).

Three phosphorylation sites are present, one of them (site I) being located on the part of the respiratory chain that is common to both pathways. It should be noted that the alternative pathway *per se* has no phosphorylation site. Therefore, its participation in electron transport decreases the efficiency of oxidative phosphorylation (Fig. 1). Theoretically, P/O ratios can vary from 3 to 1 for malate and from 2 to 0 for succinate, depending on the fraction of the electron flux (0 to 100%) that is mediated by the alternative pathway.

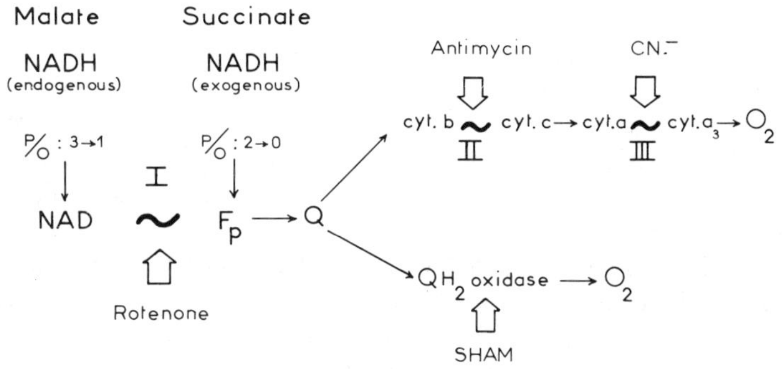

FIG. 2. Structure of the electron transport system in plant mitochondria. Fp: flavoproteins; Q: ubiquinone; Cyt.: cytochromes.

The scheme in Fig. 2 also shows the sites of action of the specific inhibitors (cyanide, antimycin, rotenone, SHAM) of electron transport in plant mitochondria. It shows too, that endogenous NADH produced by various NAD-linked dehydrogenases, such as malate dehydrogenase, donates its electrons to the respiratory chain before the first phosphorylation site (P/O: 3 → 1). On the other hand, exogenous NADH, experimentally added or produced in the cytosol, does so after phosphorylation site I (P/O: 2 → 0).

C. BASIC PROBLEMS

Although electron transport in plant mitochondria can be fairly well explained according to the scheme in Fig. 2, nevertheless two major questions arise when considering its actual operation in electron transport. One question pertains to the electron carrier composition of cyanide-insensitive and cyanide-sensitive mitochondria; the other one to the distribution of electrons between the two pathways.

1. *Presence of the Alternative Pathway*

The difference between cyanide-sensitive and cyanide-resistant mitochondria could be easily explained by assuming that the alternative pathway is absent in cyanide-sensitive mitochondria.

Many researches have been conducted to find some difference in the electron carrier composition of both types of mitochondria. The conclusion has been that the composition is much the same in both instances, although flavoproteins are generally more abundant in cyanide-insensitive tissues or mitochondria (van Herk, 1937; James and Beevers, 1950; Lance and Bonner, 1968). There is no difference in content of iron–sulphur protein (Dizengremel *et al.*, 1973) or in the nature of the various cytochrome components (Lance and Bonner, 1968). Nevertheless, it should be pointed out that cytochrome *b* is present in rather high concentrations in *Arum* mitochondria (Bendall and Hill, 1956). No data are available at the present time on the distribution of ubiquinone in both types of mitochondria. Examples of electron carrier distributions between cyanide-sensitive and cyanide-resistant mitochondria are given in Tables I and II.

From such results it was concluded that electron carrier compositions are very similar in all plant mitochondria. The implication is that the alternative pathway is present in all of them, with the provision that it is functional only under particular situations or circumstances.

Two arguments are in favour of such a conclusion. During the ageing of potato tissue slices one observes no change in the composition of mitochondrial membranes (Dizengremel, 1975; Dizengremel and Kader, 1980), yet the mitochondrial electron transport becomes 50% resistant to cyanide. The action of inhibitors of protein synthesis (cycloheximide, chloramphenicol) indicates that a protein component, which is not an electron carrier, is synthesized in the cytoplasm and makes the alternative pathway, formerly present but not functional, accessible to oxygen (Dizengremel and Lance, 1976).

TABLE I
Electron carriers in cyanide-sensitive and cyanide-resistant plant mitochondria

Material	cyt. aa_3	cyt. *b*	cyt. *c*	Fp
	(nmol/mg mitochondrial protein)			
A. *Cyanide-sensitive*				
Jerusalem artichoke	0·10	0·10	0·15	0·38
Mung bean	0·11	0·12	0·17	0·58
Cauliflower	0·12	0·15	0·21	0·69
White potato (fresh)	0·26	0·29	0·35	0·60
B. *Cyanide-resistant*				
White potato (aged)	0·21	0·27	0·33	0·55
Skunk cabbage	0·11	0·23	0·36	2·02
Sauromatum guttatum	0·12	0·36	0·32	0·95
Arum maculatum	0·18	0·75	0·44	1·86
Arum italicum	0·18	0·56	0·45	1·20

From Lance and Bonner (1968), Chauveau (1980) and Dizengremel (1980).

TABLE II
Iron–sulphur proteins in cyanide-sensitive and in cyanide-resistant plant mitochondria

Material	Inhibition by KCN (%)	Total Fe	Non-haem Fe	Labile S
		(nmol/mg mitochondrial protein)		
A. *Cyanide-sensitive*				
Jerusalem artichoke	98	9·95	8·65	6·12
Mung bean	80	10·20	9·05	5·55
Cauliflower	90	10·17	9·25	7·06
White potato (fresh)	96	8·80	7·70	5·80
B. *Cyanide-resistant*				
White potato (aged)	50	8·40	8·10	5·60
Sauromatum guttatum	0	8·40	7·10	6·68
Arum maculatum	20	8·75	7·66	6·70
Arum italicum	5	8·00	6·70	6·47

From Chauveau (1980) and Dizengremel (1980).

Another more convincing observation is that a cyanide-resistant electron transport can be experimentally induced even in the most cyanide-sensitive plant mitochondria. Depending on the experimental conditions, the oxidation of malate via malate dehydrogenase is strongly cyanide-sensitive, whereas the same oxidation carried out through malic enzyme is strongly cyanide-resistant (Rustin and Moreau, 1979; Rustin *et al.*, 1980).

Another point that should be noted is the relative instability of the alternative pathway. *Arum* mitochondria, which are normally strongly cyanide-insensitive, become cyanide-sensitive upon standing for two or three days at cold room temperature. In *Arum* and white potato mitochondria, this process can be greatly accelerated by exposure of mitochondria to temperatures of 35° to 45°C (Chauveau *et al.*, 1978).

2. *Distribution of electrons between the two pathways*

Assuming that both pathways are present and functional, the distribution of electrons between the two pathways has then to be considered. The actual electron flux depends on the availability and the nature of the substrates being oxidized. On the other hand, each one of the three parts of the respiratory chain—initial segment, cytochrome pathway, alternative pathway (Fig. 2)—has a maximal capacity for electron transfer. The problem then is to assess how the actual fluxes are distributed with regard to these parameters. Without going into details, three methods can be used, which give an estimation of the electron distribution.

(a) *The method based on phosphorylation efficiency.* This method is based on the recognition that the participation of the alternative pathway will lower the efficiency of oxidative phosphorylation in such mitochondria. All one has to do is to measure ADP/O ratios in the presence of various substrates and then alter the distribution of the electron flux by adding specific inhibitors (cyanide, SHAM) in order to obtain an estimation of the efficiency of the phosphorylation process in the cytochrome pathway. Then, very simple equations allow the determination of the distribution of the electrons between the two pathways. Examples of such determinations can be found in the literature (Lambowitz *et al.*, 1972; Lance, 1974; Passam, 1974).

The drawback of this method is that it requires mitochondria that are able to phosphorylate and produce ATP. Unfortunately, this is not the case in mitochondria that are uncoupled endogenously or in the presence of an uncoupler, or in mitochondria that do not actually phosphorylate (state 4, in the absence of ADP).

(b) *The method of Bahr and Bonner.* This method, proposed by Bahr and Bonner (1973a, b), overcomes this difficulty. It is based on the assumption that the cytochrome pathway operates at full capacity under any experimental condition. In addition, a fraction of the alternative pathway, measured by a factor ρ $(0<\rho<1)$, eventually contributes to the total electron flux. This method requires the measurement of the maximal electron transfer capacity in each pathway (V_{cyt} and V_{alt}) in the presence of appropriate inhibitors (SHAM and cyanide). The distribution of the actual electron flux (v_T) can be described by the following equation:

$$v_T = V_{cyt} + \rho \cdot V_{alt}$$

The titration by increasing concentrations of SHAM of the electron transfer rate in mitochondria oxidizing various substrates in the presence or absence of cyanide allows the expression of the results according to the above equation, from which a ρ value can be obtained (Bahr and Bonner, 1973a, b).

This method *a priori* favours the cytochrome pathway and requires that both inhibitors, SHAM and cyanide, remain strongly specific for their own pathway under any circumstances. This is not always the case (Chauveau, 1976b).

(c) *The method of de Troostembergh and Nyns.* This method was proposed by de Troostembergh and Nyns (1978) as an extension to cyanide-insensitive mitochondria of the model of Kröger and Klingenberg (1973) to account for the role of ubiquinone in animal mitochondria. It states that the assignment of the electrons takes place at the level of ubiquinone proportionally to the maximal electron transfer capacity of the cytochrome and the alternative pathways. In that case, no preference is given to the cytochrome pathway.

III. Participation of the Alternative Pathway

Rather than considering various plant organs exhaustively, this section will mainly deal with two selected examples: the situation in the white potato tuber and the measure of the contribution of the alternative pathway to the respiration of slices of various storage organs and fruits.

A. THE ALTERNATIVE PATHWAY IN POTATO TUBER

As in most storage organs the respiratory rate in white potato tuber is rather low (3 to 10 $\mu l\ O_2\ h^{-1}\ g^{-1}$ fresh weight). However, if the sensitivity of this respiration to cyanide is assayed, it is found that cyanide does not inhibit oxygen uptake but rather stimulates it five- to six-fold, as shown in Fig. 3 (Solomos and Laties, 1975). Ethylene has the same effect as cyanide on the respiration of white potato tuber (Fig. 3). On a fruit tissue, the avocado pear, cyanide and ethylene have also been found to exert similar actions (Solomos and Laties, 1974). The increase in the rate of oxidation results from an enhancement of the glycolytic pathway, and it is assumed that the increased electron transfer in the presence of cyanide is diverted to the alternative pathway. Both ethylene and cyanide would activate or effect the link between the conventional electron transport chain and the cyanide-insensitive pathway (Solomos and Laties, 1975). The existence of a cyanide-insensitive pathway is a prerequisite for the stimulation of respiration by ethylene (Solomos and Laties, 1976a, b), since ethylene has no effect on tissues whose respiration is strongly inhibited by cyanide.

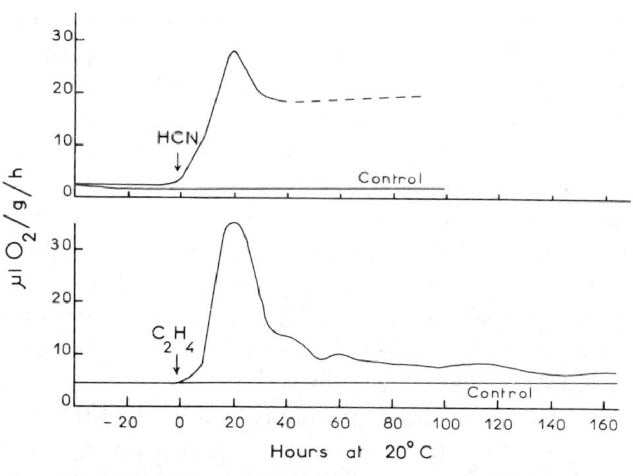

FIG. 3. Effect of cyanide and ethylene on the respiratory rate of white potato tubers. Adapted from Solomos and Laties (1975).

Such a stimulation of respiration by cyanide is not uncommon in storage organs; it occurs in cassava (Passam, 1976) and in sweet potato, beetroot, rutabaga as well as in many fruits (Solomos and Laties, 1976a).

Cyanide-insensitive tissues generally yield mitochondria that are also cyanide-insensitive. This is not only the case for the mitochondria isolated from the Araceae spadices but also for those from sweet potato tubers (Tomlinson and Moreland, 1975; Solomos and Laties, 1976a). However, in the case of white potato, mitochondria isolated from the tuber tissue or from fresh slices are strongly cyanide-sensitive (Fig. 1) (Hackett *et al.*, 1960; Dizengremel, 1975).

In white potato, but also in beetroot (Robertson *et al.*, 1947), slicing allows the tissue to become cyanide-sensitive where it was originally cyanide-insensitive. Generally, it takes 24 h of ageing in an aerated medium for the tissue and mitochondria isolated therefrom to recover their cyanide insensitivity (Hackett *et al.*, 1960).

The loss of mitochondrial cyanide resistance upon slicing was attributed to a particular sensitivity of mitochondrial membranes to physical injury. In white potato, cutting elicits a large phospholipid breakdown, with the result of a loss of cyanide resistance (Theologis and Laties, 1976; Laties, 1978). The recovery of cyanide resistance would be attributable to the synthesis of new phospholipid molecules during the ageing process. For the sweet potato tuber, the slicing would not impair the phospholipid composition and, consequently, the mitochondria would retain their cyanide resistance.

Such an explanation would therefore assume that the operation of the alternative pathway is highly dependent on the phospholipid composition of the mitochondrial membranes. It seems that this assumption should be re-evaluated. First, the phospholipid compositions of mitochondria isolated from fresh or aged potato tissue slices are very similar on a qualitative and a quantitative basis (Dizengremel and Kader, 1980). It was also found that treatment of potato tuber with ethylene induces the development of a cyanide-insensitive pathway that is retained in fresh slices and in mitochondria isolated therefrom (Rychter *et al.*, 1979). The same result is observed if the tubers are treated with ethanol, acetaldehyde or acetic acid (Janes *et al.*, 1979). The effect of these compounds is reinforced in the presence of pure oxygen.

The experiments which have been reported show the strong effect of some small molecules such as ethylene and cyanide or of some volatile end products of the glycolytic pathway in promoting the induction or activation of an alternative pathway in storage organs. This activation accompanies a rise in the respiratory rate. The correlation between the two events is quite clear. The alternative pathway probably accounts for the increased respiration. However, the significance of this should probably be different for fruits and storage organs. In one case, it could be associated with the onset of the climacteric brought about by ethylene or, indirectly, by cyanide (Solomos and Laties, 1974). Such an explanation cannot hold for storage organs.

B. CONTRIBUTION OF BOTH PATHWAYS

Since ethylene and cyanide have similar effects on both storage organs and fruits, resulting in a rise of respiration and in an activation of the alternative pathway, the knowledge of the contributions of the two pathways to the total electron transfer could shed some light on the significance of the alternative pathway.

Using the Bahr and Bonner (1973a) method with slices of such tissues, Theologis and Laties (1978a, b, c, d) have attempted to answer this question. Only two examples will be given. One deals with respiration in fresh and aged potato tuber slices (Theologis and Laties, 1978a, b), but the same results are also obtained with sweet potato slices (Theologis and Laties, 1978c). The other example will be taken from ripening avocado fruits; a similar situation also occurs in banana fruits (Theologis and Laties, 1978d).

As shown in Table III, the respiration of fresh potato slices (v_T) consists of two parts: one (V_{cyt}) is mediated by the cytochrome pathway and the other one, resistant to all inhibitors (cyanide, SHAM) added simultaneously, is called residual respiration (V_{res}). No alternative pathway is present in fresh potato slices. When the rate of electron transport is raised in the presence of the uncoupler CCCP, the increment of respiration is mediated by the cytochrome pathway. This indicates that in the control slices the cytochrome pathway is not used at its maximum capacity, the latter being expressed only in the presence of the uncoupler.

In aged slices the alternative pathway (V_{alt}) is present but normally does not participate in electron transport, as shown by a ρ value of 0. The cytochrome pathway and residual respiration can fully account for the total rate of electron transport in aged slices. In the presence of a low concentration of

TABLE III
Contribution of the cytochrome and the alternative pathways to the electron transport in fresh or aged potato slices

	Condition	v_T	V_{cyt}	V_{alt}	ρ	V_{res}
			(μl O_2 g fresh weight^{-1} h^{-1})			
Fresh	Control	35	24	0	0	11
	+CCCP	85	75	0	0	10
Aged	Control	112	80	65	0	32
	+KCN	116	48	65	0·55	32
	+CCCP	190	116	46	1·00	28

CCCP: carbonyl-cyanide *m*-chlorophenylhydrazone (10 μM); KCN: 10 μM cyanide; v_T: respiration rate; V_{cyt}, V_{alt}: maximal capacity of the cytochrome and alternative pathway, respectively; V_{res}: residual respiration (resistant to cyanide and SHAM). From Theologis and Laties (1978a).

cyanide (10 μM) part of the cytochrome pathway is curtailed, and the compensation comes from the alternative pathway, which operates at half its maximal capacity ($\rho = 0{\cdot}55$). In the presence of an uncoupler, the rate of oxidation is increased beyond the capacity of the cytochrome pathway. V_{cyt} is only multiplied by 1·5 in aged slices, while the factor is 3 in fresh slices. This means that in the control condition a large part of the cytochrome pathway was expressed. It is clear that in uncoupled slices all pathways (cytochrome, alternative and residual) operate at full capacity and limit the level of electron transfer in aged slices.

In avocado slices (Table IV) the picture is very similar. Although the alternative pathway is present at all stages of ripening, it does not normally participate in electron transport in slices of pre-climacteric or mid-climacteric fruits. It does so only in the presence of an uncoupler which nearly doubles the rate of electron transport. On the other hand, in post-climacteric slices, the uncoupler exerts but a very slight action. In that case, one observes no participation of the alternative pathway.

TABLE IV
Contribution of the cytochrome and the alternative pathways to the electron transport in avocado slices at different stages of ripening

Condition	v_T	V_{cyt}	V_{alt}	ρ	V_{res}
	(μl O_2 g fresh weight^{-1} h^{-1})				
Pre-climacteric					
Control	147	90	121	0	55
+CCCP	271	107	109	1	51
Mid-climacteric					
Control	151	85	109	0	64
+CCCP	249	105	120	0·9	41
Post-climacteric					
Control	165	80	102	0	77
+CCCP	192	102	57	0	85

See footnote to Table III. From Theologis and Laties (1978d).

One of the striking aspects of these results is the observation that in many instances the activity of the cytochrome pathway is not fully expressed under normal conditions, as shown by the effects of uncouplers. The other important fact is that, in the situations where the alternative pathway is present, it nevertheless does not participate in electron transport. It does so only in the presence of cyanide or, eventually, of an uncoupler.

These observations raise the question of the physiological significance of the alternative pathway. One criticism that can be opposed to these conclusions is the assumption that the cytochrome pathway is always working at maximal capacity (Bahr and Bonner, 1973a). Obviously this is not the case in fresh potato slices. At any rate, this hypothesis theoretically lessens the actual participation of the alternative pathway and could eventually account for the observed results.

IV. Physiological Significance of the Alternative Pathway

Despite the amount of work that has been invested in the study of the alternative pathway, progress has been slow and even today the number of unanswered questions overwhelmingly exceeds the number of points of agreement. The nature of the alternative pathway itself is still unclear. Its terminal oxidase is ill-defined and the number and nature of its electron carriers are still open questions. One does not even know if the scheme of cyanide insensitivity in plants proposed in Fig. 2 is the only pathway, or if there is more than one way to achieve cyanide resistance, as the many hypotheses that have been put forward would tend to suggest. The only points of agreement are on its non-phosphorylating nature, on its branching from the main pathway at the level of ubiquinone and on its widely acknowledged distribution in the plant kingdom.

These observations indicate that the alternative pathway must have some general significance in plant respiration. In some instances, its role seems to be rather clear, but most of the time one has to speculate on its significance. Some possible roles are listed below.

A. Thermogenesis in the Araceae

This is probably the only instance in which a clear role can be attributed to the alternative pathway (Meeuse, 1975). At some stage in the reproductive phase, the complex inflorescence of the plants belonging to the Araceae family displays a huge respiratory crisis. The site of this crisis is generally the sterile part of the inflorescence. Starch, which has been accumulated up to this stage, is burned up in a few hours due to an extremely high rate of glycolysis and respiration (Lance, 1972; ap Rees *et al.*, 1977).

Although the alternative pathway is present and functional at all stages of the inflorescence development, its participation in electron transport becomes predominant at this stage. By the operation of a non-phosphorylating alternative pathway and that of an endogenously uncoupled cytochrome pathway, the efficiency of the electron transport system becomes very low, although it still provides enough ATP for the economy of the cell, due to the high rate of respiration (Lance and Chauveau, 1975). The excess of energy that is not used for ATP synthesis is liberated as heat. As a consequence, the temperature of the sterile part of the spadix is raised 10–15°C above that of the environment. The warming up of the spadix tissue allows the emission of a great number of volatile amines (Smith and Meeuse, 1966) which attract pollinator insects.

In this instance, the operation of the alternative pathway can be closely related to thermogenesis in these tissues. However, it should be mentioned that thermogenesis is also observed in some animal tissues, particularly in the brown adipose tissue of new-born or cold-adapted mammals. In that case, the mechanism of thermogenesis is completely different from the one in the Araceae spadices (Nicholls, 1979).

B. CYANOGENIC PLANT SPECIES

A wide variety of plant tissues have the ability to produce toxic levels of cyanide. Many plant organs—tubers, leaves or seeds—contain compounds known as cyanogenic glycosides, such as amygdalin in the seeds of numerous Rosaceae, from which cyanide can be released. Cyanogenic glycosides are widely distributed in the plant kingdom (Seigler, 1977; Conn, 1980).

The problem is of importance since one of these cyanogenic plants, the root tuber of cassava (*Manihot esculenta*), is a major staple food in the tropics. Passam (1976) has made a study of the respiration of root tubers of cassava. Not surprisingly, the respiration of discs of tissue prepared from the tubers is stimulated by cyanide, while the respiratory rates of cultivars with high endogenous cyanide contents are usually lower than those of cultivars with low cyanide contents.

Similarly, mitochondria isolated from these tubers are resistant to cyanide and antimycin and their cyanide resistant electron transport can be inhibited by SHAM. Also, their degrees of cyanide resistance vary according to the cultivars from which they are isolated. From this, one could conclude that the alternative pathway could play a specific role in the respiration of organs containing cyanogenic glycosides.

However, one should be extremely cautious about such a conclusion. The alternative pathway could well be present in these tissues, but not for the above reasons. As a matter of fact, it is extremely difficult to measure endogenous concentrations of cyanide. Cyanide is produced as a result of an enzymic breakdown of the glycoside, first by a β-glucosidase and then by a hydroxynitrile lyase. Generally, some compartmentation exists between enzymes and glycosides within the cell or the tissue (Saunders *et al.*, 1977; Saunders and Conn, 1978). Cyanide is produced only when the two partners come into contact, during cutting, peeling or crushing of the plant organ. Although the amounts of cyanide that are released from cyanogenic plants can be quite high (75 mg/100 g fresh weight in cassava, 400 mg/100 g fresh weight in lima bean), it has to be expected that the endogenous concentrations of cyanide are probably very much lower in the intact tissue (Passam, 1976; Conn, 1979a).

As a matter of fact, endogenous concentrations of cyanide, even in cyanogenic species, could well be very close to nil. Two arguments are in favour of this view. First, cyanide is not a reactant of the biosynthetic pathways of the cyanogenic glycosides. Instead, the nitrile function present in cyanogenic glycosides is derived from the amino group and carbon-2 of amino acids such as tyrosine, phenylalanine, valine or leucine (Conn, 1979b, 1980), which are the precursors of the aglycone moiety of most glycosides.

Furthermore, defence mechanisms against endogenous production of cyanide seem to exist in cyanogenic species. If, for some reason, glycosides and enzymes come into contact, the cyanide that is produced can be removed by specific enzymes such as β-cyanoalanine synthase or rhodanese (Miller and Conn, 1980). A good correlation is observed between the ability to produce cyanide and cyanide metabolizing activity.

It should also be noted that the enzyme β-cyanoalanine synthase, coupled to a hydrolase, can easily convert cyanide, administered as a gas, into asparagine, using cysteine as an acceptor. This process occurs not only in cyanogenic plants but also in non-cyanogenic plants. It should be borne in mind that such a mechanism for cyanide removal could perhaps explain some cases of cyanide-resistant respiration in plant tissues.

C. ELIMINATION OF REDUCING POWER

Although the operation of the alternative pathway could be implicated in the two particular instances that have just been mentioned, one has to recognize that its distribution is much more general in plants. Its components are potentially present in all plant mitochondria, even though it is not always functional. Therefore, the significance of this pathway should eventually be traced to the characteristic properties of plant mitochondria and, in no case, confined to some particular situations.

Compared to animal mitochondria, plant mitochondria are endowed with unique properties: their electron transport can be cyanide-insensitive, but they are also able to oxidize malate at a high rate by two different mechanisms and to oxidize NADH in a number of different ways (Palmer, 1976).

Regarding this particular point, plant mitochondria oxidize endogenously produced NADH through a rotenone-sensitive NADH dehydrogenase. On the other hand, they can also oxidize exogenous (externally added or cytoplasmically produced) NADH by two different pathways. One involves a rotenone-insensitive NADH dehydrogenase located on the outer face of the inner membrane (Douce *et al.*, 1973). The second involves the participation of a NADH dehydrogenase located on the outer membrane, acting in con-

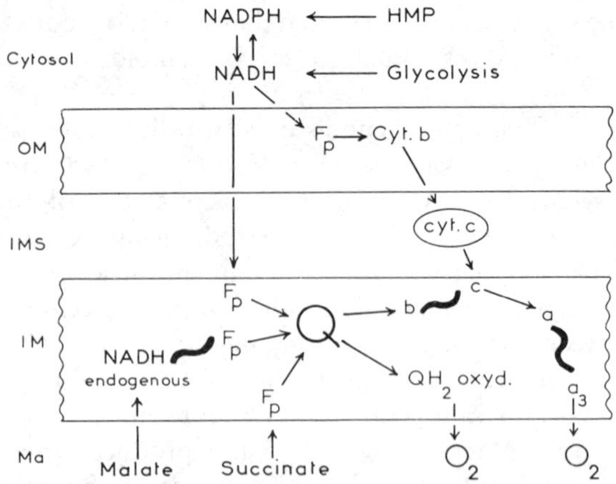

FIG. 4. Pathways for the oxidation of various substrates in plant mitochondria. OM: outer membrane; IMS: inner membrane space; IM: inner membrane; Ma: matrix space; HMP: hexose monophosphate pathway; Fp: various NADH dehydrogenases or succinate dehydrogenase; a, a_3, b, c: cytochromes.

junction with soluble cytochrome *c* in the intermembrane space (Moreau, 1976). This is shown in Fig. 4.

In cyanide-sensitive mitochondria, depending on the pathway used, between one and three molecules of ATP can be produced for each NADH oxidized. It is therefore clear that plant mitochondria can regulate the oxidation of NADH with respect to the production of ATP, by switching the electron flux from one pathway to another. However, endogenous NADH produced in the mitochondrial matrix is bound to three phosphorylation sites.

The striking feature of the alternative pathway is that it is a non-phosphorylating pathway. At first sight, it could appear as a wasteful process. When it is functional, endogenous NADH is oxidized with the production of only one ATP, whereas no ATP at all is produced from the oxidation of exogenous NADH through the inner membrane dehydrogenase. By these various ways, the oxidation of NADH, which directly or indirectly represents the main reducing power of the plant cell, can be carried out with as little production of ATP that is suitable for the cell.

This could be one of the finalities of the alternative pathway. When the level of ATP (energy charge or phosphate potential) is high within the cell, regulatory mechanisms, mainly ones involving allosteric effects, tend to reduce the intensity of cell metabolism. However, one of the major roles of the Krebs cycle is not only to produce reducing equivalents that will generate ATP through the process of oxidative phosphorylation, but also to metabolize products going to or coming from various metabolic pathways.

A high energy charge would tend to slow down the production of ATP and also that of metabolites, if there were a strong control of the energy charge on the rate of NADH oxidation. The alternative pathway seems to be perfectly suited to overcome or decrease this control.

The corollary of this is the idea that the primary need of the cell is for the removal of an excess of reducing power rather than for having a very efficient electron transport process. This could well be the case, since it is generally the capacity to produce work or use energy that controls the cell respiratory rate through the utilization of ATP. Therefore, if for some reason the reducing power generated by biosynthetic pathways and the production of ATP linked to electron transport in the respiratory chain happen to become somewhat unbalanced, the alternative pathway could be used as a well adapted regulatory mechanism.

D. ROLE IN MALATE OXIDATION

The immediate reducing power of plant cells is found in various reduced coenzymes (NADH, NADPH, $FADH_2$, etc.). However, the ultimate source of reducing power is to be found in the organic carbon compounds that are produced from the reduction of CO_2 in the photosynthetic process. All substrates are potential sources of reducing power, and among them malate ($\simeq$ reduced oxaloacetate) appears to play a major role in the physiology of plant cells.

There are numerous physiological situations in which malate appears as the main factor involved, as well as the enzymes linked to its metabolism (malate dehydrogenase and malic enzyme). Moreover, as these enzymes are found in various cell compartments such as cytosol, plastids, peroxisomes and mitochondria, all conditions are met for making malate a regulatory component of the cell metabolism.

Just to give a few examples, malate metabolism and enzymes related to it can show large variations of activity in a variety of processes. This is the case in diseased plants (Millerd and Scott, 1962), in climacteric or senescent fruits (Hulme *et al.*, 1967, 1968; Lance *et al.*, 1967; Hartman, 1973; Sacher, 1973), in ageing tissue slices (Dizengremel, 1975) or in the *Arum* spadix (Wedding *et al.*, 1975, 1978). However, the most spectacular example is probably found in CAM (crassulacean acid metabolism) plants, where CO_2 is stored at night as malate, which is decarboxylated in the morning to yield CO_2, NADH and pyruvate, through the action of malic enzyme (Osmond, 1978).

On the other hand, in plant mitochondria, it has been recently demonstrated that the oxidation of malate by malate dehydrogenase is strongly linked to the phosphorylating, cyanide-sensitive cytochrome pathway, while the oxidation of this substrate by malic enzyme is linked to the non-phosphorylating, cyanide-insensitive alternative pathway (Rustin and Moreau, 1979; Rustin *et al.*, 1980). This can be easily shown by adding NAD^+ to mitochondria oxidizing malate in the presence of cyanide (Fig. 5). The effect of adding NAD^+ is to activate malic enzyme (Macrae and Moorhouse, 1970).

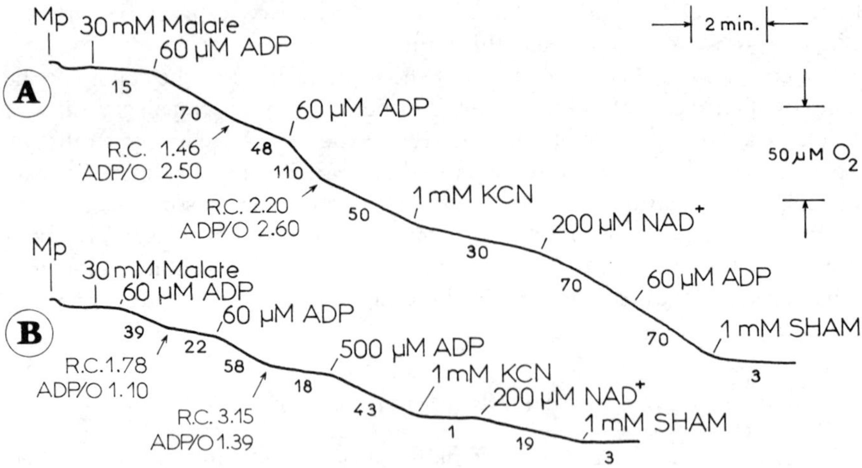

FIG. 5. Oxidation of malate in plant mitochondria via malate dehydrogenase and malic enzyme. A: cauliflower mitochondria; B: white potato mitochondria; Mp: purified mitochondria. From Rustin *et al.* (1980).

Therefore, as for NADH, plant mitochondria have the ability to oxidize malate with various degrees of efficiency since the number of ATP produced can vary from 0 to 3. In other words, malate, too, can be oxidized in a way that is independent of an eventual control by the energy charge. This represents an evident advantage for plants that have to metabolize large amounts of malate over a short period of time, such as CAM plants, or in response to rapid changes in their physiological status.

E. INTEGRATION OF THE ALTERNATIVE PATHWAY IN CELL METABOLISM

From what has been said above, it seems that the alternative pathway can be of utility to the plant cell in two different ways: for carrying an active electron transport resulting in thermogenesis or, more generally, for channelling a non-phosphorylating electron transport in the presence of a high energy charge. In the latter case, there is a release from control by the cell energy charge; the reducing power can be removed and substrates necessary for metabolic pathways can be produced according to demand.

It is probably in such directions, by linking the alternative pathway to the general metabolism of the plant cell, that this pathway can find some significance.

An example of such an integration is depicted in Fig. 6, which is half experimental and half speculative, in that all the individual steps have been experimentally proved, only the links between them having been established as a result of the above considerations.

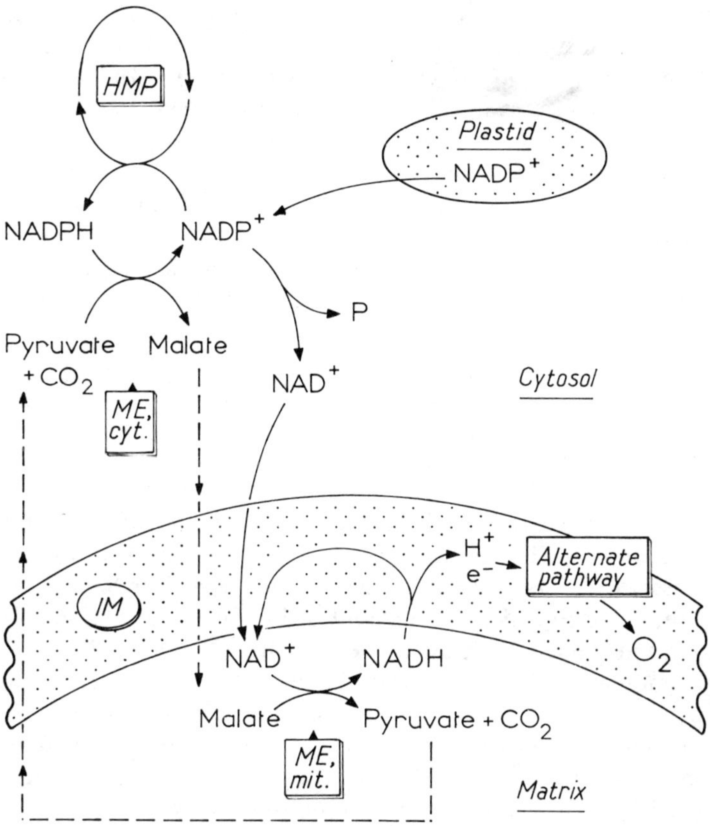

FIG. 6. Example of the integration of the alternative pathway and malate oxidation by malic enzyme in the removal of the reducing power generated in the cytosol. HMP: hexose monophosphate pathway; IM: inner mitochondrial membrane; ME: malic enzyme (cytosolic or mitochondrial).

The original facts are taken from various works dealing with the infection of plant leaves by the powdery mildew (Millerd and Scott, 1962; Edwards and Allen, 1966; Ryrie and Scott, 1968; Magyarosy *et al.*, 1976).

Upon infection by the parasite, one observes a decrease in the rate of photosynthesis. The compartmentation of $NADP^+$ is altered. Initially concentrated in the plastid compartment, $NADP^+$ is found in greater amounts in the cytosol. The activity of the hexose monophosphate pathway is increased. At the same time, the NAD^+ concentration increases in the cytoplasm, as well as the concentration of malate and malic enzyme activity. The infection is also accompanied by a considerable rise in the respiratory rate with a decrease in phosphorylative efficiency (Millerd and Scott, 1962).

All these facts can be integrated in the following way. The decompartmentalization of $NADP^+$ stimulates both the hexose monophosphate pathway and the cytosolic $NADP^+$-dependent malic enzyme. It also raises the level of cytosolic NAD^+. The increased concentration in NAD^+ favours the oxidation of malate via a NAD^+-dependent mitochondrial malic enzyme. If there is a malate–pyruvate shuttle operating between the cytosol and the mitochondrial matrix, the reducing equivalents of the NADPH produced in the cytosol by the hexose monophosphate pathway can be reoxidized in the mitochondria with no production of ATP through the mitochondrial malic enzyme linked to the alternative pathway. Such a scheme can account for most of the facts reported above and could be taken as an example of the involvement of the alternative pathway in a physiological process.

V. Conclusions

Despite a large amount of work, the nature of the alternative pathway is still elusive. Nevertheless, this pathway is so widely distributed in plants that it must play a role of general significance for the physiology of plants. Up to the present, the emphasis has been put on its effects on electron transport and energy conservation. Its ability to oxidize substrates such as NADH or malate in a way that is independent of the control by the energy charge of the cell draws attention to its relationship to the biosynthetic activity of the cell. There is a large amount of data scattered in the literature on fruit physiology that deal with cyanide-insensitive respiration, distribution of malate and NAD^+, changes in enzyme activity or compartmentation, etc. The above considerations could provide a link between these dispersed facts and allow their full integration in the cell or tissue metabolism.

REFERENCES

ap Rees, T., Wright, B. W. and Fuller, W. A. (1977). *Planta* **134**, 53–56.
Bahr, J. T. and Bonner, W. D. (1973a). *J. Biol. Chem.* **248**, 3441–3445.
Bahr, J. T. and Bonner, W. D. (1973b). *J. Biol. Chem.* **248**, 3446–3450.
Bendall, D. S. and Bonner, W. D. (1971). *Plant Physiol.* **47**, 236–245.
Bendall, D. S. and Hill, R. (1956). *New Phytol.* **55**, 206–212.
Bonner, W. D. (1965), *In* "Plant Biochemistry" (J. Bonner and J. E. Varner, Eds), pp. 89–123. Academic Press, New York and London.
Cammack, R. and Palmer, J. M. (1973). *Ann. N.Y. Acad. Sci.* **222**, 816–823.
Chance, B., Bonner, W. D. and Storey, B. T. (1968). *Ann. Rev. Plant Physiol.* **19**, 293–320.
Chauveau, M. (1976a). *Physiol. Vég.* **14**, 309–323.
Chauveau, M. (1976b). *Physiol. Vég.* **14**, 325–337.
Chauveau, M. (1980). Ph.D. thesis, Paris, pp. 277.
Chauveau, M., Dizengremel, P. and Lance, C. (1978). *Physiol. Plant.* **42**, 214–220.
Conn, E. E. (1979a). *Intern. Rev. Biochemistry* **27**, 21–43.
Conn, E. E. (1979b). *Naturwissenschaften* **66**, 28–34.

Conn, E. E. (1980). *In* "Encyclopedia of Plant Physiology" (E. A. Bell and B. V. Charlwood, Eds), Vol. 8, pp. 461–492. Springer-Verlag, Berlin.
de Troostembergh, J. C. and Nyns, E. J. (1978). *Eur. J. Biochem.* **85**, 423–432.
Dizengremel, P. (1975). *Physiol. Vég.* **13**, 39–54.
Dizengremel, P. (1980). Ph.D. thesis, Paris, pp. 272.
Dizengremel, P. and Kader, J. C. (1980). *Phytochem.* **19**, 211–214.
Dizengremel, P. and Lance, C. (1976). *Plant Physiol.* **58**, 147–151.
Dizengremel, P., Chauveau, M. and Lance, C. (1973). *C. R. Acad. Sci.* **277**, série D, 239–242.
Douce, R., Mannella, C. A. and Bonner, W. D. (1973). *Biochim. Biophys. Acta* **292**, 105–116.
Ducet, G. and Rosenberg, A. J. (1962). *Ann. Rev. Plant Physiol.* **13**, 171–200.
Edwards, D. L. (1978). *In* "Functions of Alternative Terminal Oxidases" (H. Degn, D. Lloyd and G. C. Hill, Eds), pp. 21–29. Pergamon Press, Oxford.
Edwards, D. L., Chalmers, J. H., Guzig, H. J. and Warden, J. T. (1976). *In* "Genetics and Biogenesis of Mitochondria and Chloroplasts" (T. Bücher, Ed.), pp. 865–879. Elsevier/North Holland, Amsterdam.
Edwards, H. H. and Allen, P. J. (1966). *Plant Physiol.* **41**, 683–688.
Hackett, D. P. (1957). *J. Exp. Bot.* **8**, 157–171.
Hackett, D. P. (1959). *Ann. Rev. Plant Physiol.* **10**, 113–146.
Hackett, D. P., Haas, D. W., Griffiths, S. K. and Niederpruem, D. J. (1960). *Plant Physiol.* **35**, 8–19.
Haddock, B. A. and Garland, P. B. (1971). *Biochem. J.* **124**, 155–170.
Hartman, C. (1973). *C. R. Acad. Sci.* **276**, série D, 1851–1853.
Henry, M. F. and Nyns, E. J. (1975). *Sub-Cell. Biochem.* **4**, 1–65.
Henry, M. F., Bonner, W. D. and Nyns, E. J. (1977). *Biochim. Biophys. Acta* **460**, 94–100.
Hulme, A. C., Rhodes, M. J. C. and Wooltorton, L. S. C. (1967). *Phytochem.* **6**, 1343–1351.
Hulme, A. C., Rhodes, M. J. C., Galliard, T. and Wooltorton, L. S. C. (1968). *Plant Physiol.* **43**, 1154–1161.
Huq, S. and Palmer, J. M. (1978). *FEBS Lett.* **95**, 217–220.
Ikuma, H. (1972). *Ann. Rev. Plant Physiol.* **23**, 419–436.
James, W. O. and Beevers, H. (1950). *New Phytol.* **49**, 353–374.
James, W. O. and Elliott, D. C. (1955). *Nature* **175**, 89.
Janes, H. W., Rychter, A. and Frenkel, C. (1979). *Plant Physiol.* **63**, 837–840.
Kröger, A. and Klingenberg, M. (1973). *Eur. J. Biochem.* **34**, 358–368.
Lambowitz, A. M. and Zannoni, D. (1978). *In* "Plant Mitochondria" (G. Ducet and C. Lance, Eds), pp. 283–291. Elsevier/North Holland, Amsterdam.
Lambowitz, A. M., Smith, E. W. and Slayman, C. W. (1972). *J. Biol. Chem.* **247**, 4859–4865.
Lance, C. (1972). *Ann. Sci. Nat. Bot.* **13**, 477–495.
Lance, C. (1974). *Plant Sc. Lett.* **2**, 165–171.
Lance, C. and Bonner, W. D. (1968). *Plant Physiol.* **43**, 756–766.
Lance, C. and Chauveau, M. (1975). *Physiol. Vég.* **13**, 83–94.
Lance, C., Hobson, G. E., Young, R. E. and Biale, J. B. (1967). *Plant Physiol.* **42**, 471–478.
Laties, G. G. (1978). *In* "Biochemistry of Wounded Plant Tissues" (G. Kahl, Ed.), pp. 421–466. Walter de Gruyter, Berlin.
Macrae, A. R. and Moorhouse, R. (1970). *Eur. J. Biochem.* **16**, 96–102.
Magyarosy, A. C., Schürmann, P. and Buchanan, B. B. (1976). *Plant Physiol.* **57**, 486–489.

Meeuse, B. J. D. (1975). *Ann. Rev. Plant Physiol.* **26**, 117–126.
Miller, J. M. and Conn, E. E. (1980). *Plant Physiol.* **65**, 1199–1202.
Millerd, A. and Scott, K. J. (1962). *Ann. Rev. Plant Physiol.* **13**, 559–574.
Moore, A. L., Rich, P. R., Bonner, W. D. and Ingledew, W. J. (1976). *Biochem. Biophys. Res. Commun.* **72**, 1099–1107.
Moreau, F. (1976). *Plant Sc. Lett.* **6**, 215–221.
Nakano, M. and Asahi, T. (1970). *Plant Cell Physiol.* **11**, 499–502.
Nicholls, D. G. (1979). *Biochim. Biophys. Acta* **549**, 1–29.
Osmond, C. B. (1978). *Ann. Rev. Plant Physiol.* **29**, 379–414.
Palmer, J. M. (1976). *Ann. Rev. Plant Physiol.* **27**, 133–157.
Parrish, D. P. and Leopold, A. C. (1978). *Plant Physiol.* **62**, 470–472.
Passam, H. C. (1974). *J. Exp. Bot.* **25**, 653–657.
Passam, H. C. (1976). *Plant Sc. Lett.* **7**, 211–218.
Rich, P. R. (1978). *FEBS Lett.* **96**, 252–256.
Rich, P. R. and Bonner, W. D. (1978a). *In* "Functions of Alternative Terminal Oxidases" (H. Degn, D. Lloyd and G. C. Hill, Eds), pp. 149–158. Pergamon Press, Oxford.
Rich, P. R. and Bonner, W. D. (1978b). *Biochim. Biophys. Acta* **501**, 381–395.
Rich, P. R. and Bonner, W. D. (1978c). *Biochim. Biophys. Acta* **504**, 345–363.
Rich, P. R. and Moore, A. L. (1976). *FEBS Lett.* **65**, 339–344.
Rich, P. R., Boveris, A., Bonner, W. D. and Moore, A. L. (1976). *Biochem. Biophys. Res. Commun.* **71**, 695–703.
Robertson, R. N., Turner, J. S. and Wilkins, M. T. (1947). *Aust. J. Exp. Biol. Med. Sci.* **25**, 1–8.
Rustin, P. and Moreau, F. (1979). *Biochem. Biophys. Res. Commun.* **88**, 1125–1131.
Rustin, P., Moreau, F. and Lance, C. (1980). *Plant Physiol.* **66**, 456–462.
Rychter, A., Janes, H. W. and Frenkel, C. (1978). *Plant Physiol.* **61**, 667–668.
Rychter, A., Janes, H. W. and Frenkel, C. (1979). *Plant Physiol.* **63**, 149–151.
Ryrie, I. J. and Scott, K. J. (1968). *Plant Physiol.* **43**, 687–692.
Sacher, J. (1973). *Ann. Rev. Plant Physiol.* **24**, 197–224.
Saunders, J. A. and Conn, E. E. (1978). *Plant Physiol.* **61**, 154–157.
Saunders, J. A., Conn, E. E., Lin, C. H. and Stocking, C. R. (1977). *Plant Physiol.* **59**, 647–652.
Schonbaum, G. R., Bonner, W. D., Storey, B. T. and Bahr, J. T. (1971). *Plant Physiol.* **47**, 124–128.
Seigler, D. S. (1977). *In* "Progress in Phytochemistry" (L. Reinhold, J. B. Harborne and T. Swain, Eds), Vol. 4, pp. 83–120. Pergamon Press, New York and Oxford.
Slayman, C. W. (1978). *In* "Functions of Alternative Terminal Oxidases" (H. Degn, D. Lloyd and G. C. Hill, Eds), pp. 159–168. Pergamon Press, Oxford.
Smith, B. N. and Meeuse, B. J. D. (1966). *Plant Physiol.* **41**, 343–347.
Solomos, T. (1977). *Ann. Rev. Plant Physiol.* **28**, 279–297.
Solomos, T. and Laties, G. G. (1974). *Plant Physiol.* **54**, 506–511.
Solomos, T. and Laties, G. G. (1975). *Plant Physiol.* **55**, 73–78.
Solomos, T. and Laties, G. G. (1976a). *Biochem. Biophys. Res. Commun.* **70**, 663–671.
Solomos, T. and Laties, G. G. (1976b). *Plant Physiol.* **58**, 47–50.
Storey, B. T. (1976). *Plant Physiol.* **58**, 521–525.
Theologis, A. and Laties, G. G. (1976). *Plant Physiol.* **57**, S. 93.
Theologis, A. and Laties, G. G. (1978a). *Plant Physiol.* **62**, 232–237.
Theologis, A. and Laties, G. G. (1978b). *Plant Physiol.* **62**, 238–242.
Theologis, A. and Laties, G. G. (1978c). *Plant Physiol.* **62**, 243–248.
Theologis, A. and Laties, G. G. (1978d). *Plant Physiol.* **62**, 249–255.
Tomlinson, P. F. and Moreland, D. E. (1975). *Plant Physiol.* **55**, 365–369.
van Herk, A. W. H. (1937). *Rec. Trav. Bot. Néerl.* **34**, 69–156.

van Herk, A. W. H. and Badenhuizen, N. P. (1934). *Proc. Kon. Akad. Wettensch.* **37**, 99–105.
Wedding, R. T., McCready, C. C. and Harley, J. L. (1975). *New Phytol.* **74**, 1–17.
Wedding, R. T., Black, M. K. and Harley, J. L. (1978). *New Phytol.* **81**, 211–222.

CHAPTER 4

Biosynthesis of Ethylene and Its Regulation

S. F. YANG

Department of Vegetable Crops, University of California, Davis, California, USA

I.	Introduction	89
II.	Pathway of Ethylene Biosynthesis	90
III.	Mechanism of Auxin-induced Ethylene Production in Vegetative Tissue	95
IV.	Regulation of Ethylene Production in Ripening Fruits	95
V.	Regulation of Stress-induced Ethylene Production	97
VI.	Inhibition of Ethylene Production	101
VII.	Conclusion	102
	Acknowledgement	104
	References	104

I. INTRODUCTION

The ability of certain gaseous agents to stimulate fruits to ripen has been known for many years. The Chinese knew for hundreds of years that picked fruits would ripen more quickly if placed in a chamber containing burning incense (Salisbury and Ross, 1969). This effect we now attribute to the action of ethylene. Another historical practice implicating still a different role for ethylene was the application of olive oil to hasten the enlargement and maturation of fig fruits, which has been practised pre-dating Christianity, and it is now clear that this effect is due to the action of ethylene derived from the degradation of olive oil (Saad *et al.*, 1969). The use of smoke to induce pineapple plants to bloom out of season was discovered in 1893 as a result of an accidental fire (Grabham, 1903); to the surprise of the grower, the plants burst into flower instead of being damaged. Interest in ethylene as a plant growth regulator has extended over many years. Although it was clear even in the mid-nineteenth century that the presence of gaseous materials in the air could modify the growth of plants (Girardin, 1864), the identity of ethylene as an active agent was not established until the turn of the century by Neljubow

(1901). He reported that ethylene was the component of illuminating gas causing reduction of elongation, increased radial expansion, and horizontal growth in pea seedlings.

The first suggestion that biological material releases a gas which affects the growth of nearby plant material was reported by Cousins (1910), who observed that gases from oranges caused premature ripening of bananas in mixed commercial shipments. However, it is now certain that the ethylene produced did not come from the oranges but from fungus growing on the oranges, since uninfected oranges produced very little ethylene. This copious production of ethylene by infected oranges has led Biale (1940) and Miller *et al.* (1940) to discover that the green mould, *Penicillium digitatum*, produces ethylene at high rates. Elmer (1932) showed that the presence of ripe apples and pears caused abnormal growth of potato sprouts. Independently, Kidd and West (1932) observed that a gaseous emanation from ripe apples could stimulate the ripening of unripe apples, an effect likewise caused by ethylene. Two years later Gane (1934) provided chemical proof that ethylene was indeed produced by ripe apples. Since then it has been shown that ethylene is produced from nearly all parts of higher plants including leaves, stems, roots, flowers, fruits, tubers, and seedlings, although the production rate is normally low (Burg, 1962). As part of the normal life of a plant, ethylene production is induced during certain stages of growth, including seed germination, ripening of fruit, leaf and flower senescence and abscission (Abeles, 1973). In vegetative tissue, ethylene production is known to be regulated by another class of hormone, the auxins. Ethylene production can also be induced by many external factors, including application of auxins, physical wounding and cutting, chilling injury, drought and water flooding (Yang and Pratt, 1978). It has been recognized that this increased ethylene production in turn brings about many important physiological consequences (Abeles, 1973).

Many reviews on general ethylene physiology and fruit ripening are available (Burg, 1962; Pratt and Goeschl, 1969; Abeles, 1973; Lieberman, 1979). In this chapter I shall focus my discussion on the biosynthesis of ethylene and its regulation in plant tissue.

II. Pathway of Ethylene Biosynthesis

Methionine was first suggested as a possible precursor of ethylene by Lieberman and Mapson (1964) based on the observation that methionine was rapidly converted into ethylene in a model system consisting of Cu^{2+} and ascorbic acid. Methionine was later shown to be the biological precursor in apple tissue (Lieberman *et al.*, 1966). These findings were subsequently extended to other plant tissues including banana and IAA-treated pea stems (Burg and Clagett, 1967); tomato and cauliflower floret (Mapson *et al.*, 1970); avocado (Baur *et al.*, 1971); chemical-stressed bean and tobacco leaves

(Abeles and Abeles, 1972); IAA-treated mung bean seedling (Sakai and Imaseki, 1972); and senescent morning-glory flowers (Hanson and Kende, 1976). In these tissues the ability to convert methionine to ethylene parallels the ability to produce ethylene. In apple tissue, it was shown that C-1 of methionine is converted to CO_2, C-2 to formic acid, and C-3,4 to ethylene. Sulphur, however, is retained in the tissue, since no volatile radioactivity was detected after feeding S-labelled methionine (Yang, 1974). Although apple is known to maintain a substantial rate of ethylene production for extended periods, its methionine level is quite low. This led Baur and Yang (1972) to suggest that in order for the apple to continue to produce ethylene, the sulphur of methionine must be recycled to provide an adequate supply for continued ethylene production. Adams and Yang (1977) have presented experimental evidence showing that 5′-methylthioadenosine (MTA) is the primary product derived from the CH_3S group of methionine during its conversion to ethylene. These results indicate that S-adenosylmethionine (SAM) is an intermediate in the biosynthesis of ethylene from methionine and suggest that MTA must be cycled back to form methionine to maintain ethylene production. This prediction was fully substantiated experimentally and the results are summarized in Fig. 1. Methionine is first activated by ATP to form SAM, which is

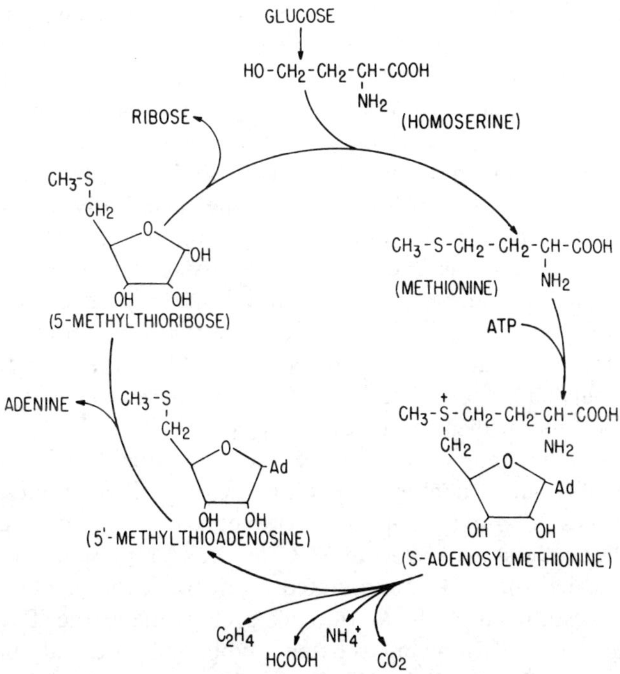

FIG. 1. Methionine–sulphur cycle in relation to the biosynthesis of ethylene in apple tissue. From Adams and Yang (1977).

fragmented to give ethylene, MTA and other products. MTA is then rapidly hydrolysed to methylthioribose which reforms methionine. The recycling of 5-methylthioribose back to methionine has also been recognized in some micro-organisms (Schwartz and Shapiro, 1954; Sugimoto *et al.*, 1976). Comparing the chemical structure of SAM and ethylene, it is evident that SAM cannot be converted into ethylene in a single step.

It has been known for some time that endogenous ethylene production, as well as the conversion of methionine to ethylene, ceases in plant tissue placed in an anaerobic atmosphere, but that a surge of ethylene production occurs upon re-exposure of the tissue to air. These observations are interpreted to indicate that an intermediate accumulates during anaerobic incubation and is subsequently converted to ethylene upon exposure to oxygen (Burg and Thimann, 1959). Adams and Yang (1979) have recently examined the metabolism of methionine under a nitrogen atmosphere and found that methionine is converted to 1-aminocyclopropane-1-carboxylic acid (ACC) and 5-methylthioribose, but not into ethylene. In the presence of air, ACC was rapidly converted into ethylene. These data indicate the following sequence for the pathway of ethylene biosynthesis in apple tissue:

$$\text{Methionine} \rightarrow \text{SAM} \rightarrow \text{ACC} \rightarrow \text{ethylene}$$

Aminoethoxyvinylglycine (AVG, 2-amino-4-aminoethoxy-*trans*-3-butenoic acid), a known inhibitor of pyridoxal-phosphate mediated enzyme reactions (Rando, 1974), greatly inhibited the conversion of SAM to ACC but did not inhibit the conversion of ACC to ethylene, suggesting that the enzyme catalysing the conversion of SAM to ACC is a pyridoxal enzyme (Adams and Yang, 1979). ACC synthase, which catalyses the conversion of SAM to ACC, was recently isolated from tomato fruit tissue and characterized by Boller *et al.* (1979) and Yu *et al.* (1979b). This enzyme specifically utilized SAM as substrate and had a molecular weight of approximately 55,000. As predicted from *in vivo* data (Adams and Yang, 1979), ACC synthase was markedly inhibited by AVG (Boller *et al.*, 1979) and aminooxyacetic acid, which are known inhibitors of pyridoxal enzymes. The enzyme required pyridoxal phosphate for maximal activity (Yu *et al.*, 1979b). Although these observations strongly support the view that ACC synthase is a pyridoxal enzyme, final proof awaits purification of the enzyme.

It is well known that pyridoxal enzymes are capable of catalysing the elimination of γ-substituent groups along with β-H of amino acids (1,2-elimination), resulting in the formation of olefins (Davis and Metzler, 1972). However, the conversion of SAM to ACC and MTA represents a new enzymic reaction in which the α-H is eliminated along with the γ-substituent (1,3-elimination), resulting in the formation of a cyclopropane ring. To account for such a reaction, the following scheme involving pyridoxal phosphate is suggested, where Y^+ stands for the methylthioadenosine sulphonium function and Py·E for the pyridoxal enzyme.

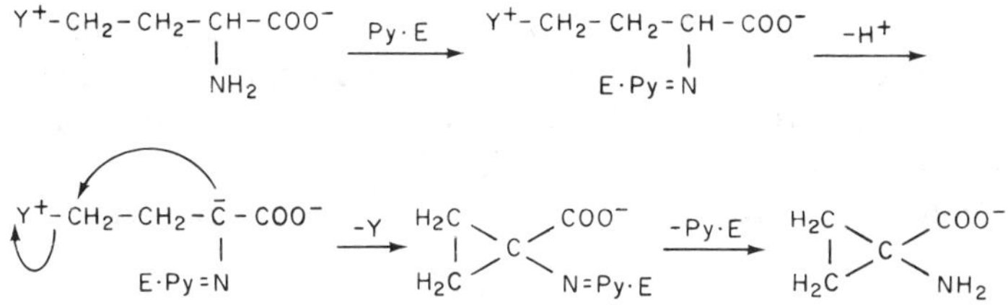

The first step is the formation of a Schiff base between SAM and the pyridoxal phosphate of the enzyme. The elimination of a proton from the α-carbon of an amino acid, catalysed by the pyridoxal enzyme, is well documented (Davis and Metzler, 1972). Once a carbanion is formed, an internal nucleophilic displacement reaction occurs, resulting in the elimination of methylthioadenosine. The elimination mechanism depicted above is a typical γ-ElcB reaction (Saunders and Cockerill, 1973), involving the intermediacy of a carbanion.

Based on the enzymic reaction catalysed by ACC synthase as described above, this enzyme can be classified as SAM methylthioadenosine-lyase (α,γ-eliminating).

From the above discussion it can be reasoned that the biological significance of SAM serving as an intermediate in the biosynthesis of ethylene from methionine is two-fold: (a) a positive sulphonium function of SAM is an excellent leaving group which facilitates the γ-elimination reaction forming ACC; (b) the eliminated product, MTA, is readily recycled back to methionine (Fig. 1) and thus conserves the sulphur atom of methionine.

It should be noted that ACC was isolated from perry pears and cider apples by Burroughs (1957). Although Burroughs (1960) observed that the amount of ACC in perry pears increased during storage, he could not detect ACC in any variety of apple or pear he examined. Undoubtedly this was due to the insensitive method he employed at that time.

Although the enzymic conversion of ACC to ethylene has been recently reported by Konze and Kende (1979), it is doubtful if their enzyme system is of any physiological significance. Based on the characteristics of the conversion of ACC to ethylene *in vivo*, it has been proposed that the enzyme system converting ACC to ethylene is membrane-associated, labile, and subject to disruption by various chemicals and treatments (Lieberman, 1979; Yu *et al.*, 1980). Until such an enzyme system is isolated and characterized, it is premature to discuss its nature. Nevertheless, it is clear from the data *in vivo* that the reaction is oxygen-dependent (Adams and Yang, 1979). It is relevant to note that the chemical oxidation of substituted cyclopropylamines to ethylene via nitrenium ion intermediates has been documented (Hiyama *et al.*, 1975). Analogous to the chemical oxidation, we can suggest that ACC is oxidized *in vivo* by an enzyme, ACC hydroxylase, forming N-hydroxy-ACC, which is then fragmented spontaneously into ethylene and cyanocarboxylic acid by a base-catalysed concerted elimination process as depicted in Fig. 2. Cyanocarboxylic acid is then degraded to CO_2 (which is derived from the carboxyl group of ACC) and HCN (which is derived from C-1 of ACC). HCN is then metabolized into formic acid, CO_2 and other products.

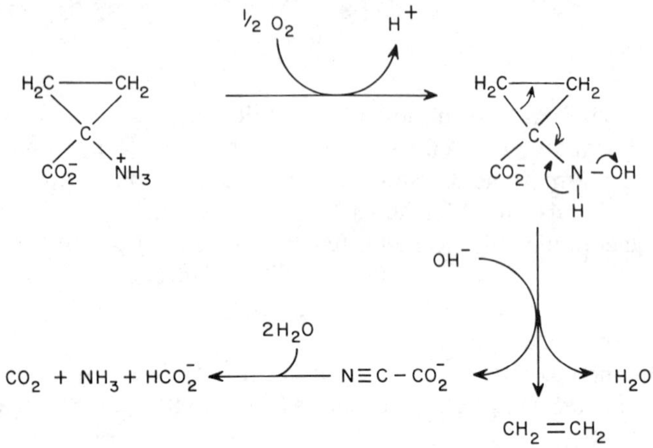

FIG. 2. A proposed reaction scheme for the conversion of ACC to ethylene catalysed by ACC hydroxylase.

When ACC was applied to various plant organs including root, stem, leaf, inflorescence, and fruit, a remarkable increase in ethylene production was observed (Cameron *et al.*, 1979). This suggests that the formation of ACC is the rate-limiting reaction in ethylene biosynthesis. Independently, Lürssen and co-workers (1979) have also discovered from screening tests that ACC is able to enhance ethylene production in a number of plants or plant organs.

Because of the important role of ACC in ethylene biosynthesis, a sensitive and simple assay for ACC in plant tissues is essential. Lizada and Yang (1979) have developed such an assay, based on the conversion of ACC to ethylene with NaOCl (a commercial bleach) in the presence of Hg^{2+}, according to the following reactions:

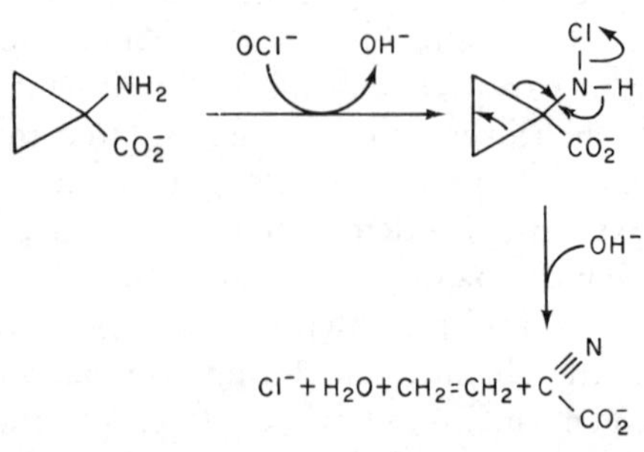

The ethylene evolved is then assayed by gas chromatography. The method is highly specific to ACC and can detect as little as 5 ρmol.

III. Mechanism of Auxin-induced Ethylene Production in Vegetative Tissue

Auxins are known to stimulate ethylene production in a wide variety of plant tissues (Abeles, 1973). Many of the effects of auxin on growth are now attributed to auxin-induced ethylene production (Abeles, 1973). The rate of ethylene production in vegetative tissues is thought to be regulated by the internal level of free auxin (Abeles, 1973). Yu *et al.* (1979a) and Yu and Yang (1979) have examined the mechanism of auxin-induced ethylene synthesis using the hypocotyls of mung bean seedlings. Although the conversion of methionine to ethylene requires the presence of auxin, the tissue is capable of converting ACC to ethylene in the absence of auxin. Even though IAA was found to stimulate ethylene production about 500 times over that of the control, the incorporation of label from methionine into SAM was little influenced by IAA treatment. Since auxin did not influence the conversion of methionine to SAM or of ACC to ethylene, but caused active conversion of methionine to ACC, auxin must exert its regulatory function on a step involved in the conversion of SAM to ACC. Indeed, both ACC content and ACC synthase activity were non-detectable in the control tissue but increased markedly upon treatment with IAA. These data indicate that the conversion of SAM to ACC is the rate-controlling step in ethylene production, and that it is subject to regulation by IAA. This view was further supported by the observation that AVG, a known inhibitor of ACC synthase, completely abolished the accumulation of ACC and the stimulation of ethylene production by auxin (Yu and Yang, 1979). Ethylene production in mung bean hypocotyls was also greatly stimulated by cytokinin + Ca^{2+} (Lau and Yang, 1974). The mechanism of this response is apparently similar to that induced by IAA, because cytokinin + Ca^{2+} treatment also caused massive accumulation of ACC (Y. B. Yu, unpublished).

IV. Regulation of Ethylene Production in Ripening Fruits

Climacteric fruits are characterized by a surge of ethylene production at the onset of ripening, and it is recognized that ethylene plays an essential role in the ripening process (Pratt and Goeschl, 1969). To study the regulation of ethylene production during the ripening of fruits, Hoffman and Yang (1980) examined the changes in internal ACC content during ripening, as well as the effect of exogenous ACC on ethylene production in pre-climacteric fruits. In pre-climacteric fruits of avocado, banana, and tomato, the ACC content was

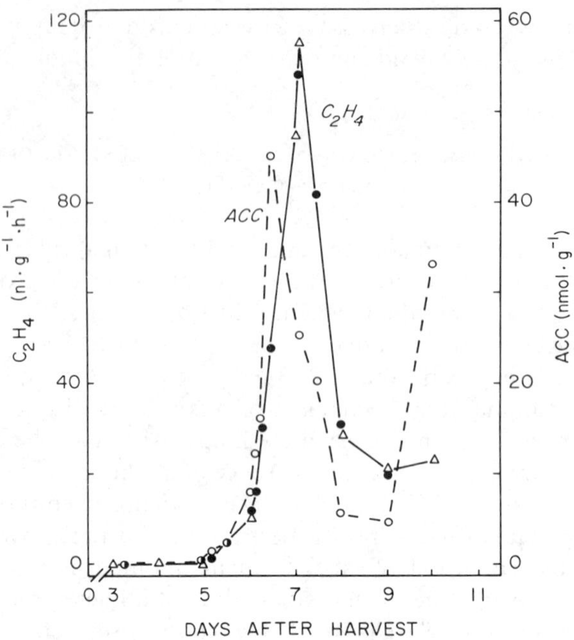

FIG. 3. Change in ACC content of avocado fruit at various stages of ripeness. Each ACC value is from a single fruit which had been monitored for ethylene production and assigned an arbitrary stage of ripeness by comparison with the established climacteric pattern of ethylene production. From Hoffman and Yang (1980).

very low (less than 0·1 nmol/g), but a massive increase occurred at the time vigorous ethylene production commenced. The relationship between the change in ACC content and the change in the ethylene production rate with ripening of avocado is illustrated in Fig. 3. ACC increased to 45 nmol/g in the later stage of the climacteric rise of respiration, then declined to 5 nmol/g, and later increased again when ethylene production declined to a low rate. The ability to convert ACC to ethylene is obviously impaired in these over-ripe fruit tissues, and this is thought to be the cause for the second increase in ACC. These data indicate that the inability of pre-climacteric fruit tissue to produce ethylene is due to its lack of ability to form ACC. However, if ACC synthase is the only enzyme in the pathway which is missing in the pre-climacteric fruits, administration of ACC should greatly increase ethylene production, as was observed in vegetative tissues. Pre-climacteric apple and cantaloup tissues, however, exhibit only a slight (less than five-fold) increase in ethylene production following the application of ACC (D. O. Adams and M. L. Bliss, unpublished observations), while the increase in ethylene production at the climacteric peak is several hundred-fold. Thus, pre-climacteric fruits are able to convert methionine to SAM (Adams and Yang, 1977, 1979), but have little ability to convert SAM to ACC and to convert ACC to ethylene. It should be noted, however, that application of ACC to intact pre-climacteric tomato fruit significantly enhanced the ripening process. These observations suggest that the slight increase in ethylene production caused by the application of ACC may be enough to trigger the ripening process. The onset of ethylene production in climacteric fruit tissues, unlike vegetative tissues, is not regulated by auxin. Thus, fruit tissues and vegetative tissues share a common pathway for ethylene biosynthesis, although the mechanism of regulation is different.

V. Regulation of Stress-induced Ethylene Production

It is well documented that various kinds of stressing of plant tissues result in increased ethylene production. This trauma-induced ethylene can be caused by chemicals, temperature extremes, drought, waterlogging, radiation, insect damage, disease or mechanical wounding, and does not appear to be associated with the normal auxin-mediated ethylene production exhibited by young developing plant parts (Abeles, 1973; Yang and Pratt, 1978). The biochemistry of ethylene formation in wounded tissues has been studied in a number of plant materials, and in all of them methionine appears to be the precursor (Yang and Pratt, 1978). To determine whether wound ethylene is also synthesized through the ACC pathway, Yu and Yang (1980) studied stress-ethylene production induced by Cu^{2+} treatment of mung bean hypocotyls, or by slicing citrus albedo tissue, immature cantaloup or tomato fruits. Both ethylene production and ACC content were very low before or immediately following wounding, but a rapid increase in ACC occurred during subsequent incubation, preceding and paralleling the rise in ethylene production. Stress did not cause a change in the quantity of endogenous SAM, however, suggesting that the conversion of SAM to ACC is a key reaction controlling production of stress-ethylene. This possibility was further supported by the observation that application of AVG, a potent inhibitor of ACC synthase, eliminated the increase in ACC formation and the production of stress-ethylene. In excised unripe tomato tissue a marked increase in ACC synthase activity was associated with an increase in ACC formation and stress-ethylene production, as shown in Fig. 4. Infiltration with cycloheximide, a protein synthesis inhibitor, completely blocked all wound-induced development of ACC synthase activity, ACC formation and ethylene production. These data indicate that wounding induces the synthesis of ACC synthase, which in turn causes accumulation of ACC and the onset of stress-ethylene production. Wound-ethylene is known to hasten the onset of ripening in climacteric-type fruits, to cause chlorophyll loss, and to reduce storage life and the quality of fruits and vegetables. In preparation for the mechanical harvesting of oranges, chemically induced wound-ethylene has been employed to hasten the abscission of the fruit.

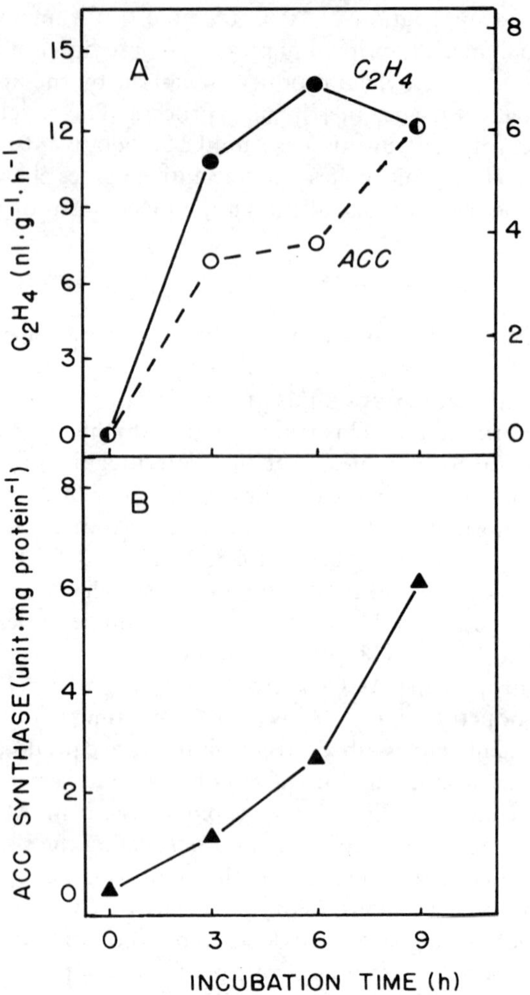

FIG. 4. Time courses of ethylene production, ACC formation (A) and development of ACC synthase (B) by excised columella tissue of unripe tomato fruit. One unit of enzyme catalyses the formation of 1 nmol ACC in 3 h at 30° C. Data from Yu and Yang (1980).

Drought is known to cause plants to increase ethylene production which in turn promotes abscission and thereby reduces water loss (McMichael *et al.*, 1972; El-Beltagy and Hall, 1974). The pathway and regulation of water deficit-induced ethylene production is similar to that of stress-ethylene discussed above; the increase in ethylene production is closely associated with the increase in ACC accumulation in response to water loss. Figure 5 shows the relationship of water loss, ACC accumulation and ethylene production in excised wheat leaves (Apelbaum and Yang, 1981).

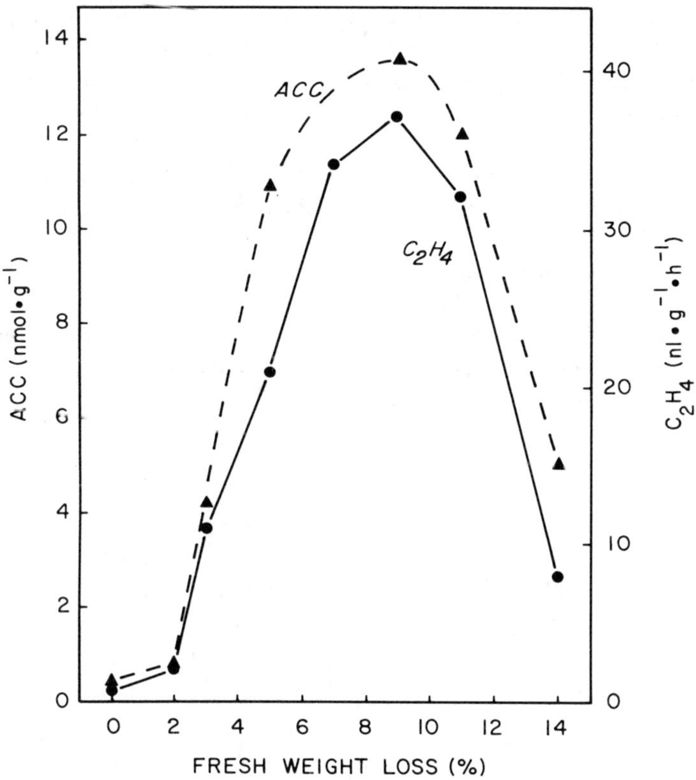

FIG. 5. Effect of the severity of water stress on ethylene production and ACC content in wheat leaves. Ethylene production and ACC content were determined 4 h after the end of the water stress treatment by which they reached the indicated water loss. From Apelbaum and Yang (1981).

It has been long recognized that waterlogging creates anaerobic conditions in the root zone and thus causes elevated synthesis of ethylene by the shoot in many plants. This elevated internal ethylene content in turn causes adventitious root formation, leaf senescence, and epinastic growth of the petioles (Jackson and Campbell, 1976; Bradford and Dilley, 1978). Since ethylene is not synthesized in the anaerobic root, and cannot be transported to the shoot in significant quantity due to its gaseous nature (Zeroni *et al.*, 1977), it has been suggested that a "signal" is synthesized in the anaerobic root and transported to the shoot, where it stimulates ethylene synthesis (Jackson and Campbell, 1976). Since ACC accumulates under anaerobic conditions and is readily converted to ethylene in aerobic tissues, ACC is a logical candidate to serve as the "signal". Bradford and Yang (1980a, b) collected the xylem sap from de-topped tomato plants and determined the ACC content of the sap. ACC content was very low in xylem sap of the control plants, but increased markedly in response to waterlogging and root anaerobiosis. This appearance of ACC in the xylem sap from the roots of de-topped flooded plants preceded both the increase in ethylene production by excised petioles and epinastic growth in intact flooded plants (Fig. 6). When ACC was supplied through the base of the cut stem of tomato shoots at concentrations comparable to those found in xylem sap, increased ethylene production and epinasty of the shoot were observed. Since the amount of ACC exported from anaerobic roots exceeded the amount of ethylene synthesized by roots maintained continuously in aerobic condition, a mere blockage of ethylene synthesis in the root by anaerobiosis is insufficient to account for the increased quantity of ACC transported. Anaerobic stress must therefore not only block conversion of ACC to ethylene in the root, but also stimulate ACC synthesis.

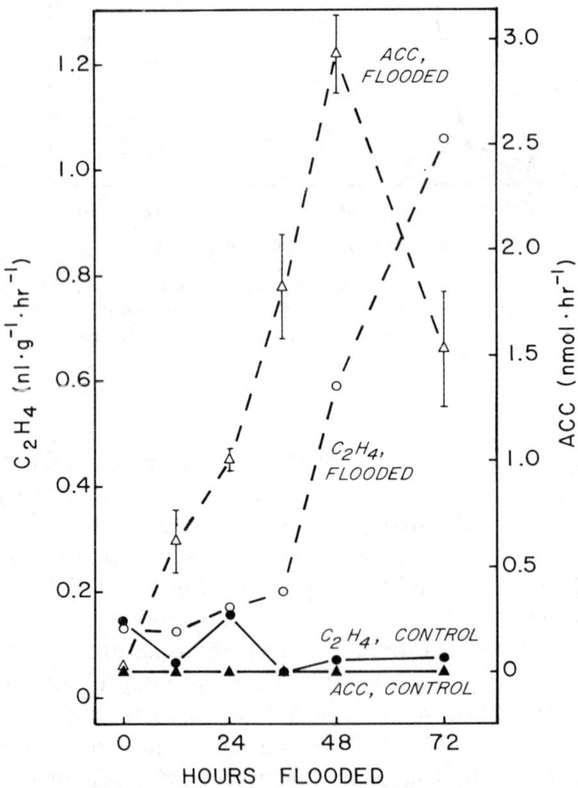

FIG. 6. Time course of changes in the quantity of ACC exuded in the xylem sap and of changes in petiolar ethylene production following flooding of tomato plants. From Bradford and Yang (1980a).

By classical definition, a hormone is an endogenous compound which is synthesized at one site and is transported to another site where it exerts a physiological effect. Due to its gaseous nature, ethylene exerts a physiological effect only at or near a site where it is synthesized. For this reason, the classical definition of a hormone does not apply to ethylene. This work has shown, however, that ACC may be synthesized in one part of the plant, be readily transported, and exert its effect through conversion to ethylene in another part of the plant.

VI. Inhibition of Ethylene Production

The inhibition of ethylene production under anaerobic or low oxygen conditions has been observed by many workers. It is now clear that oxygen participates directly in the conversion of ACC to ethylene. Temperature is another environmental factor which greatly influences ethylene production. As the temperature is increased above 30° C the rate of production falls until it ceases at 40° C. This inhibition by high temperature appears to be reversible. By examining the effect of temperature on the levels of the metabolic intermediates and on ACC-dependent ethylene production, Yu *et al.* (1980) determined that the step of the biosynthesis which is sensitive to high temperature is the conversion of ACC to ethylene.

In the above discussion, it has been noted that the reaction converting SAM to ACC is the rate-controlling step in the pathway of ethylene biosynthesis. Ethylene production can be effectively inhibited by interfering with either ACC synthase or the subsequent step, the conversion of ACC to ethylene. As discussed above, ACC synthase is thought to be a pyridoxal enzyme (Adams and Yang, 1979; Yu *et al.*, 1979b). Two types of pyridoxal enzyme inhibitors have been shown to be potent inhibitors of ACC synthase activity *in vitro* as well as *in vivo*. The first includes the vinylglycine analogues (L-2-amino-4-alkoxy-*trans*-3-butenoic acid), among which the most effective and commonly used is AVG. AVG has been shown to be a very effective inhibitor of ethylene production systems *in vivo* (Lieberman, 1979); *in vitro* it effectively inhibits ACC synthase with a *Ki* value of 0·2 μM (Boller *et al.*, 1979). Recently Bangerth (1978) and Williams (1980) found that spraying apple trees with AVG before harvest reduced pre-harvest drop, reduced loss of flesh firmness during storage and increased vegetative growth in the following growth season. The use of AVG in agricultural practice appears to have great promise and warrants further investigation. The second type of inhibitors is composed of the hydroxylamine analogues, which react with pyridoxal phosphate to form stable oximes. Among them the most effective and commonly used analogue is AOA. Murr and Yang (1975) first reported that the conversion of methionine to ethylene by apple tissue was effectively inhibited by L-canaline (2-amino-4-aminoxybutanoic acid). Later, Yu *et al.* (1979a, b) showed that

AOA inhibited ACC synthase isolated from tomato with a *Ki* value of 0·8 μM. Amrhein and Wenker (1980) have examined a number of hydoxylamine analogues and concluded that AOA and N-benzyloxycarbonyl-L-2-aminoxypropionic acid are the most potent inhibitors. When tomato plants were sprayed simultaneously with AOA and naphyl-1-acetic acid (NAA) the AOA effectively inhibited the NAA-induced ethylene production and epinastic growth of leaf petioles (Amrhein and Schneebeck, 1980). When AOA was included in the solution in which cut carnation flowers were stored, it significantly inhibited ACC formation and ethylene production, and prolonged the longevity of the flowers (Fujino *et al.*, 1980).

As noted by Adams and Yang (1979), Yu *et al.* (1979a, b) and Boller *et al.* (1979) either AVG or AOA is able to inhibit ethylene production by blocking ACC formation, but once an adequate quantity of ACC is formed these inhibitors are no longer effective. Thus, the sensitivity of ethylene-producing systems to ACC synthase inhibitors depends on the quantity of pre-formed ACC initially available and the rate at which it is consumed in ethylene production. It can be predicted that a tissue which contains a large quantity of ACC, and/or has a low ethylene production rate, is relatively insensitive to AVG or AOA inhibition.

A number of inhibitors are known to interfere with the conversion of ACC to ethylene. These include Co^{2+} and uncouplers of oxidative phosphorylation, 2,4-dinitrophenol (DNP) and carbonylcyanide *m*-chlorophenylhydrazone (CCCP) (Yu and Yang, 1979; Yu *et al.*, 1980). It has been postulated that the enzyme converting ACC to ethylene is membrane-associated, and the above-mentioned inhibitors, as well as other factors such as high temperature, osmotic shock and cold shock, inhibit ethylene production by disrupting the membrane integrity essential for the conversion of ACC to ethylene (Yu *et al.*, 1980).

VII. Conclusion

Ethylene regulates many aspects of plant growth and development, whether it is of endogenous origin or is applied exogenously. Depending upon where and when it occurs, it may have important beneficial or detrimental effects upon the growth or development of plants. Ethylene action has long been recognized to be regulated by the level of oxygen and CO_2 in the atmosphere in which harvested fruits, vegetables or flowers are stored. Such principles have been successfully applied to controlled atmosphere storage (Burg and Burg, 1969). Ag^+ has been recently shown to be effective as an anti-ethylene agent (Beyer, 1976a, b; Veen, 1979). In addition to the modification of ethylene action, manipulation of the level of ethylene in plants by either stimulation or inhibition of ethylene production is certain to be of great potential value in agriculture. Such manipulation is possible through the

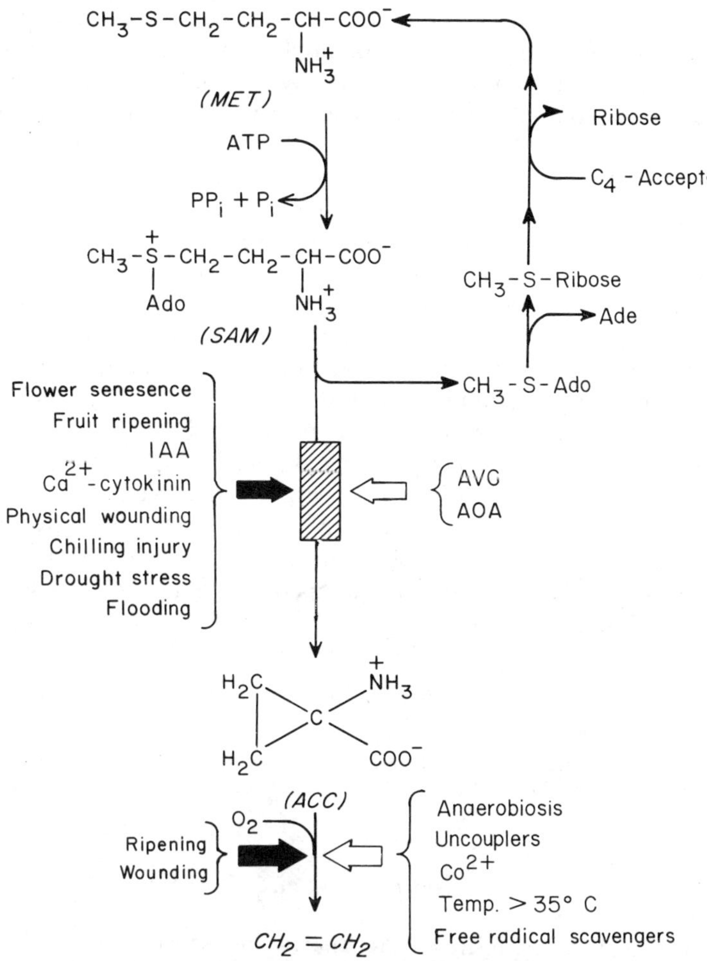

FIG. 7. Regulation of ethylene biosynthesis. ▨: this reaction is normally suppressed and is the rate-limiting step in the pathway; (➡): induction of synthesis of the enzyme; (⇦): inhibition of the reaction. Met. Ado and Ade stand for methionine, adenosine, and adenine, respectively.

application of knowledge obtained from studies of the biosynthesis of ethylene and its regulation.

The recent discovery of ACC as the key intermediate in the pathway of ethylene biosynthesis (Adams and Yang, 1979) has opened the door for many fundamental as well as applied studies. The detailed pathway of ethylene biosynthesis has been defined, and our understanding of the regulation of ethylene production has been greatly clarified. The current status of knowledge with respect to the pathway and regulation of ethylene biosythesis is summarized in Fig. 7. An important conclusion emerging from these studies is that ACC synthase, which converts SAM to ACC, is the rate-controlling enzyme in ethylene biosynthesis. Upon application of ACC to various plant tissues which normally produce little ethylene, a spectacular increase in ethylene production was observed. Since ACC is the natural precursor of ethylene, application of ACC to stimulate ethylene production appears to have advantages over either those synthetic growth regulators which induce ethylene production or, compounds which release ethylene directly without biological mediation. Many internal and external factors, including fruit ripening, flower senescence, application of auxins, physical wounding, chemical injury, root anaerobiosis, drought, and chilling injury are all known to increase ethylene production as the result of the induction of the synthesis of the same enzyme, ACC synthase. The biochemical mechanism by which these different forms of stimulus result in synthesis of ACC synthase remains to be elucidated. Conversely ethylene production can be suppressed by inhibitors of ACC synthase, a pyridoxal enzyme. Examples of such inhibitors are AVG and AOA, both of which have been tested on plants and found to be effective and useful. Ethylene production may also be inhibited by inhibitors or factors which interfere directly with conversion of ACC to ethylene. Examples are Co^{2+}, DNP and high temperature treatment (35–40° C).

ACKNOWLEDGEMENT

Our work cited herein has been supported by a Research Grant PCM 78-09278 from the National Science Foundation.

REFERENCES

Abeles, A. L. and Abeles, F. B. (1972). *Plant Physiol.* **50**, 496–498.

Abeles, F. B. (1973). "Ethylene in Plant Biology". Academic Press, New York and London.

Adams, D. O. and Yang, S. F. (1977). *Plant Physiol.* **60**, 892–896.

Adams, D. O. and Yang, S. F. (1979). *Proc. Nat. Acad. Sci. USA* **76**, 170–174.

Amrhein, N. and Schneebeck, D. (1980). *Physiol. Plant.* **49**, 62–64.

Amrhein, N. and Wenker, D. (1980). *Plant Cell Physiol.* **20**, 1635–1642.

Apelbaum, A. and Yang, S. F. (1981). *Plant Physiol.* (in press).

Bangerth, F. (1978). *J. Amer. Soc. Hort. Sci.* **103**, 401–404.

Baur, A. H. and Yang, S. F. (1972). *Phytochemistry* **11**, 3207–3214.

Baur, A. H., Yang, S. F., Pratt, H. K., and Biale, J. B. (1971). *Plant Physiol.* **47**, 696–699.

Beyer, E. M. (1976a). *Plant Physiol.* **58**, 268–271.

Beyer, E. M. (1976b). *HortScience* **11**, 195–196.

Biale, J. B. (1940). *Science* **91**, 458.

Boller, T., Herner, R. C. and Kende, H. (1979). *Planta* **145**, 293–303.

Bradford, K. J. and Dilley, D. R. (1978). *Plant Physiol.* **61**, 506–509.

Bradford, K. J. and Yang, S. F. (1980a). *Plant Physiol.* **65**, 322–326.

Bradford, K. J. and Yang, S. F. (1980b). *Plant Physiol.* **65**, 327–330.
Burg, S. P. (1962). *Ann. Rev. Plant Physiol.* **13**, 265–302.
Burg, S. P. and Burg, E. A. (1969). *Qual. Plant Mater. Veg.* **19**, 185–200.
Burg, S. P. and Clagett, C. O. (1967). *Biochem. Biophys. Res. Commun.* **27**, 125–130.
Burg, S. P. and Thimann, K. V. (1959). *Proc. Nat. Acad. Sci. U.S.* **45**, 335–344.
Burroughs, L. F. (1957). *Nature* **179**, 360–361.
Burroughs, L. F. (1960). *J. Sci. Food Agric.* **11**, 14–18.
Cameron, A. C., Fenton, C. A. L., Yu, Y., Adams, D. O. and Yang, S. F. (1979). *HortScience* **14**, 178–180.
Cousins, H. H. (1910). *Ann. Rep. Dept. Agri. Jamaica*, pp. 7–15.
Davis, L. and Metzler, D. E. (1972). *In* "The Enzymes" (P. D. Boyer, Ed.), Vol. 7, pp. 33–74. Academic Press, New York and London.
El-Beltagy, A. S. and Hall, M. A. (1974). *New Phytol.* **73**, 47–59.
Elmer, D. H. (1932). *Science* **75**, 193.
Fujino, D. W., Reid, M. S. and Yang, S. F. (1980). *Acta Hortic.* (in press).
Gane, R. (1934). *Nature (Lond.)* **134**, 1008.
Girardin, J. P. L. (1864). *Jahresber. Agr. Chem.* **7**, 199–200.
Grabham, M. (1903). *Gard. Chron.* 1903, p. 36.
Hanson, A. D. and Kende, H. (1976). *Plant Physiol.* **57**, 528–541.
Hiyama, T., Koide, H. and Nozaki, H. (1975). *Bull. Chem. Soc. Japan* **48**, 2918–2921.
Hoffman, N. E. and Yang, S. F. (1980). *J. Amer. Soc. Hort. Sci.* **105**, 492–495.
Jackson, M. B. and Campbell, D. J. (1976). *New Phytol.* **76**, 21–29.
Kidd, F. and West, C. (1932). *Gt. Brit. Dept. Sci. Ind. Res. Food Invest. Board Rep.*, pp. 55–58.
Konze, J. R. and Kende, H. (1979). *Planta* **146**, 293–302.
Lau, O. and Yang, S. F. (1974). *Planta* **118**, 1–6.
Lieberman, M. (1979). *Ann. Rev. Plant Physiol.* **30**, 533–591.
Lieberman, M. and Mapson, L. W. (1964). *Nature* **204**, 343–345.
Lieberman, M., Kunishi, A. T., Mapson, L. W. and Wardale, D. A. (1966). *Plant Physiol.* **41**, 376–382.
Lizada, M. C. C. and Yang, S. F. (1979). *Anal. Biochem.* **100**, 140–145.
Lürssen, K., Naumann, K. and Schröder, R. (1979). *Z. Pflanzensphysiol.* **92**, 285–294.
Mapson, L. W., March, J. F., Rhodes, M. J. C. and Wooltorton, L. S. C. (1970). *Biochem. J.* **117**, 473–479.
McMichael, B. L., Jordan, W. R. and Powell, R. D. (1972). *Plant Physiol.* **49**, 658–660.
Miller, E. V., Winston, J. R. and Fisher, D. F. (1940). *J. Agr. Res.* **60**, 269–278.
Murr, D. P. and Yang, S. F. (1975). *Plant Physiol.* **55**, 79–82.
Neljubow, D. (1901). *Bot. Zentralbl* **10**, 128–138.
Pratt, H. K. and Goeschl, J. D. (1969). *Ann. Rev. Plant Physiol.* **20**, 541–581.
Rando, R. R. (1974). *Science* **185**, 320–324.
Saad, F. A., Crane, J. C. and Maxie, E. C. (1969). *J. Amer. Soc. Hort. Sci.* **94,** 335–337.
Sakai, S. and Imaseki, H. (1972). *Planta* **105**, 165–173.
Salisbury, F. B. and Ross, C. (1969). *In* "Plant Physiology", p. 467. Wadsworth Publishing Co., Belmont, California.
Saunders, W. H. and Cockerill, A. F. (1973). "Mechanisms of Elimination Reactions", pp. 566–586. Wiley-Interscience, New York.
Schwartz, M. and Shapiro, K. (1954). *J. Bacteriol.* **67**, 98–102.
Sugimoto, Y., Toraya, T. and Fukui, S. (1976). *Arch. Microbiol.* **108**, 175–182.
Veen, H. (1979). *Planta* **145**, 467–470.
Williams, M. W. (1980). *HortScience* **15**, 76–77.
Yang, S. F. (1974). *Recent Adv. Phytochem.* **7**, 131–164.
Yang, S. F. and Pratt, H. K. (1978). *In* "Biochemistry of Wounded Plant Storage Tissues" (G. Kahl, Ed.), pp. 595–622. de Gruyter, Berlin.

Yu, Y. B. and Yang, S. F. (1979). *Plant Physiol.* **64**, 1074–1077.
Yu, Y. B. and Yang, S. F. (1980). *Plant Physiol.* **66**, 281–285.
Yu, Y. B., Adams, D. O. and Yang, S. F. (1979a). *Plant Physiol.* **63**, 589–590.
Yu, Y. B., Adams, D. O. and Yang, S. F. (1979b). *Arch. Biochem. Biophys.* **198**, 280–286.
Yu, Y. B., Adams, D. O. and Yang, S. F. (1980). *Plant Physiol.* **66**, 286–290.
Zeroni, M., Jerie, P. H. and Hall, M. A. (1977). *Planta* **134**, 119–125.

CHAPTER 5

Ethylene Action and Metabolism

E. M. BEYER JR

Biochemicals Department, E. I. du Pont de Nemours and Company, Experimental Station, Wilmington, Delaware, USA

I.	Introduction	107
II.	Early Studies of Ethylene Action	108
III.	Early Studies of Ethylene Metabolism	109
IV.	^{14}C-Ethylene Purification	110
V.	Tissue Preparation and Treatment	111
VI.	^{14}C-Ethylene Metabolism in Peas	112
	A. ^{14}C-Ethylene Conversion to $^{14}CO_2$	112
	B. ^{14}C-Ethylene Conversion to ^{14}C-Ethylene Oxide	113
	C. Changes in Metabolism during Pea Seedling Growth	115
	D. Factors Influencing Ethylene Metabolism in Peas	115
	E. ^{14}C-Ethylene Tissue Metabolites	116
	F. ^{14}C-Ethylene Tissue Metabolism in Other Tissues	117
VII.	Physiological Significance	117
	A. Ethylene Inactivation	117
	B. Mechanism of Action	118
References		120

I. Introduction

Ethylene is unique among the plant hormones because of its structural simplicity and its gaseous nature. This latter characteristic has led some people to even doubt its hormonal status (Abeles, 1972). Nevertheless, ethylene is now recognized as an important part of the hormonal make-up of most, if not all, plants. Essentially all phases of plant growth are affected by ethylene with concentrations of 0·01, 0·1, and 10·0 μl/l generally representing threshold, half-maximal and saturating doses, respectively (Abeles, 1973). While at times the physiological effects of ethylene are subtle, often they are spectacular as in the case of ripening, abscission, breaking of dormancy and modification of sex expression. As with the other plant hormones, the mechanism by which ethylene induces such changes is still unknown. However, recent studies with ^{14}C-labelled ethylene suggest that ethylene metabolism may be required. This chapter reviews the evidence for ethylene metabolism in plants and its possible involvement in ethylene action.

II. Early Studies of Ethylene Action

The primary mechanism of ethylene action encompasses the biochemistry involved in the initial event in the ethylene action sequence. While a considerable amount is known about the secondary or downstream events associated with ethylene action, very little is known about this primary event. Prior to 1970, the work of Burg and Burg (1965, 1967) represented the only significant attempt to probe ethylene action at this level. Their analysis of the structural requirements for ethylene action indicated that (a) the ability of a compound to mimic ethylene requires a terminal carbon atom adjacent to an unsaturated bond; (b) biological activity is inversely related to molecular size; and (c) substituents which lower the electron density in the unsaturated position reduce activity. Since these requirements for ethylene-like activity closely resembled those for metal-olefin binding, they postulated that, *in vivo*, ethylene binds to a metallic receptor site. Supporting this view was the quantitative relationship they found between the biological activity of several selected olefins and the tendency of these compounds to form complexes with silver.

Using Michaelis-Menten kinetics, they also analysed the apparent sensitivity of ethylene action to O_2 and the antagonistic relationship between CO_2 and ethylene that had been noted much earlier (see Abeles, 1973). Lineweaver-Burk plots of the effect of increasing ethylene concentrations on the percent inhibition of pea stem elongation in the presence of ambient, 1·8 or 7·1% CO_2 yielded curves identical to classical Michaelis-Menten competitive inhibitor enzyme kinetics. Based on these results and the structural similarity of CO_2 to the weak ethylene mimic allene, it was proposed that CO_2 acts as a competitive inhibitor of ethylene action. A similar type of kinetic analysis of the effect of O_2 on ethylene action in the pea stem bio-assay resulted in the additional suggestion that O_2 was an essential component in receptor site activation.

Collectively, these results formed the basis for one of the few models for ethylene action that has appeared in the literature. In the Burg and Burg (1965) model, ethylene is viewed as a dissociable activator, binding to the metal of a metalloenzyme whose receptivity to ethylene is dependent on a direct or indirect interaction with oxygen. The binding of ethylene is postulated to activate this enzyme, thereby setting into motion the chain of biochemical events that ultimately culminate in the complete ethylene response. Although based on indirect evidence, this model markedly influenced subsequent thinking on ethylene action and served as the basis for later work in this area.

Both Abeles *et al.* (1972) and Beyer (1972) attempted to test the metallic receptor concept using deuterated ethylene. If a reversible metal ion complex was involved in ethylene action, it was reasoned that this interaction might lead to isomerization and hydrogen exchange (Coates *et al.*, 1968). To test this idea, *cis*-dideuteroethylene ($C_2H_2D_2$) was applied to pea seedlings for several days and the gas phase then analysed for the presence of the *trans*-isomer. In a similar system, hydrogen exchange was evaluated by applying tetradeuteroethylene (C_2D_4) to pea seedlings and then analysing the gas phase for the presence of any intermediate deuterated ethylene species (e.g. C_2HD_3, $C_2H_2D_2$, C_2H_3D). As reported by Abeles *et al.* (1972) and Beyer (1972), these and other experiments with deuterated ethylene were all negative. No evidence could be found for isomerization or hydrogen exchange and since C_2D_4 and C_2H_4 were found to have identical biological activity, no evidence was obtained for an isotopic effect.

Essential to the success of these experiments was the ability of ethylene to act strictly as a dissociable activator molecule. Hydrogen exchange and isomerization could only have been observed if the C_2D_4 and *cis*-$C_2H_2D_2$ were free to diffuse out of the tissue after interaction with the ethylene metallic receptor site. If metabolism occurred during the course of receptor site activation, this would have eliminated any possibility of observing these effects. Based on the Burg and Burg (1965) scheme, metabolism did not seem out of the question since in organometallic chemistry the metal catalysed oxidation of ethylene to ethylene oxide is a well-known and commercially important process (Kirk-Othmer, 1980). Although non-physiological temperatures are required for this oxidation, it seemed possible that a similar oxidative mechanism could exist in plants with the appropriate enzymatic catalyst. The metalloenzyme proposed by Burg and Burg (1965) could have potentially fulfilled this role with the O_2 requirement being for the direct oxidative metabolism of ethylene. Alternatively, ethylene might be covalently bound to tissue components in an ethylation-type reaction. Unfortunately, even as late as 1972, it was impossible to even speculate on the answer to these questions since, surprisingly, the metabolic fate of ethylene in plants had not yet been resolved.

III. Early Studies of Ethylene Metabolism

The early uncertainty that surrounded the metabolic fate of ethylene in plants occurred despite the fact that several studies with ^{14}C- and ^{3}H-labelled ethylene had already shown that a small but significant amount of radioactivity from ethylene was incorporated into both fruit and vegetative tissues. For example, Buhler *et al.* (1957), found that green pear fruit and ripe avocadoes incorporated radioactivity from ethylene, and Jansen (1963, 1964) later obtained similar levels of incorporation using mature green avocadoes. However, in other similar experiments, Buhler *et al.* (1957) failed to observe any incorporation in a number of different fruit and Behmer (1958) had a similar experience with apple tissue. In contrast to the negative results obtained by some workers with fruit tissues, all of the early studies with vegetative tissues were positive. These included studies with cotton, *Coleus* (Varner and Ho, 1976), and Japanese morning glory (Shimokawa and Kasai, 1968; Shimokawa *et al.*, 1969).

Unfortunately, these studies did not prove very convincing because adequate precautions had not been taken with regard to the radiochemical purity of the applied ethylene or to microbial contamination. A number of these studies also used ^{14}C-ethylene regenerated from mercuric perchlorate, a reagent that was later shown to cause the formation of labelled impurities readily incorporated into plant tissues (Jansen, 1969). These criticisms, plus the failure of some fruit tissues to incorporate ^{14}C-ethylene, ultimately led Abeles (1973), McGlasson (1970) and Varner and Ho (1976) to conclude in the early 1970s that ethylene was probably not metabolized by plants.

This conclusion, however, was clearly premature considering that up until this time no critical study had even been conducted where highly purified labelled ethylene was applied to plant tissue under aseptic conditions. Because of this and the question concerning the possible involvement of metabolism in the action of the gas, a study was undertaken in this laboratory in 1973 to critically examine the suspected metabolic inertness of ethylene. The results of this study are summarized below.

IV. ^{14}C-Ethylene Purification

The early concerns about the radiochemical purity of labelled ethylene appear to have been well founded since all of the ^{14}C-labelled ethylene samples that have been purchased and critically analysed in this laboratory by flame ionization chromatography have been found to be contaminated. Since some of these impurities are labelled and therefore a potential source of artefacts, it is essential that the labelled ethylene be purified prior to its use. For this purpose, a two-step preparative gas chromatographic procedure was developed (Beyer, 1975a). In this procedure (Fig. 1) the labelled ethylene is first separated from impurities on a silver nitrate-ethylene glycol coated Gas Chrom R column at 0° C. The ^{14}C-ethylene fraction is collected and the once purified material is repurified on a Porapak T column which has very different separation characteristics. The ^{14}C-ethylene purified by this procedure has been shown by high sensitivity gas chromatographic analysis and specific trapping techniques (Beyer, 1975a, b) to be of ultrapure quality. Although radiolysis eventually leads to the formation of new impurities (Beyer, 1975a), it has been found that the freshly purified ^{14}C-ethylene can be safely used for two to three months, provided it is stored in an oxygen-free atmosphere over dilute alkaline solutions (e.g. 0·1 N NaOH).

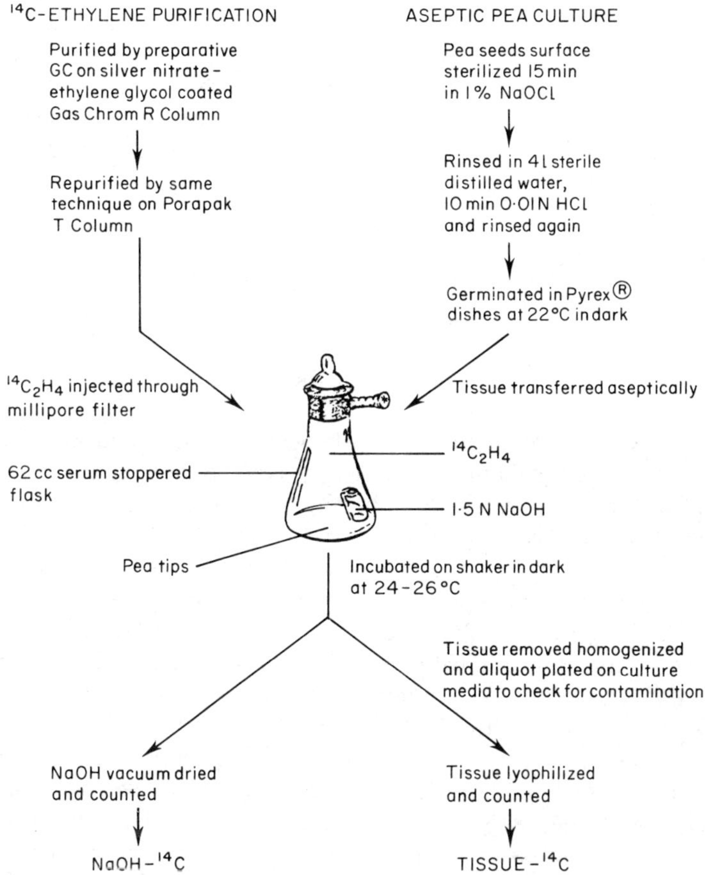

FIG. 1. Experimental procedure for treating aseptically grown etiolated pea tissue with purified ^{14}C-labelled ethylene and for analysing the tissue and NaOH for radioactivity after incubation.

V. TISSUE PREPARATION AND TREATMENT

Etiolated pea seedlings (*Pisum sativum* L. cv. Alaska) were selected for the initial studies with purified ^{14}C-ethylene, since (a) pea tissue was used in the mechanistic studies conducted by Burg and Burg (1965, 1967), (b) considerable evidence exists that ethylene is a natural regulator in this tissue (Abeles, 1973), and (c) peas have been found to be relatively easy to grow aseptically. The general procedure developed and routinely used for growing, treating and analysing the tissue and NaOH trap for radioactivity is presented in Fig. 1. Two types of data are generated by this procedure. The radioactivity remaining in the NaOH after dissolved ^{14}C-ethylene has been removed by vacuum drying is referred to as NaOH-^{14}C and is expressed in specific activity units as DPM/tissue dry weight/exposure period. This radioactivity was initially thought to contain only $^{14}CO_2$, but recently it has been shown to also contain another ^{14}C component (see below). The radioactivity remaining in the tissue after lyophilization to remove dissolved ethylene is referred to as tissue-^{14}C or tissue incorporation and is expressed in the same specific activity units as the NaOH-^{14}C data.

VI. ^{14}C-Ethylene Metabolism in Peas

A. ^{14}C-ETHYLENE CONVERSION TO $^{14}CO_2$

In the first experiments conducted with peas in 1973, 1 cm pea tips from aseptically grown four-day-old etiolated peas were incubated for six hours with 11 μl/l of freshly purified ^{14}C-ethylene. The results, reported here for the first time (Table I) clearly indicated that living pea tissue was capable of converting ^{14}C-ethylene to a volatile metabolite(s) trapped in the NaOH trap and other non-volatile components recovered in the tissue. In these initial experiments, ^{14}C-ethylene was used immediately after purification to avoid artefacts that might result from radiolysis. As a further precaution, the exposure time was limited to six hours to minimiize any build-up of radiolysis products. In retrospect, this decision to use short exposure times was unfortunate since it resulted in overlooking the importance of ethylene oxide in this system until only recently (see discussion below).

Very soon after the results in Table I were verified, the NaOH radioactivity was analysed to check to see if this radioactivity was associated with $^{14}CO_2$. To accomplish this, NaOH was used from tissue (four-day-old pea tips) or blank flasks (without tissue) that had been exposed to 14 μl/l of ^{14}C-ethylene for six hours (Table II). After vacuum drying to remove dissolved ^{14}C-ethylene, one-half of the NaOH samples were counted and the other half were acidified with 3 N H_2SO_4, placed on a shaker for 15 min at 50° C, and then counted. The results, as presented in Table II, strongly suggested that the principal radioactive component in the NaOH was $^{14}CO_2$, since upon acidification 89% of the radioactivity was released. Positive proof was obtained by cochromatography of the released radioactivity with authentic CO_2. As indicated in Table III, when the CO_2 fraction was collected from the column and counted, 95% of the ^{14}C-injected was recovered with the CO_2 peak. Based on similar results in repeated six-hour experiments with pea tips

TABLE I
Radioactivity recovered in NaOH trap or tissue following a six-hour exposure of four-day-old pea tips (1 cm) to 11 μl/l of purified $^{14}C_2H_4$ (22 mCi/mmol)

Experiment—year	Flask contents	Radioactivity recovered in NaOH trap (dpm/3 reps)	Radioactivity recovered in tissue (dpm/3 reps)
I—1973	No tissue (blanks)	60	—
	With tissue	1914	693
	Heat killed	90	81

TABLE II
Effect of acidification on radioactivity in NaOH traps following a six-hour exposure of four-day-old pea tips (1 cm) to 14 μl/l of purified $^{14}C_2H_4$ (22 mCi/mmol)

Flask contents	NaOH radioactivity after removing dissolved $^{14}C_2H_4$ (dpm/3 reps)	NaOH radioactivity after acidification (dpm/3 reps)	Percent loss
With tissue	2460	271	89
No tissue	76	60	9

TABLE III
Gas chromatographic analysis of radioactivity released into gas phase following NaOH acidification. (Radioactivity trapped during a six-hour exposure of pea tips to 10 μl/l of $^{14}C_2H_4$, 22 mCi/mmol. Samples acidified in glass syringes and 1 cc gas samples analysed on a Porapak Q column and collected in NaOH trap)

Initial radioactivity in NaOH samples (dpm/3 reps)	Percent released by acidification	Radioactivity in 1 cc gas sample	Radioactivity collected with CO_2 peak	Percent recovery from column as $^{14}CO_2$
1799	91	533	508	95

and also carnation flowers, it was concluded in 1973 that the radioactivity in the NaOH was indeed $^{14}CO_2$.

B. ^{14}C-ETHYLENE CONVERSION TO ^{14}C-ETHYLENE OXIDE

In 1978, Jerie and Hall identified ethylene oxide as the major metabolite of ethylene in *Vicia faba* cotyledons, and in this laboratory we recently found

(Blomstrom and Beyer, 1980) that ethylene is converted to ethylene glycol in pea tissue (see below). Since the logical sequence from ethylene to ethylene glycol is via an ethylene oxide intermediate, it seemed likely that ethylene oxide should also be a product of ethylene metabolism in peas. If this were true, the labelled ethylene oxide would be expected to appear in the NaOH trap since it would diffuse out of the tissue and upon coming in contact with the NaOH in the traps would readily undergo hydrolysis to ethylene glycol, a compound that is not lost from the NaOH during lyophilization. Since a small amount of radioactivity still remained in the NaOH after acidification (Table II), it seemed possible that this radioactivity might reflect such a conversion. To investigate this, the NaOH from a 24-hour exposure of four-day-old pea tips to 5 μl/l of ^{14}C-ethylene was examined. In marked contrast to the earlier six-hour experiments (Table II), 35% rather than 89% of the NaOH radioactivity was found to be released as $^{14}CO_2$ upon acidification (Table IV, Expt I). Because of the alarming discrepancy between these and our initial results, the earlier six-hour experiments were quickly repeated. However, as before (Table IV, Expt II), the NaOH radioactivity was found to contain about 90% $^{14}CO_2$.

TABLE IV
Comparison of percent $^{14}CO_2$ in NaOH trap following six- or 24-hour exposure of pea tips to 5 μl/l of purified $^{14}C_2H_4$ (110 mCi/mmol)

Experiment—year	Flask contents	Exposure time (h)	NaOH Radioactivity after removing dissolved $^{14}C_2H_4$ (dpm/rep)	NaOH radioactivity after acidification (dpm/rep)	Percent $^{14}CO_2$
I—1979	Tissue	24	5859	3817	35
	No tissue	24	260	249	4
II—1979	Tissue	6	1495	149	90
	No tissue	6	85	80	6

An explanation for these differences has recently been obtained by sampling the NaOH trap for $^{14}CO_2$ at frequent intervals during a 24-hour exposure period. In these experiments, pea seedlings were found continuously to convert ^{14}C-ethylene to $^{14}CO_2$, but after a lag period of six to seven hours, another ^{14}C-ethylene metabolite appeared that was also trapped in the NaOH. The rate of production of this metabolite increased so rapidly after six hours that by the end of the 24-hour exposure period it accounted for the majority of the radioactivity in the NaOH trap. HPLC analysis has demonstrated that this component is ^{14}C-ethylene glycol formed by the alkaline hydrolysis of ethylene oxide that was biologically derived from labelled ethylene. Thus, ethylene oxide is also a metabolite of ethylene in pea seedlings, but in marked contrast with *Vicia faba* cotyledons (Jerie and Hall, 1978), its detection requires relatively long exposure periods and the presence of a NaOH trap.

C. CHANGES IN METABOLISM DURING PEA SEEDLING GROWTH

Ethylene metabolism, like ethylene biosynthesis and ethylene tissue sensitivity, varies considerably during growth and development. Such changes were first recognized in studies with intact pea seedlings (Beyer, 1975b). While during the first day after imbibition the seedlings had only a limited capacity to metabolize ethylene, this capacity increased many fold over the next two days. By day 3, tissue and NaOH radioactivity were up 11- and 55-fold, respectively. Although the NaOH radioactivity in this study was initially reported as $^{14}CO_2$, it is now recognized that a significant amount of the ^{14}C in the NaOH was derived from ^{14}C-ethylene oxide (see discussion above).

D. FACTORS INFLUENCING ETHYLENE METABOLISM IN PEAS

In many respects, ethylene metabolism resembles ethylene biosynthesis since both are dependent on a high degree of tissue integrity (Beyer and Blomstrom, 1980). Heat, tissue homogenization, metabolic inhibitors (e.g. cycloheximide, NaCN, 2,4-dinitrophenol, N-ethylmaleimide), anaerobic conditions, COS and CS_2 either inhibit or destroy activity. Similarly, certain chelating agents, especially strong copper chelators (e.g. diethyldithiocarbamate) have been found to be strongly inhibitory.

Other treatments stimulate metabolism. For example, increasing the temperature from 15 to 25° C has been shown to increase tissue and NaOH radioactivity by approximately 2- and 4·5-fold, respectively (Beyer and Blomstrom, 1980). Above 28 to 30° C, both activities rapidly decline. Increasing the ^{14}C-ethylene concentration will also increase ethylene metabolism (Beyer, 1975c). Between 0·4 and 100 μl/l of ^{14}C-ethylene, there is a relatively linear relationship between ^{14}C-ethylene concentration and metabolism. Saturation does not occur until well beyond 100 μl/l. Below 0·4 μl/l, the amount of radioactivity recovered in the NaOH is less than that recovered in the tissue whereas the reverse is true at higher concentrations. With some tissues, additional stimulation of ethylene metabolism can be achieved by increasing the ambient oxygen concentration (Beyer, 1975c; Beyer and Blomstrom, 1980). For example, with intact pea seedlings, tissue-^{14}C and the ^{14}C recovered in the NaOH trap were increased about 40% by elevating the O_2 concentration to 50%. No increase, however, was obtained in similar experiments with pea tips, indicating that in this tissue ethylene metabolism is saturated at about 20% O_2. Presumably, such differences are related to the denser cotyledonary tissue present in intact seedlings where oxygen diffusion may become limiting. Unlike the above inhibitory or promotive treatments, light and the other plant hormones appear to have no direct effect on ethylene metabolism (Beyer, 1975c). Significant differences were not found when intact etiolated seedlings were incubated with ^{14}C-ethylene in either light (fluorescent or incandescent) or dark. Similarly, indoleacetic acid, gibberellic acid, benzyladenine and abscisic acid were without effect. While such results have indicated the absence of a direct effect of light and hormones on ethylene metabolism, the longer-term, more indirect effects have not been thoroughly studied with the exception of the effect of auxin and abscisic acid on ethylene metabolism during leaf abscission in cotton (Beyer, 1979b).

One of the interesting phenomenon associated with ethylene metabolism in intact peas is the dramatic effect that removal of the cotyledons has on metabolism (Beyer, 1975c). Using latex rubber membranes to isolate the root–shoot axis from the cotyledons during ^{14}C-ethylene exposure of intact seedlings (Beyer, unpublished data), it has been found that both the axis and cotyledonary tissues metabolize ethylene. However, detaching the cotyledons from the root–shoot axis just prior to exposing either part to ^{14}C-ethylene greatly reduces their ability to metabolize ethylene even though other metabolic functions are unaffected (e.g. respiration, ethylene biosynthesis). It would appear that some factor is required from the root–shoot axis for normal ethylene metabolism, especially in the cotyledons.

E. ^{14}C-ETHYLENE TISSUE METABOLITES

Following ^{14}C-ethylene exposure and lyophilization, approximately 80% of the ^{14}C remaining in the tissue can be readily removed by extraction with 50–80% ethanol. When the ethanolic extract is subjected to ion exchange chromatography, over 90% of the radioactivity appears in the neutral and basic fractions (Giaquinta and Beyer, 1977). These fractions can be further separated into several basic and two major neutral metabolites using thin layer or paper chromatography. Using a variety of different analytical procedures (e.g. HPLC, GC-MS), the two neutral components have recently been identified as ethylene glycol and its glucose conjugate (Blomstrom and Beyer, 1980). Interestingly, the well known ethylene mimic, propylene, has also been found to be converted to the analogous products, 1,2-propanediol and its glucose conjugate (Blomstrom and Beyer, 1980). Based on the recent finding that ethylene oxide is also a major metabolite in peas (see discussion above), it now appears likely that ethylene glycol and propanediol are directly derived from ethylene oxide and propylene oxide, respectively.

F. ^{14}C-ETHYLENE METABOLISM IN OTHER TISSUES

In addition to peas, a number of other tissues, including tomato fruit (Blomstrom and Beyer, 1980), carnation flowers (Beyer, 1977), morning glory flowers (Beyer and Sundin, 1978), cotton abscission and petiole tissue (Beyer, 1979b), and *Vicia faba* cotyledons (Jerie and Hall, 1978) have also been found to metabolize ethylene. Several important general features of ethylene metabolism have emerged from these studies. First, the rate of ethylene metabolism varies enormously in different tissues. *Vicia faba* cotyledons, for example, metabolize ethylene about 20 times faster than pea tips. In fact, as reported by Jerie and Hall (1978), metabolism is so rapid in *Vicia faba* cotyledons that it can easily be monitored by following the disappearance of unlabelled ethylene using gas chromatography. This is very similar to the rapid metabolism of propylene by pea seedlings (Beyer, 1978). Secondly, although rates vary considerably, the same basic metabolic system seems to be operating. This is suggested by the fact that the primary metabolites of ethylene in all tissues so far examined are the same simple oxidation products (e.g. CO_2, ethylene oxide, ethylene glycol). Thirdly, changes in NaOH and tissue radioactivity do not always parallel each other during changes in growth and development. For example, in cut carnation flowers (Beyer, 1977), tissue-^{14}C in the reproductive and receptacle tissue increased dramatically over a five-day period but during the same period there was only a slight increase in the amount of radioactivity recovered in the NaOH trap. A similar interdependency has also been noted during tomato fruit ripening (Beyer and Blomstrom, 1980) and cotton leaf abscission (Beyer, 1979b).

VII. Physiological Significance

A. ETHYLENE INACTIVATION

There are several reasons for believing that ethylene metabolism does not represent a system for regulating endogenous ethylene levels. First, based solely on the amount of ethylene metabolized by most tissues, this system would appear totally ineffective for this purpose. In all tissues so far examined, with the exception of *Vicia faba* cotyledons, the amount of ethylene metabolized represents only a few percent of that naturally produced. In morning glory flowers for example, it has been estimated that less than 0·2% of the ethylene formed during periods of maximum production would be metabolized (Beyer and Sundin, 1978).

Secondly, when ethylene metabolism is selectively inhibited in pea tissue with carbon disulphide (CS_2), there is no concomitant increase in the amounts of ethylene evolved from the tissue. At 8 μl/l, CS_2 has been found to have no effect on ethylene biosynthesis in a variety of plant tissues (Beyer, unpublished data). However, at this same concentration, it markedly inhibits ethylene metabolism. If metabolism were removing even 20% of the ethylene being produced by the tissue, such a treatment would be expected to result in a detectable increase in the amount of ethylene evolved from the tissue. Since this does not occur, it must be concluded that ethylene metabolism does not normally remove a significant amount of the ethylene produced.

Thirdly, during development, peak rates in ethylene biosynthesis do not always correspond with peak rates in ethylene metabolism (Jerie and Hall, 1978; Beyer, unpublished data). If the purpose of metabolism was to reduce ethylene tissue levels, it would seem that a parallel relationship should exist between the activities of these two systems. Finally, since ethylene rapidly diffuses out of most tissues, an ethylene inactivation system seems entirely unnecessary.

B. MECHANISM OF ACTION

What then might be the function of this metabolic system if it is not a means of regulating internal ethylene levels? It certainly is difficult to imagine that such a system would exist without some essential function. One possibility that has been suggested is that ethylene metabolism is an integral part of the ethylene action mechanism (Beyer, 1975b, c). While the evidence in favour of this hypothesis is still somewhat rudimentary, a sufficient number of parallel relationships have been observed between ethylene action and metabolism to give some credence to this idea.

One type of evidence is the relatively close correlation that has been found to exist between changes in ethylene tissue sensitivity and changes in ethylene metabolism. This has been especially apparent in studies of flower development in morning glory (Beyer and Sundin, 1978) and abscission in cotton (Beyer, 1979b). As reported by Kende and Hanson (1976), morning glory flower buds develop very rapidly and within a two-day period they pass from a state of insensitivity to ethylene to a very responsive condition. Ethylene metabolism studies (Beyer and Sundin, 1978) conducted during this period have revealed that during the insensitive period the tissue also does not have the ability to metabolize ethylene (i.e. NaOH-^{14}C). However, just as the buds become responsive, this ability appears and then increases very rapidly.

A similar parallel relationship has been noted during cotton leaf abscission (Beyer, 1979b). Abscission zone tissue taken daily from the second true leaf of cotton plants was found to metabolize ^{14}C-ethylene at a fairly constant rate during 5·5-hour incubation periods. A rapid increase, however, occurred when the leaf blade was excised to induce abscission. Excision resulted in more than a six-fold increase in metabolism and this increase preceded by one day the first signs of abscission. Moreover, hormone treatments that delayed or stimulated abscission had similar effects on ethylene metabolism.

Results such as these have provided indirect evidence for the idea that ethylene metabolism may be closely related to ethylene action. Other more convincing evidence (Beyer, 1979a) has been obtained using the anti-ethylene agents, CO_2 and Ag^+. Both have been found to inhibit or reduce ethylene action and metabolism. With Ag^+, a remarkable quantitative relationship was found to exist between the ability of Ag^+ to reduce ethylene action and metabolism in pea epicotyl tissue. At 0·2 μl/l of ethylene, Ag^+ (100 mg/l) negated 48% of the growth inhibition caused by ethylene and it inhibited ethylene metabolism (tissue-^{14}C) by exactly the same extent. As the concentration of ethylene was increased to 20 μl/l, these inhibitory effects of Ag^+ on ethylene action and metabolism were found to progressively disappear in a remarkably parallel fashion.

Correlations of the type discussed above strongly suggest that ethylene action and metabolism may be closely interrelated and have led to the suggestion that ethylene metabolism is actually a reflection of ethylene action, i.e. ethylene is metabolized at the ethylene receptor site. This idea can best be envisaged by recalling the Burg and Burg (1965) scheme for ethylene action. Based on their mechanism of action studies, they proposed that ethylene action involves the binding of ethylene and oxygen to the metal of a metalloenzyme. In this scheme, ethylene was viewed strictly as a dissociable activator molecule being released unchanged after enzyme activation. If this scheme were now modified so that during receptor site activation ethylene actually reacted with oxygen to form ethylene oxide, the basic concepts that have emerged from this work on ethylene metabolism would be integrated into one scheme. Although there is no direct evidence that the ethylene receptor site is an oxygenase, the proposed involvement of copper (Beyer, 1976) would certainly be consistent with this idea. The known tendency of ethylene to form complexes with copper, the sensitivity of metabolism to copper chelators, COS and CS_2, and the ability of copper to catalyse the oxidation of ethylene to ethylene oxide in non-biological aqueous systems all support this notion (see Beyer and Blomstrom, 1980). A tentative working model incorporating these ideas into a general scheme for ethylene metabolism and action is presented in Fig. 2.

Interest in this laboratory in the possible role of ethylene oxide in ethylene action has been stimulated by the realization that ethylene oxide may be a general product of ethylene metabolism. This interest has been intensified by the recent discovery that ethylene oxide acts synergistically with ethylene in the triple response of peas (Table V). Ethylene oxide, when applied at the highest concentration tested (110 μl/l) reduced growth by less than 10% (12·9 *vs* 11·9 mm). However, when combined with 0·25 μl/l of ethylene, it increased ethylene induced growth inhibition from 44% to 71%. Interestingly, this synergism was not restricted to just growth inhibition since all other ethylene effects were also amplified (e.g. epicotyl swelling). This increase in ethylene effectiveness was not due to a stimulation of endogenous ethylene production

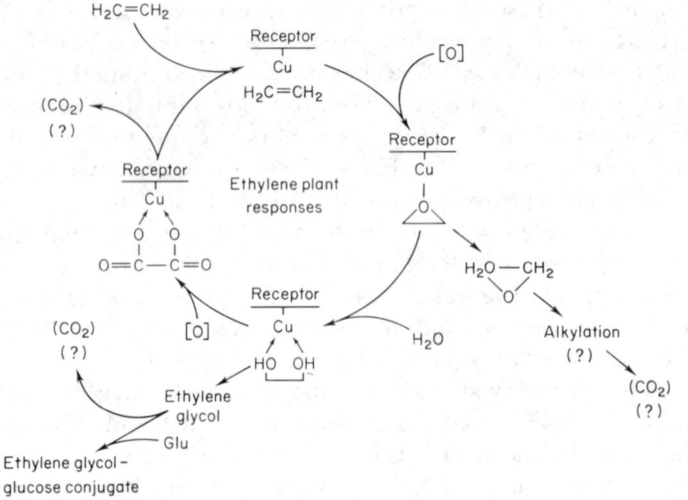

FIG. 2. Tentative working hypothesis for initial events involved in ethylene metabolism and action.

since in other experiments ethylene oxide was found to have no significant effect on ethylene production over this concentration range. As suggested (Fig. 2), these studies raise the interesting possibility that ethylene oxide may be directly involved in ethylene action. Studies are under way to further explore this possibility.

TABLE V

Effect of various concentrations of ethylene oxide with and without ethylene (0·25 μl/l) on the growth of two-day-old etiolated pea seedlings during a three-day exposure period. (Seedlings were grown in six-litre desiccators that were continuously purged with either air, C_2H_4O, C_2H_4, or $C_2H_4O+C_2H_4$)

C_2H_4O Conc. (μl/l)	Pea epicotyl growth $-C_2H_4$ (mm)	Pea epicotyl growth $+C_2H_4$ (mm)	% Reduction
0	12·9	7·2	44
28	12·5	6·2	50
55	12·1	4·6	62
110	11·9	3·4	71

REFERENCES

Abeles, F. B. (1972). *Ann. Rev. Plant Physiol.* **23**, 259.
Abeles, F. B. (1973). "Ethylene in Plant Biology." Academic Press, New York and London.

Abeles, F. B., Ruth, J. M., Forrence, L. E. and Leather, G. R. (1972). *Plant Physiol.* **49**, 669.

Behmer, M. (1958). *Klosterneuderg, Aust. Hoehere Bundeslehr-und Versuchanstalt fuer Wein, Obst-Und Gartenbau, Serie B. Obst-und Garten, Mitt.* **8**, 257.

Beyer, E. M., Jr (1972). *Plant Physiol.* **49**, 672.

Beyer, E. M., Jr (1975a). *Plant Physiol.* **55**, 845.

Beyer, E. M., Jr (1975b). *Nature* **255**, 144.

Beyer, E. M., Jr (1975c). *Plant Physiol.* **56**, 273.

Beyer, E. M., Jr (1976). *Plant Physiol.* **58**, 268.

Beyer, E. M., Jr (1977). *Plant Physiol.* **60**, 203.

Beyer, E. M., Jr (1978). *Plant Physiol.* **61**, 893.

Beyer, E. M., Jr (1979a). *Plant Physiol.* **63**, 169.

Beyer, E. M., Jr (1979b). *Plant Physiol.* **64**, 971.

Beyer, E. M., Jr and Blomstrom, D. C. (1980). *Proc. Tenth International Conf. on Plant Growth Substances* (in press).

Beyer, E. M., Jr and Sundin, O. (1978). *Plant Physiol.* **61**, 896.

Blomstrom, D. C. and Beyer, E. M., Jr (1980). *Nature* **283**, 66.

Buhler, D. R., Hansen, E. and Wang, C. H. (1957). *Nature* **179**, 48.

Burg, S. P. and Burg, E. A. (1965). *Science* **148**, 1190.

Burg, S. P. and Burg, E. A. (1967). *Plant Physiol.* **42**, 144.

Coates, G. E., Green, M. L. H. and Wade, K. (1968). "Organometallic Compounds", Vol. 2. 3rd Ed. Methuen and Co. Ltd, London.

Giaquinta, R. T. and Beyer, E. M., Jr (1977). *Plant and Cell Physiol.* **18**, 141.

Hall, W. C., Miller, C. S. and Herrero, F. A. (1961). *Proc. Fourth Int. Pl. Growth Regul. Conf.* **4**, 751.

Jansen, E. F. (1963). *J. Biol. Chem.* **238**, 1552.

Jansen, E. F. (1964). *J. Biol. Chem.* **239**, 1664.

Jansen, E. F. (1969). *First Food Sci. Technol. Proc. Int. Congr.* **1**, 475.

Jerie, P. H. and Hall, M. A. (1978). *Proc. R. Soc. Lond. B.* **200**, 87.

Kende, H. and Hanson, A. D. (1976). *Plant Physiol.* **57**, 523.

Kirk-Othmer (1980). "Encyclopedia of Chem. Tech.", Vol. 9. 3rd Ed. John Wiley and Sons, New York.

McGlasson, W. B. (1970). "Biochemistry of Fruits and Their Products." Food Sci. and Tech. Monograph Series. Academic Press, New York and London.

Shimokawa, K. and Kasai, Z. (1968). *Agr. Biol. Chem.* **32**, 680.

Shimokawa, K., Yokoyama, K. and Kasai, Z. (1969). *Mem. Res. Inst. Food Sci., Kyoto* **30**, 1.

Varner, J. E. and Ho, D. T. (1976). "Plant Biochemistry." Academic Press, New York and London.

CHAPTER 6

Enzymes and Texture Changes During Ripening

G. E. HOBSON

Glasshouse Crops Research Institute, Littlehampton,
West Sussex BN16 3PU, England

The progressive increase in softness of the tissue together with a change in colour of the skin or flesh, and a production of a wide spectrum of aroma compounds are some of the most easily recognizable changes that accompany ripening in climacteric fruit. There is a radical loss of cellular organization, and it is clear even to the casual observer that far-reaching changes occur at this time to the cell walls, as, for instance, resistance to fungal attack falls away. However, the mechanisms whereby intercellular binding substances are solubilized, the exact sequence of changes that take place as the cell wall is broken down, and the reasons why some organelles disappear during ripening while others survive are still, for the most part, unanswered questions.

A typical plant cell (Fig. 1) consists of a layer of cytoplasm within a robust cell wall, the structure of which has much in common with the lightness and strength associated with glass-reinforced-plastic technology. The bulk of fruit tissue is composed of parenchyma cells which expand considerably during growth so that at maturity their walls are thin and their shape results in the formation of intercellular spaces. Functional chloroplasts and the normal array of organelles thus eventually find themselves sandwiched in a thin layer of cytoplasm, bounded on one side by a tonoplast and on the other by a plasmalemma. It is now clear that the cell wall consists of alternating layers of cellulose microfibrils embedded in a ground substance mainly consisting of hemicelluloses with some protein, and each cell is cemented to the next one by a pectin-rich middle lamella. Plasmadesmata, often or always associated with vesicles (Ben-Arie *et al.*, 1979), interconnect the cytoplasm of adjacent cells and in the ripe fruit are thought to give the tissue a finite amount of cohesion. In addition, turgor pressure keeps the protoplast pressed against the cell wall and this helps to maintain the tissue firmness.

In a typical climacteric fruit such as the tomato, increase in size is by cell

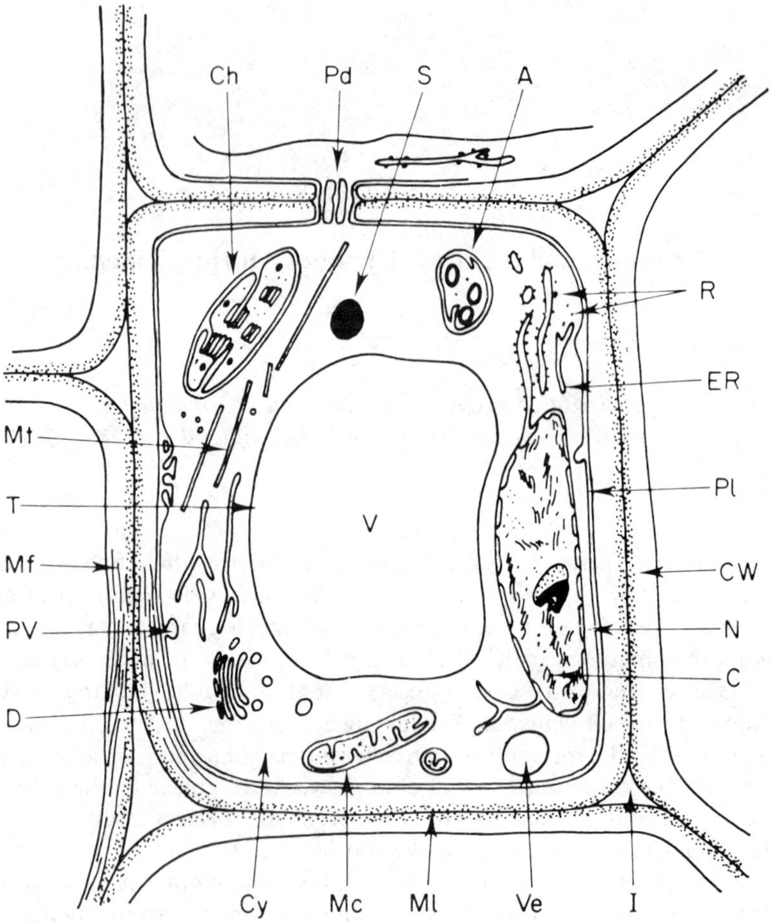

FIG. 1. A schematic representation of a plant cell (from Jewell, 1979, with permission). A, amyloplast; C, chromatin; Ch, chloroplast; CW, cell wall; Cy, cytoplasm; D, dictyosome, several of which constitute the Golgi apparatus; ER, endoplasmic reticulum with attached ribosomes, R; I, intercellular space; Mc, mitochondrion; Mf, microfibril; ML, middle lamella; Mt, microtubule; N, nucleus; Pd, plasmadesmata; Pl, plasmalemma; PV, pinocytotic vesicle, an infolding of the cell membrane; S, spherosome; T, tonoplast; V, vacuole; and Ve, vesicle.

division for the first seven (Davies and Cocking, 1965) to fourteen (Smith, 1935) days, and then by cell expansion. During the next few weeks, partial separation of the cell walls at the middle lamella occurs (Mohr and Stein, 1969), but this is probably caused by rapid cell expansion at this time (Davies and Cocking, 1965), contact being lost with adjacent cells to form intercellular spaces. This incipient separation is thought to be mechanical in origin rather than enzymic. Near ripeness, the pericarp cells become very large and cell separation much more complete, but within the cell the plasmalemma and tonoplast remain intact, both physically and physiologically (Mohr and Stein, 1969; Vickery and Bruinsma, 1973).

The linear glucan molecules in cellulose are combined into bundles to form very elongated elementary fibrils in the primary cell walls, and these crystalline structures are thought to aggregate into larger units in secondary walls. In unripe avocados, the middle lamella is quite obvious and fibrils are packed tightly in orderly array on both sides of the middle lamella (Pesis *et al.*, 1978). In post-climacteric fruit, no middle lamella was observed and many of the fibrils were missing. Observations on ripening apple and pear tissue by Ben-Arie and her associates (1979) have confirmed a progressive dissolution of the cell wall and middle lamella, but a persistence of certain organs and organelles such as the cell wall–plasmadesmata complex and the mitochondria.

From this brief summary of some of the anatomical changes that can be observed as fruit ripen, alterations in gross structure can be correlated with differences in the texture of the tissue. Moreover, these modifications are, in the main, brought about by a series of co-ordinated enzymic transformations, and to a much smaller extent by mechanical strains generated by the extensive swelling of the tissue during fruit development.

What does chemical analysis tell us about the nature of cell wall components? Knee (see Chapter 7, this volume) deals at length with this topic, and a number of recent reviews are available (Robinson, 1977; Hall, 1981). In order to clarify the probable linkages between the various components in the primary walls of sycamore callus cells, a model (Fig. 2) was constructed by Robinson (1977), based on work by Keegstra *et al.* (1973) and Albersheim (1975), which appears to be relevant to fruit cells as well. In summary, the pectic substances are thought to consist of a neutral homogalacturonan covalently attached to an acidic, polymeric rhamnogalacturonan, and perhaps surrounded by a protecting coat of arabans and galactans or a polymer containing both sugars. The hemicelluloses consist of xyloglucans and glucuronoarabinoxylans non-covalently associated with cellulose. The primary walls may well contain some 5–10% protein which is particularly rich in hydroxyproline residues; when partially digested, fragments of such proteins remain linked with arabinosyl and galactosyl residues. Cellulose fibres are closely allied to hemicellulose to which they are hydrogen bonded, and the cellulose–hemicellulose complexes are interconnected by a network of pectic polysaccharides probably with protein attachments, and the whole forms the primary wall of the cell.

Having considered a few of the changes in the characteristics of fruit cells during ripening, and briefly reviewed chemical evidence about wall structure, it is now appropriate to consider some of the properties of the enzymes involved in softening. Lewis, writing in 1978 about the avocado, thought that

> softening of the mesocarp is probably due to changes in the pectic substances. The rise in respiration is accompanied by softening, a decrease in protopectins, a rise in water-soluble pectins and a decrease in the degree of esterification.

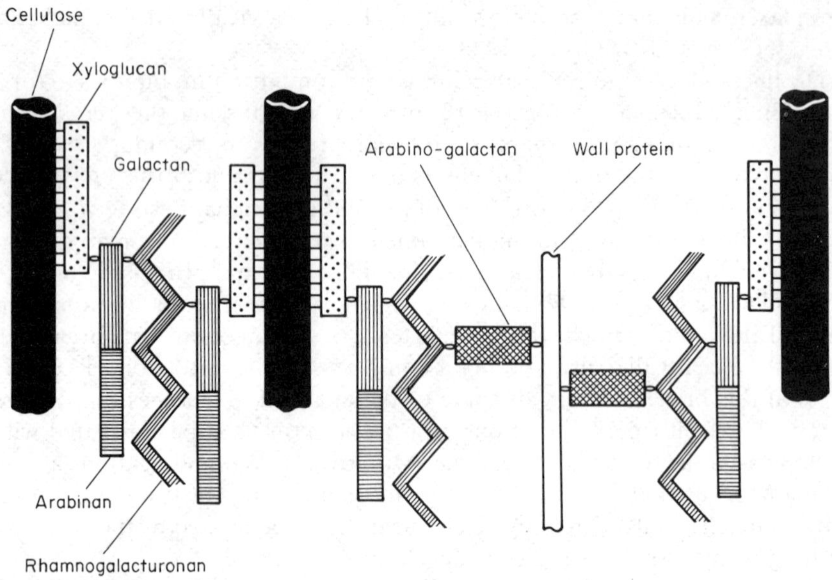

FIG. 2. Suggested scheme for the structure of the primary wall of sycamore callus cells (redrawn by Robinson (1977) from Keegstra *et al.* (1973) and Albersheim (1975) and copied with permission).

There is much evidence to suggest that these processes apply generally to climacteric fruit and are under enzymic control.

A complex of calcium with pectic substances, very often associated with unripe fruit, is known as "protopectin" (Joslyn, 1962), and it is chemically and mechanically enmeshed with other wall substances to form a type of giant molecule. An enzyme largely uncharacterized and referred to as protopectinase is thought to be involved in the initial stages of the dissolution of the pectin–cellulose–hemicellulose (PCH) compilation. It is well recognized that calcium is effective in binding homogalacturonan chains, the metal ion being the ideal size to cross-link with carboxylate side groups on the polymer to form an "egg-box model" (Grant *et al.*, 1973). The pectic substances only begin to become water-soluble after polygalacturonase activity has been initiated (Hobson, 1963; review by Pressey, 1977). In normal tomatoes, bound calcium decreased about three-fold during the later stages of development, whereas non-ripening mutant (*rin*) fruit which have no polygalacturonase activity showed an increase in calcium binding during the same period (Poovaiah, 1979). Since Grant *et al.* (1973) demonstrated that calcium binding is much stronger when the homogalacturonan chains contain many residues, the action of protopectinase in breaking up the PCH complex could actually be due to endopolygalacturonase causing chain shortening, and thereby weakening the forces holding the polymers together through calcium bridges.

Another possible explanation for the appearance of water-soluble pectin during the ripening and softening of climacteric fruit is that the release of pectin could involve cleavage of its links with other cell components through hydrolysis of terminal residues or side chains (Pressey, 1977). However, no substantial release of either galacturonic acid or rhamnose has been found during analyses of cell walls from ripening tomatoes (Gross and Wallner, 1979), but a loss of galactose and arabinose was recorded. The three major candidate enzymes involved in the mechanism of softening thus fall into two groups, pectinesterase and polygalacturonase on the one hand, and cellulase on the other.

Pectinesterase seems to be present in non-limiting quantities in much of the period of development of many fruits (Pressey, 1977). It must be the inaccessibility of the methoxy groups on the side chains of the protopectin and pectin molecules that prevents the enzyme from quickly and completely de-esterifying the substrate. It has been calculated that if the enzyme were to act unrestrictedly on all the pectin in a tomato fruit, the substrate would be completely demethoxylated in four minutes! The degree of esterification of pectin in many fruits decreases during ripening (Woodmansee *et al.*, 1959; see review by Pressey, 1977). Pectin is water-soluble whereas pectic acid, resulting from de-esterification, has much more restricted solubility. However, polygalacturonase is able to degrade pectic acid much more efficiently, hence the transformation from pectin to oligogalacturonans is extremely complex in terms of solubility, with many competing reactions. Although Lee and Macmillan (1970) demonstrated that pectinesterase acted at both the reducing end and at random along the chain, demethoxylation does not seem to go to completion. Multimolecular forms of the enzyme have been found in a number of fruits (see Rombouts *et al.*, 1979), and up to eight components have been detected in the tomato (Delincee, 1976). An explanation of this diversity might be that each isoenzyme is specifically involved in the removal of methoxyl groups from polymers of specific length, as a preliminary step for the action of polygalacturonase. The mechanisms whereby the strongly-bound enzyme is released from the cell wall (Nakagawa *et al.*, 1971) and is transported to its site of action are factors about this widely-distributed and rugged enzyme that still need to be elucidated.

In a number of climacteric fruits whose texture alters markedly during ripening, maximum loss of firmness is coincident with rapid synthesis of the enzyme polygalacturonase. Numerous studies have been directed towards establishing a relationship between firmness of the fruit and activity of the enzyme (see Pressey, 1977), and these have had a high level of success. Experimental material using F_1 hybrids between the non-ripening mutants and a normal cultivar of tomatoes has again shown a very good correlation between pectolytic enzyme activity, increased solubilization of the pectic substances and loss of firmness (Buescher *et al.*, 1976). An investigation of peaches by Pressey *et al.* (1978) led to the conclusion that fruit from the softer freestone varieties contained both exo- and endo-polygalacturonase, while fruit from the firmer clingstone varieties contained virtually only exo-polygalacturonase. Apples have been found to have only the exo-polygalacturonase enzyme (Bartley, 1978), and the extent to which fruits exhibiting this form of the enzyme soften seems to be much more limited than those containing the endo-polygalacturonase form. Many climacteric fruits (see Pressey, 1977) demonstrate a rising polygalacturonase activity (probably endo-polygalacturonase) during ripening, and the importance of this enzyme in conjunction with pectinesterase in the mechanism of softening of fruit tissue is beyond dispute. Results from examinations of the enzymes in several ripening fruits show that they consist of a number of isoenzymic forms, each of which appear to carry out specific transformations on a particular group of substrates. From the work of Grierson and his collaborators (see Chapter 8, this volume), at least two forms of endo-polygalacturonase have been isolated from the ripening tomato. In summary, the statement by Lewis *loc. cit.* about changes in the pectic substances during avocado ripening can be applied generally to many other fruit whose texture changes abruptly at this time, but there is some evidence that enzyme systems other than pectolytic ones also have a part to play.

Cellulose occurs in orderly arrangements throughout the primary and secondary walls of plant cells, and it might be expected that cellulase should play a key role in texture changes. In the tomato, cellulase activity decreases during maturation (Sobotka and Stelzig, 1974), and then increases from the onset of ripening (Hobson, 1968a); there appear to be four distinct components, two of which are endocellulases, one a non-specific β-glucosidase and the last active against cellobiose. However, the total activity in ripe tomatoes is only about 0·1% of that of polygalacturonase (Table I). In non-ripening mutants of the tomato such as "Never ripe", the fruit soften more slowly than usual, and cellulase and polygalacturonase activities are both low. A more extreme mutant, "ripening inhibitor", softens even more slowly and contains weak cellulase activity with no polygalacturonase. It has been suggested that such softening as does take place could be due to this limited cellulase activity, aided by glycosidases (Gross and Wallner, 1979) that are probably active and which can result in loss of texture even in the absence of polygalacturonase (Wallner and Walker, 1975). Cellulase has been described in many ripening fruits; perhaps the most active source is the avocado (Pesis *et al.*, 1978), but the enzyme has also been found in litchis, bananas, pineapples (Hobson, 1968b), dates (Hasegawa and Smolensky, 1971) and peaches (Hinton and Pressey, 1974). However, water-soluble carboxymethyl cellulose is generally used as a substrate to demonstrate these activities, and it has not been shown that the enzyme is capable of eroding the highly organized crystalline structure found in the microfibrils of fruit. The dissolution of the fibrils in ripened avocados (Pesis *et al.*, 1978), apples and pears (Ben-Arie *et al.*, 1979) could as well be due to disintegration of the ground substance of the

TABLE I
Effect of mutant genes on cellulase and polygalacturonase activities in ripe (or equivalent) tomato fruit (cv. Ailsa Craig)

Enzyme	Normal Ailsa Craig	*Nr Nr* Ailsa Craig	*rin rin* Ailsa Craig	*Rin rin* F_1 cross	Significance (*P*)	Least significant difference at $P=0.05$
Cellulase[a] activity (mg glucose h^{-1} 100 g^{-1} fresh tissue)	1·653	0·290	0·228	0·424	0·001	0·396
Polygalacturonase[b] activity (mg galacturonic acid h^{-1} 100 g^{-1} fresh tissue)	3·77 (5895)	2·50 (314)	— (0)	3·58 (3832)	0·001	0·16

[a] Measured viscometrically by K. Manning of this Institute and expressed as increase in reducing end groups using glucose standards.

[b] The activity figures were statistically analysed after conversion to logarithmic form, and the antilogarithm of each mean is given in parentheses.

cell wall as to direct attack upon the cellulose elements. The possibility that cellulase is acting merely as a glucanase should be borne in mind when, for example, interpreting the evidence of Ben-Arie and her collaborators (1979) where they demonstrated that only a mixture of cellulase and polygalacturonase would transform unripe pear tissue into one very close to the naturally ripened product.

Thus the involvement of other hydrolytic enzymes in addition to polygalacturonase in texture changes during the ripening of fruits seems very likely. In a survey of the glycosidases in tomato fruit, Wallner and Walker (1975) showed that the only substantial activities present were those of β-galactosidase and laminarinase. Although in the absence of polygalacturonase the remaining enzymes were unable to initiate cell wall hydrolysis (Wallner and Bloom, 1977), Wallner and Walker (1975) concluded that major modifications to the cell wall had occurred once incipient ripeness of tomato fruit had been achieved. More recently, Gross and Wallner (1979) demonstrated that a decline in galactose and arabinose in tomato cell walls is unconnected with polygalacturonase activity, as it occurs in mutant tissue in which the enzyme is not active. Hence, a mechanism concerned with the release of a galactan or an arabinogalactan could regulate polygalacturonase activity by denying access to the PCH complex until hydrolysis has taken place. Polyuronide liberation through the action of polygalacturonase without the release of neutral sugars has not been detected.

In many types of climacteric fruit, there is substantial evidence to show that without protein synthesis, normal ripening does not take place (see review by Hobson, 1979). Increases in the activities of enzymes such as glycosidases and hydrolases during the ripening of fruits are common, and evidence is accumulating to show that several enzymes have a direct or an indirect effect upon tissue firmness. The possible interrelations between the various factors that transform a hard, unripe fruit cell into one much more attractive for man to eat is illustrated in Fig. 3. Results from recent investigations (Strand *et al.*, 1976; Tigchelaar *et al.*, 1978; Ishii, 1978) suggest that as well as degrading the pectic substances in the middle lamella and cell wall, polygalacturonase is also able to release amounts of protein, some of which could be inactive while bound to the cell wall (Jansen *et al.*, 1960) and on desorption could be involved in the furtherance of ripening.

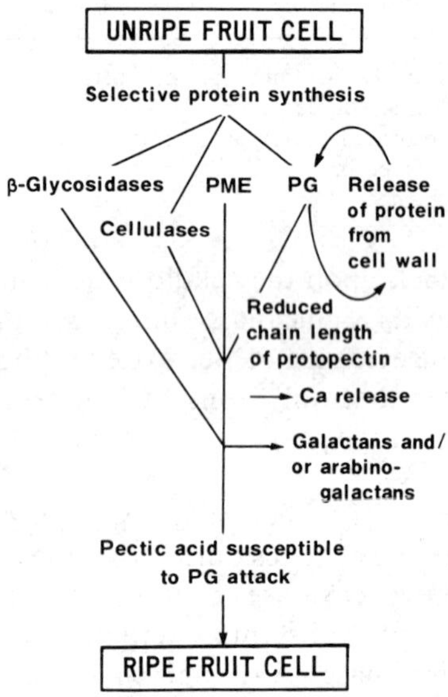

FIG. 3. Tentative scheme outlining the enzymes responsible for texture changes during the ripening of climacteric fruit.

On less controversial grounds, it is clear that a certain amount of loss of firmness by tissue prior to incipient ripeness takes place (Besford and Hobson, 1972; Awad and Young, 1979). There is every likelihood that glycosidases are able to contribute towards softening and tissue disintegration before polygalacturonase is active, especially in parts of fruit such as the locular cavities of tomatoes (Wallner and Walker, 1975). The more general changes in texture are thought to result from sequential attacks on the PCH polymer, initially by pectinesterase, glycosidases and glucanases (often referred to as cellulases). With the rise in respiration, a rapid synthesis of polygalacturonase overtakes the fruit cell, and calcium in soluble form is released at this time as are the carbohydrates galactose and arabinose (Knee, 1973). We must suppose that this then opens the way for the liberation and degradation of the water-soluble pectic substances, erosion of the middle lamella, disintegration of the primary cell wall and much more extensive loss of tissue firmness. The central role of polygalacturonase is emphasized by the behaviour of the non-ripening tomato mutants, where in the virtual absence of this enzyme ripening is extremely slow, tissue softening much more drawn out, and resistance to fungal attack is maintained for many months.

Professor Biale concluded one of his major reviews on the respiration of fruits written more than 20 years ago with the words "The field of fruit physiology is at present fruitful of thoughts on the nature of ripening and senescence". It is to be hoped that this chapter, concerning the mechanisms whereby texture alters during fruit development, gives some insight into recent results and ideas on the nature of ripening and softening.

REFERENCES

Albersheim, P. (1975). *Sci. Amer.* **232**, 80–95.
Awad, M. and Young, R. E. (1979). *Plant Physiol.* **64**, 306–308.
Bartley, I. M. (1978). *Phytochemistry* **17**, 213–216.
Ben-Arie, R., Kisler, N. and Frenkel, C. (1979). *Plant Physiol.* **64**, 197–202.
Besford, R. T. and Hobson, G. E. (1972). *Phytochemistry* **11**, 2201–2205.
Buescher, R. W., Sistrunk, W. A., Tigchelaar, E. C. and Ng, T. J. (1976). *HortScience* **11**, 603–604.
Davies, J. W. and Cocking, E. C. (1965). *Planta* **67**, 242–253.
Delincee, H. (1976). *Phytochemistry* **15**, 903–906.
Grant, G. T., Morris, E. R., Rees, D. A., Smith, P. J. C. and Thom, D. (1973). *FEBS Letters* **32** (1), 195–198.
Grierson, D., Tucker, G. A. and Robertson, N. G. (1981). *In* "Recent Advances in the Biochemistry of Fruits and Vegetables" (J. Friend and M. J. C. Rhodes, Eds), Phytochemical Society of Europe Symposium. Academic Press, London and New York.
Gross, K. C. and Wallner, S. J. (1979). *Plant Physiol.* **63**, 117–120.
Hall, M. A. (1981). *In* "Quality in Stored and Processed Vegetables and Fruit" (P. W. Goodenough and C. V. Cutting, Eds). Academic Press, London and New York.
Hasegawa, S. and Smolensky, D. C. (1971). *J. Food Sci.* **36**, 966–967.
Hinton, D. M. and Pressey, R. (1974). *J. Food Sci.* **39**, 783–785.
Hobson, G. E. (1963). *Biochem. J.* **86**, 358–365.
Hobson, G. E. (1968a). *J. Food Sci.* **33**, 588–592.
Hobson, G. E. (1968b). *Annu. Rep. Glasshouse Crops Res. Inst., 1967*, pp. 134–136.
Hobson, G. E. (1979). *Curr. Adv. Plant Sci.* **11** (10), 1–11.
Ishii, S. (1978). *Plant Physiol.* **62**, 586–589.
Jansen, E. F., Jang, R. and Bonner, J. (1960). *Food Res.* **25**, 64–72.
Jewell, G. G. (1979). *In* "Food Microscopy" (J. G. Vaughan, Ed.), pp. 1–34. Academic Press, London and New York.

Joslyn, M. A. (1962). *In* "Advances in Food Research" (C. O. Chichester, E. M. Mrak and G. F. Stewart, Eds), Vol. 11, pp. 1–107. Academic Press, New York and London.

Keegstra, K., Talmadge, K. W., Bauer, W. D. and Albersheim, P. (1973). *Plant Physiol.* **51**, 188–197.

Knee, M. (1973). *Phytochemistry* **12**, 1543–1549.

Knee, M. (1981). *In* "Recent Advances in the Biochemistry of Fruits and Vegetables" (J. Friend and M. J. C. Rhodes, Eds), Phytochemical Society of Europe Symposium. Academic Press, London and New York.

Lee, M. and Macmillan, J. D. (1970). *Biochemistry* **9**, 1930–1934.

Lewis, C. E. (1978). *J. Sci. Food Agric.* **29**, 857–866.

Mohr, W. P. and Stein, M. (1969). *Can. J. Plant Sci.* **49**, 549–553.

Nakagawa, H., Sekiguchi, K., Ogura, N. and Takehana, H. (1971). *Agr. Biol. Chem.* **35**, 301–307.

Pesis, E., Fuchs, Y. and Zauberman, G. (1978). *Plant Physiol.* **61**, 416–419.

Poovaiah, B. W. (1979). *Commun. Soil Sci. Plant Anal.* **10**, 83–88.

Pressey, R. (1977). *In* "Enzymes in Food and Beverage Processing" (R. L. Ory and A. St. Angelo, Eds), pp. 172–191.

Pressey, R., Avants, J. K. and Dull, G. G. (1978). *Abs. XXth Int. Hort. Congr., Sydney, Australia*, 1096.

Robinson, D. G. (1977). *In* "Advances in Botanical Research" (H. W. Woolhouse, Ed.), Vol. 5, pp. 89–151. Academic Press, London and New York.

Rombouts, F. M., Wissenburg, A. K. and Pilnik, W. (1979). *J. Chromatogr.* **168**, 151–161.

Smith, O. (1935). *N.Y. Agric. Exp. Stn, Ithaca*, Mem. No. 184, 1–16.

Sobotka, F. E. and Stelzig, D. A. (1974). *Plant Physiol.* **53**, 759–763.

Strand, L. L., Rechtoris, C. and Mussell, N. (1976). *Plant Physiol.* **58**, 722–725.

Tigchelaar, E. C., McGlasson, W. B. and Buescher, R. W. (1978). *HortScience* **13**, 508–513.

Vickery, R. S. and Bruinsma, J. (1973). *J. exp. Bot.* **24**, 1261–1270.

Wallner, S. J. and Bloom, H. L. (1977). *Plant Physiol.* **60**, 207–210.

Wallner, S. J. and Walker, J. E. (1975). *Plant Physiol.* **55**, 94–98.

Woodmansee, C. W., McClendon, J. H. and Somers, G. F. (1959). *Food Res.* **24**, 503–514.

CHAPTER 7

Composition and Metabolism of Cell Wall Polysaccharides in Ripening Fruits

M. KNEE AND I. M. BARTLEY

East Malling Research Station, East Malling, Maidstone, Kent, England

I. Introduction 133
II. Cell Wall Structure 134
A. General Features of Plant Primary Walls 134
B. Structure of the Middle Lamella 134
C. Special Features of Fruit Cell Walls 135
III. Wall Degradation during Fruit Ripening 139
A. Cellulose 139
B. Hemicelluloses 139
C. Loss of Galactose Residues 139
D. Increase of Soluble Pectin 141
IV. Role of Synthetic Processes in Cell Wall Changes 142
A. Synthesis versus Degradation in Ripening Fruit 142
B. Cell Expansion and Other Cell Separating Systems 145
V. Model of the Cell Separation Process 146
References 147

I. INTRODUCTION

The softening of ripening fruits is economically important, as a process affecting the edibility of the fruits as well as influencing the length of time for which they can be kept. The process is also of general biological interest as a gross example of cell wall degradation having interesting parallels with other cell separating systems.

The texture of living plant tissues is affected by their cellular anatomy, the water relations of their cells, and the composition of their cell walls. Fruit cells retain their normal osmotic properties during ripening (Simon, 1977) and there is probably little loss of turgor pressure, although this may be affected by dehydration if fruit is kept in a low humidity. Indeed, turgor pressure probably provides a driving force for cell separation following structural changes in the middle lamella and cell wall. While recognizing the importance of structure and water relations this chapter concentrates on polysaccharide composition and metabolism as determinants of fruit texture.

II. Cell Wall Structure

A. GENERAL FEATURES OF PLANT PRIMARY WALLS

The primary wall consists of cellulose microfibrils embedded in a matrix of other polysaccharides and protein. In recent years there has been considerable progress in structural studies on the matrix polysaccharides, facilitated by advances in techniques such as methylation analysis. The model structures for rhamnogalacturonan with its side branches of chains of arabinose and galactose residues and for xyloglucan, worked out in Albersheim's laboratory (Talmadge *et al.*, 1973, Bauer *et al.*, 1973, and Keegstra *et al.*, 1973) have gained general acceptance. More controversial have been the proposed covalent linkages between these polymers and others in the cell wall. Keegstra *et al.* (1973) presented evidence for a linkage of xyloglucan to rhamnogalacturonan through a galactose chain and of rhamnogalacturonan to protein through arabinogalactan linked to serine. The existence of these linkages has not been substantiated and other authors (e.g. Monro *et al.*, 1976) have preferred structures based on non-covalent bonding. Uncertainty about interpolymer bonding hinders interpretation of compositional differences: for example does the presence of neutral branches on pectin prevent interchain bonding and increase wall extensibility (Stoddart *et al.*, 1967), or do they hinder the slippage of individual chains through linkage or entanglement with other polymers, and thus reduce extensibility?

B. STRUCTURE OF THE MIDDLE LAMELLA

Histological studies of ripening fruit reveal extensive cell separation and it is natural to suppose that changes in the intercellular matrix or middle lamella are responsible for this. The Albersheim model was developed from the analysis of suspension cultured cells and has little to say about middle lamella structure, except by implication that it would be co-extensive with and similar to the matrix of the primary wall. Traditionally the middle lamella has been thought of as pectin rich and this has been confirmed by use of ferric hydroxamate staining prior to electron microscopy (Albersheim and Killias, 1963). The fact that cells of non-woody plant tissues can be dissociated using chelating agents (Letham, 1962; Ginsburg, 1961; Linehan and Hughes, 1969) suggests that cohesion of the middle lamella depends on ionic rather than covalent bonds. Divalent cations, especially calcium are the obvious candidates but as Rees (1969) pointed out single calcium ions would only form weak bonds between isolated carboxyl groups. More stable bonding occurs by cooperative effects when sequences of uronic acid lie parallel and each pair of residues encloses a calcium ion: the so-called "eggbox" junction zones (Rees, 1975).

C. SPECIAL FEATURES OF FRUIT CELL WALLS

Apart from the vascular tissue and specialized cells, for example the stone cells in pears, the cell walls of the fleshy parts of most fruits are unlignified. They also contain a low proportion of hydroxyproline rich protein and small amounts of xylose and mannose residues which are characteristic of hemicellulose polymers; on the other hand, they contain a high proportion of galacturonic acid, galactose and arabinose residues which are typical of pectic polymers (Table I).

TABLE I
Gross composition of fruit cell walls

Component monomers	% total accounted for	
	Apple	Strawberry
Rhamnose	0·4	1·1
Fucose	0·7	N.D.
Arabinose	19·5	6·5
Xylose	5·9	1·9
Mannose	1·9	0·7
Galactose	5·8	7·6
Glucose	47·5	31·1
Galacturonic acid	16·6	40·3
α-Amino acid	1·7	10·7
Hydroxyproline	0·04	0·1

N.D.—not detected.

Complexes of polygalacturonide, neutral polysaccharide and hydroxyproline rich protein can be isolated from the soluble and cell wall fractions of apple tissue (Knee, 1973a). But protease and glycanase degradation did not yield glycopeptides which would be expected from covalent attachment of pectin and protein (Knee, 1975). Thus the complexes are thought to be aggregates and, probably, artefacts arising in the extraction procedure.

TABLE II
Sequential degradation of apple fruit cell walls by purified enzymes

Substrate	Enzyme	Monomeric composition of products GalUA	Ara	Xyl	Gal	Glc
UCW	PG pI 4·6	2·07	0·30	T	0·30	0·05
	PG pI 9·7	1·16	0·74	0·07	0·40	0·12
	Glucanase	0·84	0·60	0·09	1·03	0·45
	(not released)	0·00	1·12	0·69	2·07	30·30[a]
RCW	PG pI 4·6	1·73	0·24	T	0·05	T
	PG pI 9·7	0·54	0·73	0·06	0·13	0·08
	Glucanase	0·35	0·07	0·01	0·03	0·02
	(not released)	0·00	1·23	0·80	0·59	4·30

UCW, cell walls from unripe fruit; RCW, cell walls from ripe fruit; T, trace; [a] includes starch.

Purified glycanases, mainly from fungal sources, have been used in structural studies on apple fruit cell walls (Knee *et al.*, 1975). For example, cell walls were degraded sequentially (Table II): one polygalacturonase isoenzyme (PG pI 4·6) released uronic acid with little associated neutral material, a second (PG pl 9·7) yielded uronic acid with a higher proportion of arabinose and galactose residues; a glucanase which was contaminated with polygalacturonase, could then release the remaining uronic acid associated with an even higher proportion of neutral residues. Although it causes limited degradation of cell walls, PG pI 4·6 caused separation of cells when applied to apple discs. It was concluded that this enzyme selectively attacked an unbranched polygalacturonide (or "homogalacturonan") which is important in middle lamella structure. The other isoenzyme, PG pl 9·7, could also degrade a highly branched polymer; analysis of the fragments showed that there were separate side chains of arabinose and galactose units. The arabinose was present in relatively large units (*c*. 5000 daltons) which could be degraded by α-arabinofuranosidase to release lateral 1,3 linked residues from a 1,5 linked core (Tagawa and Kaji, 1969). The arabinose rich fragments also contained most of the rhamnose released by the PG pI 9·7, suggesting an attachment of arabinan to the rhamnose residues which are known to be interpolated in the polygalacturonic acid chains (Barrett and Northcote, 1965). Galactose was present in relatively small units ($<$1000 daltons) which could be degraded by a β-1,4 galactanase (Fig. 1). These oligosaccharides were thought to be fragments of a branched polygalacturonide sharing most of the features of the rhamnogalacturonan of sycamore suspension culture cell walls (Talmadge *et al.*, 1973) and forming the main constitutent of the matrix of the primary wall (Fig. 2).

The small amount of hemicellulose present in apple cell walls is relatively inaccessible to glucanase attack. Chromatographic purification of an alkaline

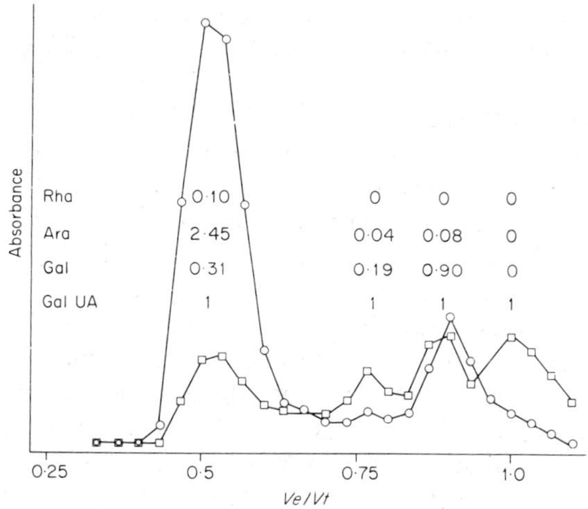

FIG. 1. Gell permeation chromatogram (Sephadex G-25) of oligosaccharides released by polygalacturonase action on apple fruit cell walls. ○—○, total carbohydrate; □—□, uronic acid; *Ve*, elution volume; *Vt*, total column volume.

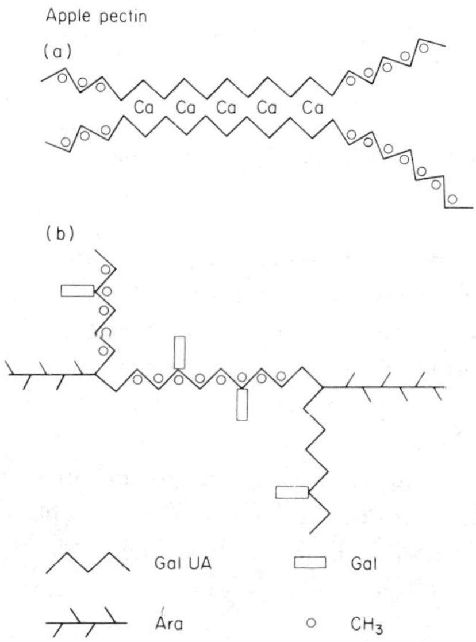

FIG. 2. Proposed structures for polygalacturonate fractions in apple fruit. (a) Sections of two "homogalacturonan" chains with linkage region. (b) Portion of branched rhamnogalacturonan. The kinks in the backbone represent rhamnose residues which are possibly the site of attachment of arabinan branches.

extract of apple cell walls yielded a polysaccharide of similar gross composition to the xyloglucan extensively studied by Bauer *et al.* (1973) (Knee, 1973a).

Subsequent chemical studies showed that the homogalacturonan in apple cell walls could be solubilized by partial methylation of its free carboxyl groups with diazomethane (Table III). Diazomethylation, or treatment with the chelating agent, sodium polyphosphate, also caused loss of cell to cell cohesion in ethanol extracted tissue (Knee, 1978a). This suggests that at least some of the 20% of non-esterified carboxyl groups in the cell wall are important in maintaining cell cohesion by cross-linkage through divalent metal ions. Accepting Rees' argument that single cross-links would be weak, a stable structure would require the alignment of non-esterified sequences of uronic acid residues in otherwise esterified chains. Unfortunately nothing is known about the distribution of methyl ester groups in polygalacturonate and the structure presented in Fig. 2 must be regarded as conjectural.

TABLE III
Chemical modification of cell walls and cell cohesion in apple fruit tissue[a]

Reagent	Tissue rupture force (N)	Monomeric composition of solubilized material (% total in cell wall) GalUA	Ara	Gal
None	7·0	26	14	15
CH_2N_2	0·5	55	2	2
$HCl-CH_3OH$	0·4	71	92	91
SPP	0·4	34	ND	ND

N.D., not determined; SPP, 2% sodium polyphosphate (pH 4·0); [a] data from Knee (1978a).

The branched rhamnogalacturonan could be solubilized by treatment of cell walls with anhydrous HCl in methanol (Table III). As well as methylating carboxyl groups this reagent hydrolysed the weak glycosidic bonds between arabinofuranosyl residues. It is unlikely that arabinan is involved in covalent links to polymers other than polygalacturonate, but it may aid binding in some way (Knee, 1978a).

Although xylose and glucose residues are often present in pectic fragments obtained by enzymic (Knee *et al.*, 1975) or chemical (Knee, 1978a) degradation of apple fruit cell walls there has been no direct evidence of linkage between hemicellulose and pectin.

III. Wall Degradation During Fruit Ripening

A. CELLULOSE

Crystalline α-cellulose is very resistant to enzymic attack and can only be degraded by the combined operation of several glycanases and other enzymes such as glucose oxidase and a peroxide utilizing system (Eriksson, 1977). The necessary enzymes have not been found in fruit and it is not surprising that degradation of cellulose during ripening has not been reported either. In view of this it is unfortunate that the β-1,4 glucanase activity which is present in some fruits and other plant parts should so often be called "cellulase".

B. HEMICELLULOSES

The small quantities of monomers characteristic of hemicelluloses, xylose, glucose and mannose do not decline in the ripening of apples (Bartley, 1976), strawberries (Neal, 1965), tomatoes (Wallner and Bloom, 1977) and pears (Jermyn and Isherwood, 1956). All of these fruits, with the exception of apples, are known to contain β-1,4 glucanase activity (Hobson, 1968; Barnes and Patchett, 1976). Labavitch and Ray (1974) have examined cell wall changes following IAA treatment of coleoptiles and suggest that although there may not be quantitative changes in xyloglucan, solubilization of this fraction is associated with increased wall plasticity. Further doubt that β-1,4 glucanase has a primary role in fruit softening arises from the observation that the activity of this enzyme is normal in the non-ripening *rin* mutant of tomatoes (Poovaiah and Nukaya, 1979).

C. LOSS OF GALACTOSE RESIDUES

The major change in monomeric composition in the cell walls of ripening apples is the loss of about 60% of their galactosyl residue content (Knee, 1973b). Fragments liberated by purified polygalacturonase (PG pI 9·7) from cell walls of ripe apples contained less galactose than those from unripe apples (Table II) indicating that the loss of this monomer occurred in the branched pectin fraction. Similarly in tomatoes (Wallner and Bloom, 1977) loss of galactose is the dominant change, but in pears (Jermyn and Isherwood, 1956) and strawberries (Neal, 1965) there is less change. However there is evidence of turnover of galactose residues during growth of strawberry fruits (Knee *et al.*, 1977) and this seems to be a feature of cell expansion in other tissues (Labavitch and Ray, 1974). Apples (Bartley, 1974), tomatoes (Wallner and Walker, 1975) and pears (Yamaki and Matsuda, 1977) all contain β-galactosidase activity. The apple enzyme was shown to be capable of

TABLE IV
Hydrolysis of cell walls of cortical tissue from apples stored in 2% oxygen by β-galactosidase[a]

Time in store (days)	Galactose anhydrosugar content (mg g^{-1} fr wt) Cell wall preparation	H_2O soluble products of incubation[b] Polymer (insoluble in EtoH)	Low MW (soluble in EtoH)
0	2·75	0	0·52
79	1·26	0	0·12
119	0·97	0	0·08

[a] Data taken from Bartley (1978). [b] 72 h at 20°C.

degrading β-1,4 linked galactan (Bartley, 1974) and of releasing galactose residues from apple cell wall preparations (Table IV; Bartley, 1978). Enzyme activity was present at all stages of ripening in the various fruits examined and activity associated with the cell wall was detectable at all stages in the apple (Bartley, 1977; Fig. 3). This leads to the suggestion that hydrolysis of terminal galactosyl residues is a constant feature of some fruit cell walls but that galactose residues decline only when the synthesis of new galacturonorhamnan carrying branches of galactose units fails to make good the losses.

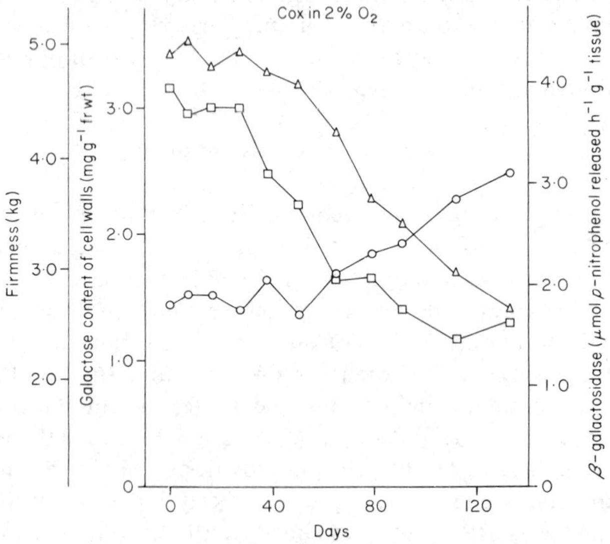

FIG. 3. Changes in fruit firmness, galactose content of cortex cell walls and β-galactosidase activity in Cox's Orange Pippin apples stored in 2% O_2 at 3°C (Bartley, 1977). △—△, Firmness; □—□, galactose; ○—○, β-galactosidase.

The significance of the loss of galactose residues is not clear. The loss took place after harvest in *rin* tomatoes but little softening occurred (Gross and Wallner, 1979). In apples storage in 2% O_2 delays other compositional changes and softening appeared to commence with the loss of galactose residues (Fig. 3).

D. INCREASE OF SOLUBLE PECTIN

It has been known for at least 60 years that soluble pectin increases in ripening apples (Carré, 1922). The process is matched by a decline in the insoluble pectin fraction and it is natural to conclude that a change in part of the insoluble fraction converted it to a soluble form. The ensuing years have seen various interpretations of what that change might be. Joslyn (1962) has reviewed the older literature and we will confine ourselves to more recent developments. The soluble pectin in ripe apples has a low proportion of neutral monomers (Knee, 1973a) and this is in keeping with its supposed origin in the middle lamella, which we believe to be rich in this type of pectin. Its physical properties and hence its molecular weight are comparable with those of the soluble pectin in the unripe fruit (Table V) and it is hard to see how it could be a degradation product. The apple is exceptional among fruits in that it appears to lack endo-polygalacturonase (PG). In other fruits this enzyme is an obvious means of achieving solubilization and the soluble pectin isolated from peaches (Pressey *et al.*, 1971), strawberries (Woodward, 1972) and pears (Knee, 1973c) shows progressive depolymerization with ripening. An interesting contrast is provided by different varieties of peach: "free stone" varieties possess high activities of both endo- and exo-PG; they show extensive softening and a marked rise in soluble pectin on ripening. "Clingstone" varieties have low activities of endo-PG, and show less softening and pectin solubilization (Pressey and Avants, 1978). Comparison with these fruits leaves the apple as a conspicuous and, we believe, significant exception. In the absence of endo-PG how does it achieve solubilization of pectin and cell separation?

An exo-PG is present (Bartley, 1978) and this must account for the loss of about 10% of total uronic acid residues during ripening. The enzyme

TABLE V
Sedimentation and viscosity data for pectin fractions from ripe and unripe apples[a]

Apples	Limiting viscosity number ml/g		Sedimentation coefficient	
	Soluble pectin	Wall bound pectin	Soluble pectin	Wall bound pectin
Unripe	337	319	1·61	1·36
Ripe	585	166	1·24	1·34

[a] Data from Knee (1978a).

preferentially attacks non-esterified polygalacturonate but is capable of degrading apple cell walls to release monomers and some polymeric material (Table VI). The viscosity of the wall-bound pectin in apples declines during ripening (Table V) although the neutral glycan branches on this polymer would be expected to block exo-PG attack.

TABLE VI
Hydrolysis of cell walls of cortical tissue from apples stored in 2% oxygen by exo-polygalacturonase[a]

Time in store (days)	Galacturonic acid anhydrosugar content ($mg\ g^{-1}$ fr wt)		
	Cell wall preparation	H_2O soluble products of incubation[b]	
		Polymer (insoluble in EtoH)	Low MW (soluble in EtoH)
0	3·48	0·14	0·43
79	2·93	0·13	0·38
119	2·75	0·09	0·34

[a] Data taken from Bartley (1978). [b] 72 h at 20°C.

IV. ROLE OF SYNTHETIC PROCESSES IN CELL WALL CHANGES

A. SYNTHESIS VERSUS DEGRADATION IN RIPENING FRUIT

Most studies of cell wall changes in ripening fruit have concentrated on degradative aspects and the possible role of synthetic processes has rarely been considered. A study of polysaccharide metabolism in developing strawberry fruits (Knee *et al.*, 1977) revealed that incorporation of ^{14}C-glucose into wall polymers ceased at the onset of ripening, and when fruits were labelled with $^{14}CO_2$ at an early stage a loss of label from insoluble polymers and a corresponding increase of label in soluble polymers occurred during ripening (Fig. 4). This was consistent with the predominance of degradative changes in ripening.

However, the restricted range of degradative enzymes in apple and the limited degradation of wall polymers prompted an investigation of synthetic processes in this fruit (Knee, 1978b).

Methyl labelled ^{14}C-methionine and ^{3}H-inositol were selected as precursors, respectively of methyl ester and galacturonic acid residues in pectin.

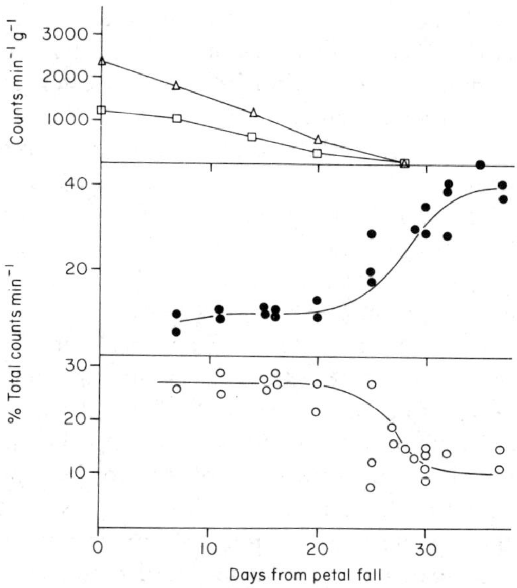

FIG. 4. Polysaccharide metabolism in developing strawberry fruits (Knee *et al.*, 1977). △—△, ^{14}C-glucose incorporation into soluble polymers; □—□, ^{14}C-glucose incorporation into cell wall; ●—●, proportion of label from $^{14}CO_2$ applied early in development in soluble polymers; ○—○, proportion of label in cell wall.

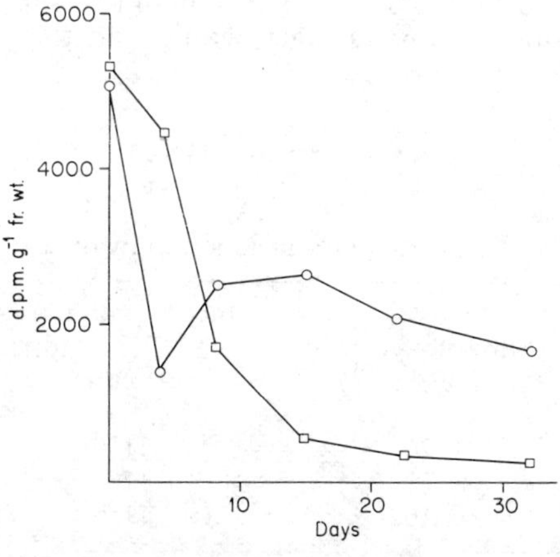

FIG. 5. Pectin metabolism in ripening apples (Knee, 1978b). ○—○, incorporation from $^{14}CH_3$ methionine into methyl ester; □—□, incorporation from ^{3}H-inositol into galacturonic acid.

Unripe fruit showed good conversion of both precursors into the expected products. On ripening inositol incorporation into galacturonic acid declined sharply but methionine incorporation into methyl ester groups was maintained (Fig. 5). There was no change in the degree of esterification of the total pectin but the esterification of the soluble fraction increased with ripening. Although esterification of wall-bound pectin had been suggested as a mechanism for solubilization and cell separation (Neal, 1965), this seemingly obvious interpretation of the results was rejected. The immediate precursor of the methyl groups, S-adenosyl methionine (SAM) would be unlikely to be exported from the cytoplasm and the only known methylating system was associated with a cytoplasmic, particulate fraction (Kauss and Hassid, 1967).

But SAM can be taken up from the surrounding medium by mammalian and yeast cells (Stramentinoli and Pezzoli, 1979) and it might, therefore, cross the plasmalemma in the opposite direction. Alternatively, the continuation of methionine incorporation in the absence of inositol incorporation could be evidence of continued pectin synthesis utilizing glucose as the precursor of galacturonic acid in preference to inositol, or it could arise from the methylation of a pre-existing cytoplasmic pool of polygalacturonate. The utilization of glucose for polysaccharide synthesis in ripe fruits is difficult to demonstrate because of the large pools of hexose sugars present. But application of large amounts of ^{3}H-glucose did give some incorporation into galacturonic acid in ripening pear slices (Knee, 1980, unpublished results). Whatever the interpretation of methionine incorporation, the process could be irrelevant to cell separation during fruit ripening. That it is relevant is suggested by various observations. Rhizobitoxine inhibits methionine incorporation into pectin by pear slices at concentrations which do not inhibit protein synthesis (Table VII). This inhibitor is better known for its effects on ethylene biosynthesis (Owens *et al.*, 1971). Application of a related compound, amino-ethoxy vinyl glycine (AVG), to apples on the tree inhibited the production of soluble pectin during ripening and this inhibition was only

TABLE VII
Effects of rhizobitoxine on methionine metabolism by discs of ripening pear tissue

Concentration of inhibitor	Ethylene synthesis (nl g^{-1} h^{-1})	Incorporation from $^{14}CH_3$-methionine (dpm g^{-1} h^{-1}) Pectin[a]	Protein[b]
0	26	3120	530
5×10^{-6}M	21	1760	600
5×10^{-5}M	16	1150	480

[a] Activity in methanol insoluble fraction lost on treatment with NH_4OH. [b] Residual activity after NH_4OH treatment. (Assay system as described by Knee, 1978b).

partially overcome by treatment with ethylene (Fig. 6). Also, the failure of pears to develop normal levels of soluble pectin and soften properly after long storage at −1° C is associated with lower incorporation from methionine into pectin; endo-PG activity develops normally in the long stored fruit (Knee and Casimir, 1980, unpublished results).

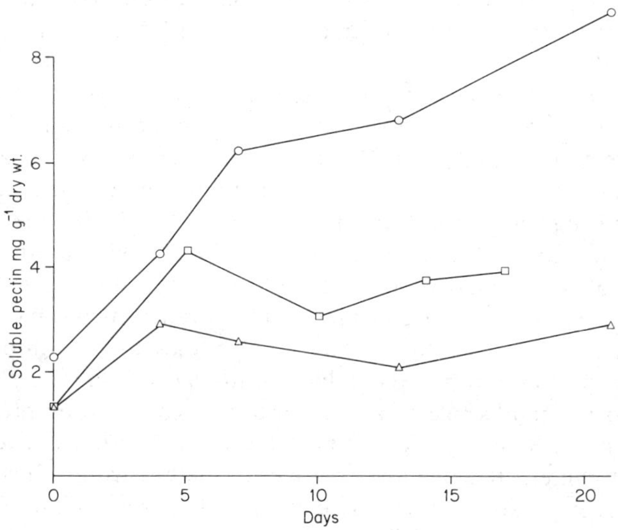

FIG. 6. Changes in soluble pectin in Golden Delicious apples treated with amino-ethoxy vinyl glycine (AVG). ○—○, control; △—△, AVG treated; □—□, AVG and ethylene treated. Fruit samples were provided by Dr F. Bangerth. AVG was applied as three sprays (0·5 g l^{-1}) 38, 30 and 18 days before harvest and fruit was ripened at 20°C in air or with ethylene in air (400 μl l^{-1}).

B. CELL EXPANSION AND CELL SEPARATING SYSTEMS

As mentioned above there is growing evidence that changes in wall properties influencing cell expansion are accomplished by turnover of cell wall components (Labavitch and Ray, 1974; Darvill *et al.*, 1978). Cell separation in senescent tissues should be regarded as an extension of these processes with the balance shifted more in favour of degradation than synthesis. From this point of view continued pectin synthesis in ripening fruit could be regarded as a (futile) gesture towards replacement of the polysaccharides being degraded in the cell wall; this would be in keeping with the concept of ripening as a disturbance of the homeostasis of the unripe fruit (Romani, 1978)

We have argued above that the incorporation from methionine is related to soluble pectin formation and fruit softening. This could involve synthesis of new highly esterified pectin with low cohesive properties or esterification of existing pectin. Newly synthesized pectin would need to be secreted into the middle lamella region in order to affect cell cohesion. There is some doubt as to whether molecules of this size can traverse the primary wall (Carpita *et al.*, 1979). In another cell separating system, the root cap, there is good evidence for the synthesis of a specialized polysaccharide which lubricates the passage of the root tip through the soil. Secretion of this polysaccharide seems to occur only after degradative enzymes have loosened wall structure (Paull and Jones, 1976).

Abscission zones are possibly more closely analogous to ripening fruit. Interestingly, endo-PG activity does not seem to have been reported although an exo-PG from citrus and bean abscission zones has been described (Riov, 1974; Berger and Reid, 1979). Furthermore, methionine treatment accelerates abscission and high incorporation of label from ^{14}C-methionine occurs in cotton abscission zones (Valdovinos and Muir, 1965). This was attributed to esterification of wall-bound pectin although the biochemical plausibility of this does not seem to have been considered. The parallel enhancement by IAA of wall plasticity and growth in *Avena* coleoptiles are accompanied by increased incorporation from ^{14}C-methionine into pectin (Jansen *et al.*, 1960). This occurred in the absence of an increase in degree of esterification and apparently without a stimulation of polysaccharide synthesis. The authors attributed it to an enhancement of turnover of methyl groups during pectin synthesis. This enigmatic finding closed an episode of research to relate IAA induced growth to pectin esterification. More recently the formulation of the Albersheim model of primary wall structure led to suggestions that increased extensibility of walls could result from turnover of interpolymer bonds (Keegstra *et al.*, 1973) or polymers themselves (Labavitch and Ray, 1974; Darvill *et al.*, 1978). These models do not envisage quantitative changes in the bonds or polymers in question. If the occurrence of extracellular methylation of polygalacturonate can be substantiated this could be a means of regulating the cohesion of the wall matrix and middle lamella.

V. Model of the Cell Separation Process

Many questions remain to be answered concerning the structure and metabolism of plant cell walls in general and fruits in particular. With this in mind the following is offered as a hypothesis to suggest further experimental approaches which may result in modification or rejection of the model. It is proposed that the middle lamella is stabilized by junction zones formed by calcium ions between free carboxyl groups in predominantly esterified pectin molecules. Methylation of the carboxyl groups disrupts these linkages, but elsewhere on the pectin chains the action of pectin esterase can form new carboxyl groups which would be available for formation of new cross-links.

Enhanced turnover of linkage zones could be the initial reaction promoting cell separation in all fruits (and abscission zones?). The apple lacks endo-PG and the exo-PG would cause a limited degradation of pectin. In other fruits endo-PG degrades the solubilized pectin; cell separation and softening are more extensive.

Substantiation of this model could aid practical control of fruit softening. Considerable effort has gone into the control of the production and action of ethylene which stimulates the cell separation process. However, the process itself might be amenable to modification. Pears lose the capacity to incorporate methionine into pectin on prolonged storage; this is a disadvantage in that the pear does not develop its attractive "melting-ripe" texture, but in other fruits, such as the apple, partial inactivation of the system could be of benefit.

REFERENCES

Albersheim, P. and Killias, U. (1963). *Am. J. Bot.* **50**, 732–745.
Barnes, M. F. and Patchett, B. J. (1976). *J. Fd Sci.* **41**, 1392–1395.
Barrett, A. J. and Northcote, D. H. (1965). *Biochem. J.* **94**, 617–627.
Bartley, I. M. (1974). *Phytochemistry* **13**, 2107–2111.
Bartley, I. M. (1976). *Phytochemistry* **15**, 625–626.
Bartley, I. M. (1977). *J. Exp. Bot.* **28**, 943–948.
Bartley, I. M. (1978). *Phytochemistry* **17**, 213–216.
Bauer, W. D., Talmadge, K. W., Keegstra, K. and Albersheim, P. (1973). *Pl. Physiol.* **51**, 188–196.
Berger, R. K. and Reid, P. D. (1979). *Pl. Physiol.* **63**, 1133–1137.
Carpita, N., Sabularse, D., Montezinos, D. and Delmer, D. P. (1979). *Science* **205**, 1146–1147.
Carré, M. H. (1922). *Biochem. J.* **16**, 704–712.
Darvill, A. G., Smith, C. J. and Hall, M. A. (1978). *New Phytol.* **80**, 503–516.
Eriksson, K. E. (1977). "Cell Wall Biochemistry Related to Specificity in Host-Plant Pathogen Interactions" (J. Raa and B. Solheim, Eds), pp. 71–83. University Tromso Press, Tromso, Norway.
Ginsburg, B. Z. (1961). *J. Exp. Bot.* **12**, 85–107.
Gross, K. C. and Wallner, S. (1979). *Pl. Physiol.* **63**, 117–120.
Hobson, G. E. (1968). *Ann. Rep. Glasshouse Crops Res. Inst. 1967*, 134–136.
Jansen, E. F., Jang, R., Albersheim, P. and Bonner, J. (1960). *Pl. Physiol.* **35**, 87–97.
Jermyn, M. A. and Isherwood, F. A. (1956). *Biochem J.* **64**, 123–132.
Joslyn, M. A. (1962). *Adv. Fd Res.* **11**, 1–107.
Kauss, H. and Hassid, W. Z. (1967). *J. Biol. Chem.* **242**, 3449–3453.
Keegstra, K., Talmadge, K. W., Bauer, W. D. and Albersheim, P. (1973). *Pl. Physiol.* **51**, 188–196.
Knee, M. (1973a). *Phytochemistry* **12**, 637–653.
Knee, M. (1973b). *Phytochemistry* **12**, 1543–1549.
Knee, M. (1973c). *J. Sc. Fd Agric.* **24**, 1137–1145.
Knee, M. (1975). *Phytochemistry* **14**, 2181–2188.
Knee, M. (1978a). *Phytochemistry* **17**, 1257–1260.
Knee, M. (1978b). *Phytochemistry* **17**, 1261–1264.

Knee, M., Fielding, A. M., Archer, S. A. and Laborda, F. (1975). *Phytochemistry* **14**, 2213–2222.
Knee, M., Sargent, J. A. and Osborne, D. J. (1977). *J. Exp. Bot.* **28**, 377–396.
Labavitch, J. M. and Ray, P. M. (1974). *Pl. Physiol.* **53**, 669–673.
Letham, D. S. (1962). *Exp. Cell Res.* **27**, 352–355.
Linehan, D. J. and Hughes, J. C. (1969). *J. Sci. Fd Agric.* **20**, 119–123.
Monro, J. A., Penny, D. and Bailey, R. W. (1976). *Phytochemistry* **15**, 1193–1198.
Neal, G. E. (1965). *J. Sci. Fd Agric.* **16**, 604–611.
Owens, L. D., Lieberman, M. and Kunishi, A. (1971). *Pl. Physiol.*, **48**, 1–4.
Paull, R. E. and Jones, R. L. (1976). *Z. Pflanzenphysiol.* **79**, 154–164.
Poovaiah, B. W. and Nukaya, A. (1979). *Pl. Physiol.* **64**, 534–537.
Pressey, R. and Avants, J. K. (1978). *J. Fd Sci.* **43**, 1415–1417.
Pressey, R., Hinton, D. M. and Avants, J. K. (1971). *J. Fd Sci.* **36**, 1070–1073.
Rees, D. A. (1969). *Adv. Carbohydr. Chem. Biochem.* **24**, 267–332.
Rees, D. A. (1975). "Biochemistry of Carbohydrates". M.T.P. Internat. Rev. Sci. Biochem. Ser. 1, Vol. 5 (W. J. Whelan, Ed.), pp. 1–42. Butterworth, London, University Park Press Baltimore.
Riov, J. (1974). *Pl. Physiol.* **53**, 312–316.
Romani, R. J. (1978). *J. Fd Biochem.* **2**, 221–228.
Simon, E. W. (1977). *J. Exp. Bot.* **28**, 1147–1152.
Stoddart, R. W., Barrett, A. J. and Northcote, D. H. (1967). *Biochem J.* **102**, 194–204.
Stramentinoli, G. and Pezzoli, C. (1978). "Transmethylation". Developments in Neuroscience Vol. 5 (E. Usdin, R. T. Borchardt and C. R. Creveling, Eds), pp. 37–47. Elsevier/North Holland, New York, Amsterdam, Oxford.
Tagawa, K. and Kaji, A. (1969). *Carbohyd. Res.* **11**, 293–301.
Talmadge, K. W., Keegstra, K., Bauer, W. D. and Albersheim, P. (1973). *Pl. Physiol.* **51**, 158–173.
Valdovinos, J. G. and Muir, R. M. (1965). *Plant Physiol.* **40**, 335–340.
Wallner, S. J. and Bloom, H. L. (1977). *Pl. Physiol.* **60**, 207–210.
Wallner, S. J. and Walker, J. E. (1975). *Pl. Physiol.* **55**, 94–98.
Woodward, J. R. (1972). *J. Sci. Fd Agric.* **23**, 465–473.
Yamaki, S. and Matsuda, K. (1977). *Pl. Cell Physiol.* **18**, 81–93.

CHAPTER 8

The Molecular Biology of Ripening

D. GRIERSON, G. A. TUCKER AND N. G. ROBERTSON

Department of Physiology and Environmental Studies, University of Nottingham, School of Agriculture, Sutton Bonington, Loughborough, England

I.	Introduction	149
II.	Tomato Ripening Mutants	150
III.	Fruit-Softening Enzymes	150
IV.	Purification and Properties of Polygalacturonase	151
V.	Conversion of PG2 → PG1	155
VI.	Regulation of Polygalacturonase Activity During Ripening	156
VII.	Messenger-RNA Synthesis During Ripening	157
VIII.	Translation of Polygalacturonase m-RNA *in vitro*	158
IX.	Conclusions and Prospects	160
Acknowledgement		160
References		160

I. Introduction

A large number of physiological and biochemical changes occur during the ripening of fruits. These include the degradation of starch or other storage carbohydrate, the production of sugars, the synthesis of pigments and volatile compounds and the modification or partial solubilization of cell walls. In climacteric fruits these changes take place over a relatively short period of time in association with an increase in CO_2 production, the respiratory climacteric. Ethylene, which is a naturally occurring plant-growth-regulator, is implicated in the control of ripening. Removal of ethylene from the gas-phase surrounding fruits retards ripening whereas an exogenous supply of ethylene frequently accelerates ripening.

An understanding of the control of ripening is important because it has implications for fruit storage and fruit quality. Ripening control is also of scientific interest since many of the changes that occur during ripening are caused by alterations in the activities of specific enzymes, some of which are active only during ripening. The problem can therefore be viewed in the wider context of gene expression and the control of metabolism.

Although there are many different types of fruit with specific characteristics it is possible to formulate general questions such as (1) how is enzyme activity controlled? and (2) what is the mechanism of action of ethylene? Such problems are common to ripening processes in many fruits. We are investigating these problems by studying ripening in tomato (*Lycopersicon esculentum* Mill). This is an important crop in its own right but it also has advantages in that there are many experimental lines and ripening mutants available and tomatoes can be conveniently produced for experimental purposes throughout most of the year.

II. Tomato Ripening Mutants

During the course of this work two glasshouse varieties have been used: Potentate—a slow ripener and Ailsa Craig—a fast ripener. Two nuclear mutations, Neverripe (*Nr*) and ripening inhibitor (*rin*) have been introduced into these varieties by repeated back-crossing to produce near-isogenic lines (Darby, 1978). The phenotypic characteristics of mature fruit homozygous for either *rin* or *Nr* are shown in Table I. Essentially the fruit grow and develop normally but they do not undergo all of the changes associated with ripening.

TABLE I
Tomato ripening mutants

Neverripe (*Nr*)	Dominant gene, chromosome 9
Phenotype:	Reduced C_2H_4 and CO_2 production
	Reduced lycopene synthesis (orange fruit)
	Softens less than wild type at a comparable age
Ripening inhibitor (*rin*)	Recessive gene, chromosome 5
Phenotype:	Only background levels of C_2H_4 and CO_2
	Only traces of lycopene (yellow fruit)
	Extremely slow softening

III. Fruit-softening Enzymes

Apart from pigment synthesis, fruit-softening is one of the most obvious changes during tomato ripening. A large part of the cell-wall pectin is solubilized (see Hobson, Chapter 6, this volume) and this is associated with changes in cell-wall degrading enzymes. At least two enzymes are implicated in this process, pectin methyl esterase and polygalacturonase (Grierson *et al.*,

1980). The latter enzyme is virtually absent from mature green tomatoes but increases greatly in activity during ripening (Fig. 1). The lower amount of polygalacturonase in Potentate compared to Ailsa Craig is probably related to the slower ripening and lower softening of the former (Hobson, 1965). Homozygous *rin* fruit produce almost no polygalacturonase at any stage and they soften extremely slowly (Table I), probably by the action of a cellulase (Hobson, 1968). *Nr* fruit, which do soften appreciably, only produce low levels of polygalacturonase.

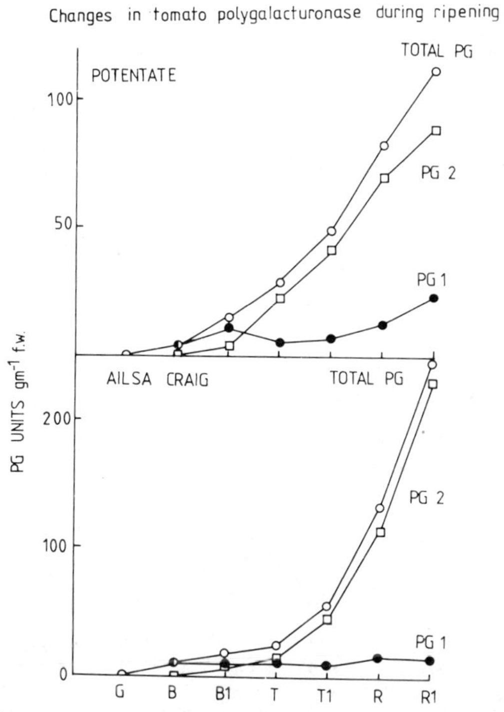

FIG. 1. Polygalacturonase activity extracted from individual fruits at various stages of ripeness. Total polygalacturonase was extracted from individual fruits of either Potentate or Ailsa Craig tomatoes and an aliquot was assayed for total activity (○). Further aliquots were heated at 65°C for 5 min before assay and the heat-stable activity taken as 90% of the original PG1 (●) (see Fig. 4). PG2 activity (□) was calculated by difference. Stages of fruit ripeness were: green (G); breaker (B); late breaker, i.e. half of the fruit coloured (B1); turning, i.e. appearance of orange colour (T); completely orange fruit (T1); firm red fruit (R); soft red fruit (R1). From Tucker *et al.* (1980).

IV. Purification and Properties of Polygalacturonase

We have detected and characterized two isoenzymes of polygalacturonase (PG) in Potentate and Ailsa Craig tomatoes (Grierson *et al.*, 1980). These enzymes, designated PG1 and PG2 have different molecular weights and can be separated by ion exchange chromatography (Fig. 2). The isoenzymes have been purified (Tucker, Robertson and Grierson, in preparation). Their properties are summarized in Table II. During ripening PG1, which has a molecular weight of approximately 100,000, is synthesized first, followed by PG2 (Fig. 1), which has a molecular weight of about 42,000. *Nr* tomatoes produce only PG1 under normal conditions and, as mentioned above, *rin* tomatoes produce virtually no PG.

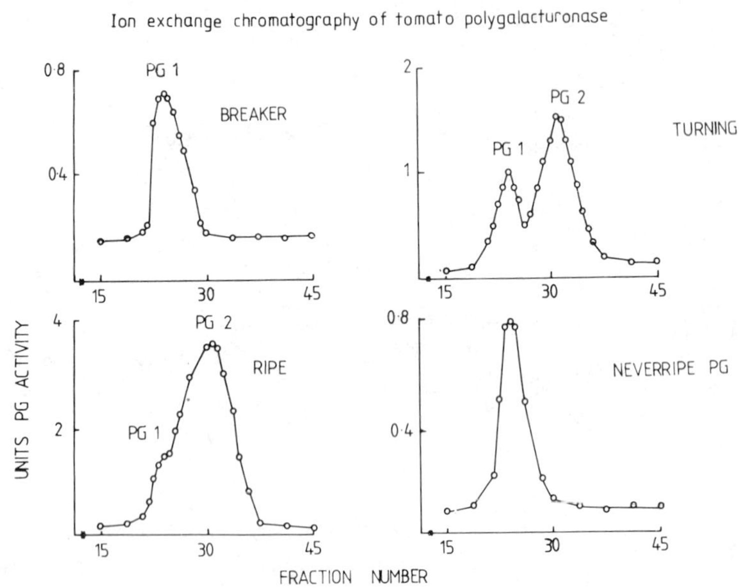

FIG. 2. Chromatography on Sephadex DEAE—A50 of polygalacturonase isoenzymes extracted from Ailsa Craig fruit at various stages of ripeness. Polygalacturonase was extracted from single fruits at breaker stage, i.e. just starting to show colour change (A); turning stage, i.e. fruit with an orange colour (B); red fruit (C); and "ripe" (=orange) fruit of the *Nr* mutant (D). From Tucker *et al.* (1980).

Despite the differences in molecular weight of native PG1 and PG2 the subunits of each isoenzyme run at identical positions on sodium dodecyl sulphate/polyacrylamide gel electrophoresis (Fig. 3). Furthermore, partial proteolysis of the proteins prior to electrophoresis yields similar polypeptides for PG1 and PG2, suggesting that the protein component of the two isoenzymes are very similar or identical (Fig. 3). However, the two enzymes have substantially different properties. The temperature for 50% inactivation of PG1 is 78° C whereas for PG2 it is 57° C (Fig. 4). Furthermore, isopycnic centrifugation of the two isoenzymes in CsCl shows that they have quite

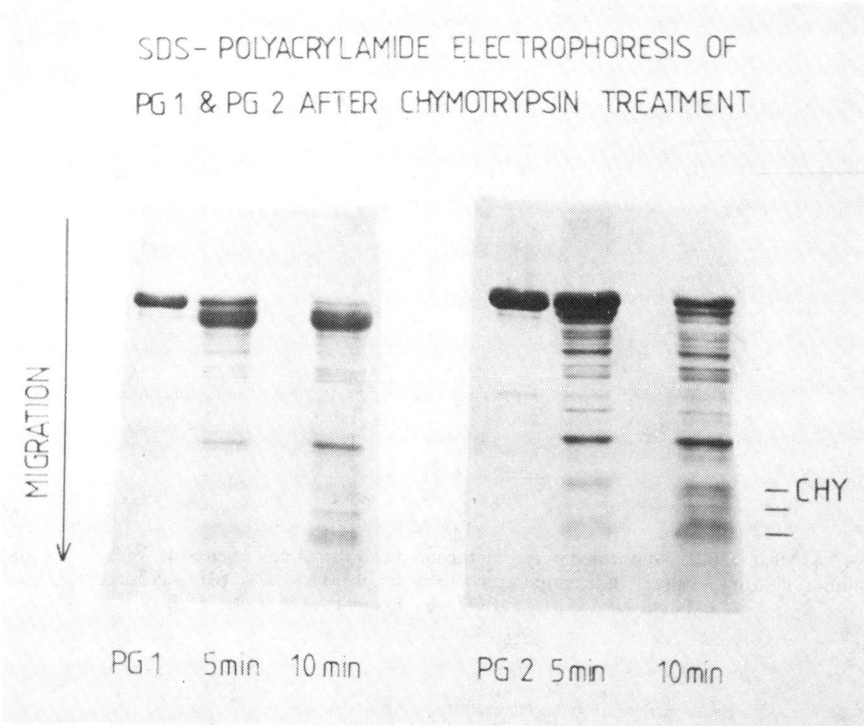

FIG. 3. SDS-polyacrylamide gel electrophoresis and finger-printing of polygalacturonases 1 and 2. This is a photograph of the gel after staining with Coomassie blue. Proteins were fractionated in a 10–15% (top–bottom) gradient of polyacrylamide in the presence of 1% sodium dodecyl sulphate. Reading from left to right: PG1 alone; PG1 plus chymotrypsin for 5 min; PG1 plus chymotrypsin for 10 min; PG2 alone; PG2 plus chymotrypsin for 5 min; PG2 plus chymotrypsin for 10 min. An amount equal to 50 μg of polygalacturonase was added to each track, plus 2 μg chymotrypsin where appropriate. The ink lines on the right of the photograph show the positions of polypeptides in the chymotrypsin preparation. From Tucker *et al.* (1980).

different densities (Fig. 5). Total polygalacturonase gives two peaks (density = 1·343 and 1·300 g ml^{-1}). After heating at 65° C (see Fig. 4) only the 1·343 g ml^{-1} peak remains, suggesting this is PG1 (Fig. 5), whereas purified PG2 labelled *in vitro* with ^{125}I bands with the 1·300 g ml^{-1} peak, suggesting this represents PG2. The difference in density between PG1 and PG2 could be due to major differences in conformation and ion-binding or by the preferential attachment of some other component, such as a carbohydrate, to PG1.

The most satisfactory explanation for these results is that PG1 is a dimer of PG2 and the difference in density and the slight discrepancy in molecular weights (see Table II) is due to binding of carbohydrate. We have no direct evidence for the presence of carbohydrate in PG1 (as suggested by its density) but this possibility is being tested.

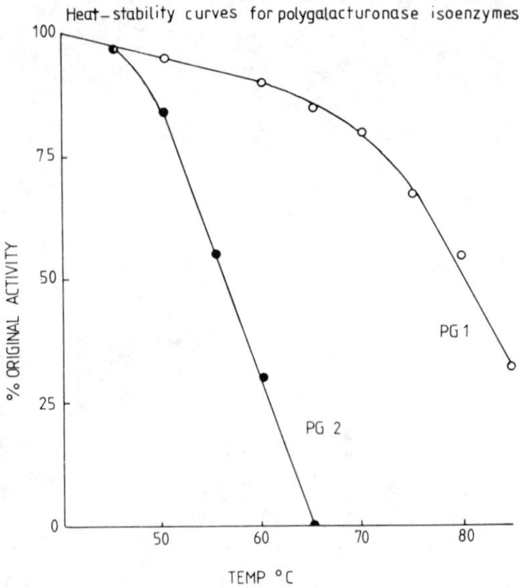

FIG. 4. Heat stability of purified polygalacturonases 1 and 2. Each isoenzyme was purified and aliquots held at the appropriate temperatures for 5 min in 0·15 M NaCl pH 6 before being cooled and assayed. PG1 (○); PG2 (●). From Tucker *et al.* (1980).

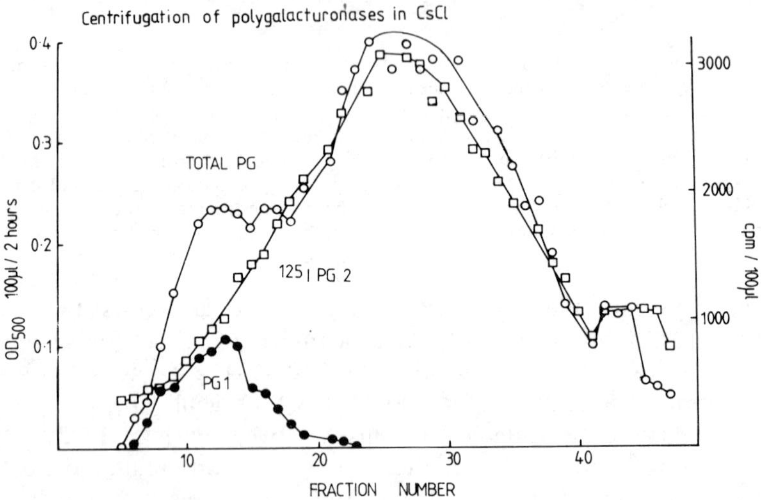

FIG. 5. Isopycnic centrifugation of polygalacturonases in CsCl. Total polygalacturonase was isolated from orange fruit and a trace of ^{125}I-labelled PG2 was added. The mixture was then centrifuged in CsCl (initial density 1·3 g cm^{-3}) at 45,000 r.p.m. for 90 h at 20°C in a Beckman type 75 Ti rotor. Exactly 100 μl fractions were collected, divided and used for (a) measurement of CsCl density using a refractometer, (b) assay of total PG activity (○), (c) assay of PG1 activity after heating for 15 min at 65°C (●), (d) measurement of radioactivity from ^{125}I-labelled PG2 (□). From Tucker *et al.* (1980).

TABLE II
Properties of polygalacturonase isoenzymes

	PG1	PG2	PG *Nr*
Elution volume from DEAE Sephadex (ml)	335	400	335
Molecular weight (Sephadex—G100)	100,000	42,000	100,000
Sub-unit molecular weight (SDS-polyacrylamide)	46,000	46,000	46,000
Temperature for half-inactivation (°C)	78	57	78
Density in CsCl (g ml^{-1})	1·343	1·300	1·343
Reaction with PG2 antibody	+	+	+

V. CONVERSION OF PG2 → PG1

The time-course for PG activity during ripening suggests that PG1 appears before PG2 (Fig. 1). The increase in activity of the latter isoenzyme is mainly responsible for the large rise in total PG activity. It is possible that the two isoenzymes are either produced separately or one is converted into the other. The basic similarity in the polypeptides of the two isoenzymes (Fig. 3) supports, but does not prove, the latter hypothesis.

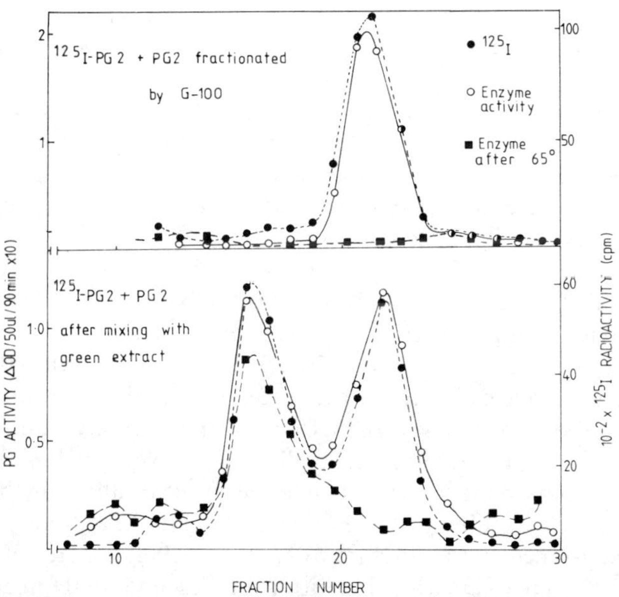

FIG. 6. *In vitro* conversion of PG2 to a form with properties of PG1. (Top panel) PG2 plus a trace of ^{125}I-labelled PG2 was run on Sephadex G-100 and the eluate assayed for ^{125}I radioactivity (●), total enzyme activity (○) and enzyme activity after heating at 65°C (■). (Bottom panel) an aliquot of the same mixture was treated with a salt-soluble, heat-stable, non diaysable extract from green tomatoes and then run on a Sephadex G-100 column and assayed as before.

We have been able to convert PG2 into a distinct isoenzyme form *in vitro* which has the properties of PG1. Figure 6 (top panel) shows the elution profile of PG2, purified from ripening tomatoes, following gel-filtration with Sephadex G-100. This isoenzyme, which can be identified by its heat-sensitivity and elution volume, co-chromatographs with a trace of PG2 previously iodinated *in vitro* with ^{125}I. The bottom panel in Fig. 6 shows the properties of the enzyme after mixing with an extract from green tomatoes (which have no PG activity, Fig. 1). The green extract converts part of the purified PG2 to a form with the elution volume and heat stability of PG1. The similar shift in position of ^{125}I radioactivity confirms that this involves the bulk transformation of PG2 isoenzyme and cannot be explained by the activation of an inactive form of PG1. These results support the suggestion that the protein components of the two isoenzymes are the same and provides evidence for a possible interconversion *in vitro*. At the present time, however, we have not been able to achieve the conversion of PG1 → PG2 *in vitro*. The factor responsible for converting PG2 → PG1 *in vitro* has not been identified. It is heat stable and non-dialysable and may be a carbohydrate.

The significance of two forms of PG is not yet clear. It is possible that different forms are required for transport across the plasmalemma and entry into the wall.

VI. Regulation of Polygalacturonase Activity During Ripening

The large increase in PG activity during ripening could be explained by the activation of a pre-existing inactive enzyme or by *de novo* synthesis. Preliminary evidence suggests the conversion of PG1 → PG2 is not a mechanism for increasing enzyme activity since purified samples of the two isoenzymes have similar specific activities.

There is evidence for *de novo* enzyme synthesis. We have raised an antibody against purified PG2 which reacts with both PG1 and PG2 (Table II). The cross-reactivity is not surprising in view of the similarity in polypeptide composition of the two isoenzymes (Fig. 3). It is possible to use the antibody to make quantitative estimates of the amount of total PG protein in tomato extracts by radioimmunoassay. In this method the antibody is allowed to react with a test sample of purified PG2 labelled *in vitro* with ^{125}I. The amount of PG present in tomato extracts can then be measured by allowing these samples to compete with the iodinated test PG for binding to the antibody. In this way it is possible to measure nanogram quantities of PG. The results show that PG1 and PG2 give similar competition curves and that green tomatoes contain very little PG protein compared to "ripe" *Nr*, which contain low levels of PG1 (Fig. 7). Further experiments (Tucker, Robertson and Grierson, in preparation) indicate that the amount of antibody-reacting-material present in tomato extracts increases in proportion to the amount of PG enzyme measured by conventional assay. If the assumption is made that the PG antibody would recognize an enzymically-inactive form of PG, then these results can be interpreted to indicate that the increase in enzyme activity is due to *de novo* synthesis and not to activation of pre-existing enzyme.

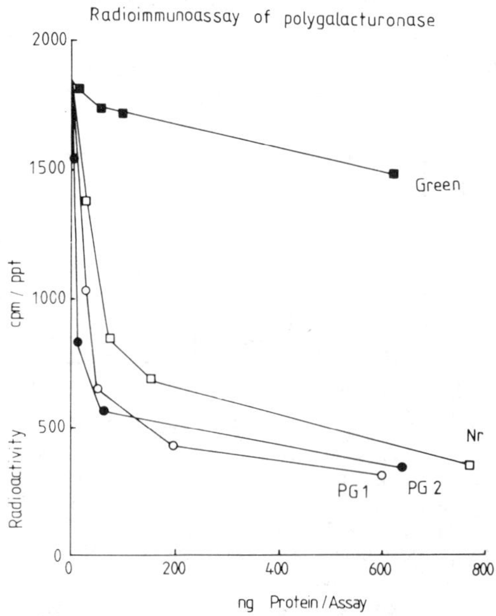

FIG. 7. Radioimmunoassay curves for polygalacturonase. Assays were performed by mixing 100 μl of antibody raised against PG2 with 100 μl of sample. After equilibration 100 μl of ^{125}I-labelled PG2 was added to compete for binding sites on the antibody. Purified PG1 (○), purified PG2 (●), green fruit extract (■), *Nr* fruit extract (□). From Tucker *et al.* (1980).

We have sought confirmatory evidence for the conclusion that the enzyme is synthesized *de novo* during ripening by radio-isotope pulse-labelling and density-labelling experiments. The preliminary results suggest there is incorporation of supplied radioactive amino acids into PG during ripening and also an increase in density of the enzyme when tomato segments are incubated in D_2O.

VII. MESSENGER-RNA SYNTHESIS DURING RIPENING

The increase in polygalacturonase activity appears to be due to enzyme synthesis. This supports the view that ripening involves a series of precisely co-ordinated synthetic events. This leads to the question of whether control of enzyme synthesis is exerted at the level of mRNA transcription or translation.

It has previously been shown that there is an increase in RNA synthesis and RNA turnover before and during ripening and that this involves the synthesis of rRNA, tRNA and mRNA (Rattanapanone *et al.*, 1977). Polydisperse polyadenylated RNA, similar to that shown in Fig. 8, is synthesized by ripening fruit, and poly(A)-containing mRNA purified from ripening tomatoes has been shown to direct the synthesis of proteins *in vitro*. Furthermore, there is evidence for a change in mRNA content during ripening (Rattanapanone *et al.*, 1978). Therefore it is quite possible that mRNA for specific enzymes is synthesized when fruits begin to ripen.

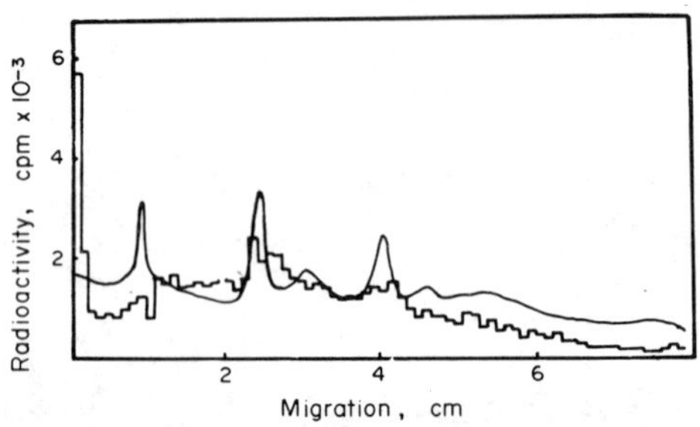

FIG. 8. Polyacrylamide gel electrophoresis of poly(A)-containing RNA from tomatoes. Tomato pericarp tissue was labelled with ^{3}H-uridine for 24 h, total nucleic acid extracted and poly(A)-containing RNA purified by oligo(dT)-cellulose chromatography. The radioactive poly(A)-containing RNA was mixed with unlabelled RNA (as carrier) and fractionated in a 2·4% gel for 3·5 h. Smooth curve, absorbance at 265 nm; histogram, ^{3}H radioactivity per mm gel slice. From Rattanapanone *et al.* (1977).

VIII. TRANSLATION OF POLYGALACTURONASE mRNA *IN VITRO*

Figure 9 shows that purified tomato RNA directs the cell-free synthesis of proteins. If the protein products synthesized *in vitro* can be identified, then the presence or absence of mRNA coding for specific proteins can be established. Figure 10 shows that a proportion of the radioactive proteins synthesized in response to mRNA from ripening tomatoes reacts with antibody to PG. The reaction appears to be specific for tomato RNA translation products: there is very little reaction with proteins synthesized in response to tobacco mosaic RNA. This strongly suggests that PG mRNA is present in the tomato extracts and is being accurately translated. Preliminary results suggest that there is mRNA for PG in green tomatoes but that there is an increase in the amount of this mRNA as the fruits ripen. However, this question requires further attention. In our experience the first sign of tomato ripening is increased ethylene evolution above a background level of about 0·02 nl g^{-1} fresh weight h^{-1}. This begins while tomatoes are still green. It is therefore important to establish whether PG mRNA is present before ethylene evolution begins and/or whether PG mRNA is synthesized in response to ethylene. We are carrying out experiments to try and answer this question.

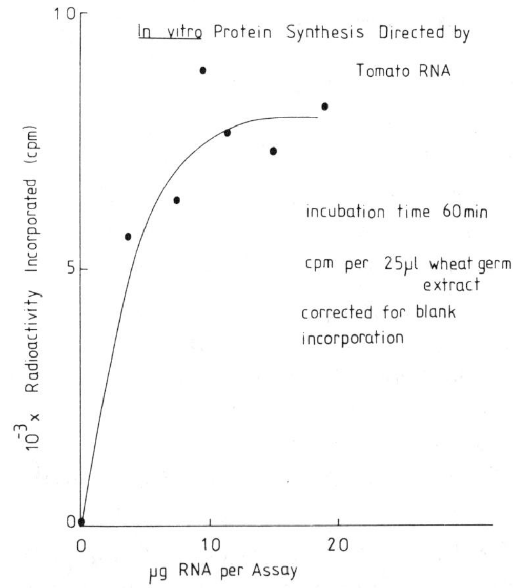

FIG. 9. *In vitro* protein synthesis directed by tomato RNA in a wheat-germ cell-free system. Different concentrations of tomato total RNA were added to a wheat-germ *in vitro* protein synthesis system and the incorporation of ^{35}S-methionine into protein was measured. The results are expressed as c.p.m. above the blank (i.e. no added RNA) incorporated into TCA-precipitable protein per 25 μl incubation mixture per hour.

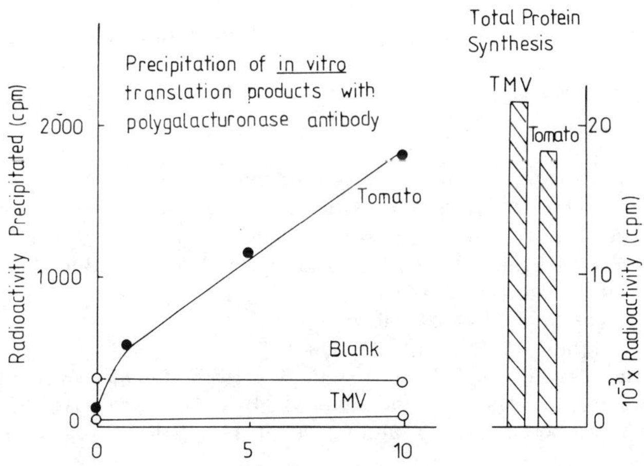

FIG. 10. Immunoprecipitation of protein synthesized *in vitro* in response to tomato RNA, using antibody raised against polygalacturonase 2. The total incorporation into protein *in vitro* in response to RNA from orange tomatoes and tobacco mosaic virus is shown on the right. The results on the left show the titration of the *in vitro* synthesized proteins against increasing concentrations of PG2-antibody.

IX. Conclusions and Prospects

1. These results show that synthesis of specific enzymes occurs during fruit ripening. It is not yet clear whether control is exerted at the level of transcription or translation but further work should settle this issue.

2. In order to understand the role of ethylene in ripening it is important to establish the timing of PG mRNA transcription and translation in relation to endogenous ethylene synthesis and in response to external ethylene.

3. The study of tomato mutants may be of assistance in establishing the mechanism of ripening. For example, further information about the role of PG1 and PG2 should come from comparative studies of normal and *Nr* tomatoes (which synthesize only PG1). In addition, in relation to transcriptional and translation control mechanisms, it will be interesting to see whether *rin* fruit (which produces no PG) synthesize mRNA for this enzyme.

4. Further studies on the *in vitro* translation of PG mRNA may also help in our understanding of the transport of enzymes across membranes. It will be interesting to establish whether there is a PG precursor-protein with additional amino acid sequences for transport of the enzyme across the plasmalemma.

ACKNOWLEDGEMENT

This work was supported by a grant from the Agricultural Research Council.

REFERENCES

Darby, L. A. (1978). *Horticultural Research* **18**, 73–84.

Grierson, D., Tucker, G. A. and Robertson, N. G. (1980). The regulation of gene expression during the ripening of tomato fruits. *In* "Quality in Stored and Processed Vegetables and Fruit" (P. W. Goodenough and R. K. Atkin, Eds). Academic Press, London and New York (in press).

Hobson, G. E. (1965). *J. Hort. Sci.* **40**, 66–72.

Hobson, G. E. (1968). *J. Food Sci.* **33**, 588–592.

Rattanapanone, N., Grierson, D. and Stein, M. (1977). *Phytochemistry* **16**, 629–633.

Rattanapanone, N., Grierson, D. and Stein, M. (1978). *Phytochemistry* **17**, 1485–1486.

Tucker, G. A., Robertson, N. G. and Grierson, D. (1980). *Europ. J. Biochem.* **112**, 119–124.

CHAPTER 9

Polyphenol Oxidases in Fruits—Changes During Ripening

A. M. MAYER AND E. HAREL

Botany Department, Hebrew University, Jerusalem, Israel

I. Introduction 161
II. Catechol Oxidase in Apples 168
III. Catechol Oxidase and Laccase in Peaches 171
IV. Catechol Oxidase in Grapes 173
V. Catechol Oxidase in Olives 175
VI. Callus and Suspension Cultures as Model Systems 176
VII. The Function of Catechol Oxidase in Fruits 178
References 179

I. Introduction

The polyphenol oxidases comprise a large group of enzymes, all of which are characterized by their ability to utilize molecular oxygen during the oxidation of phenolic substrates.

They may be divided into two main groups: the laccases (formerly *p*-diphenol: oxygen oxidoreductases E.C. 1.10.3.2) and the catechol oxidases (formerly *o*-diphenol: oxygen oxidoreductases E.C. 1.10.3.1). Both groups have recently been classed under the general name monophenol monooxygenase E.C. 1.14.18.1 The latter designation is unfortunate and misleading. The main reaction catalysed by laccase can be seen in equation I, while catechol oxidase catalyses two distinct reactions (equation II and equation III).

In general we will use the terms catechol oxidase and laccase. A detailed review of these two enzymes has recently been published (Mayer and Harel, 1979) in which most of the literature has been quoted as it relates to plants. Bacteria and fungi were included in the above-mentioned review but the voluminous literature pertaining to animals, including man was not con-

Equation I

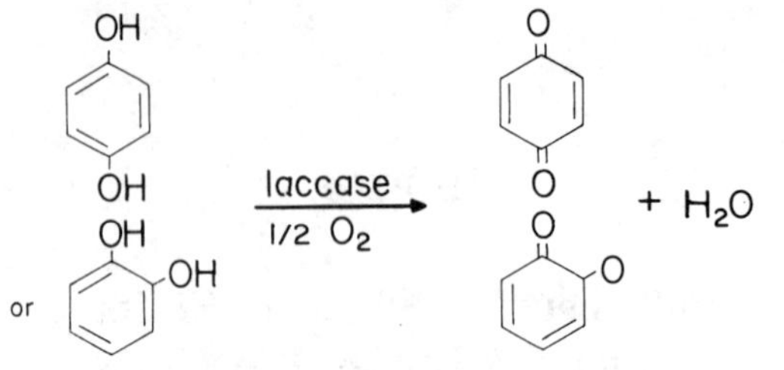

Equation II

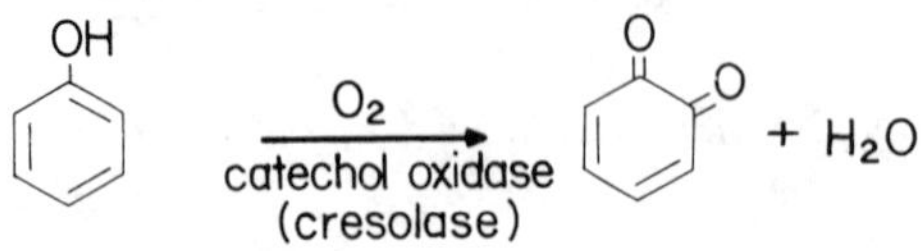

Equation III

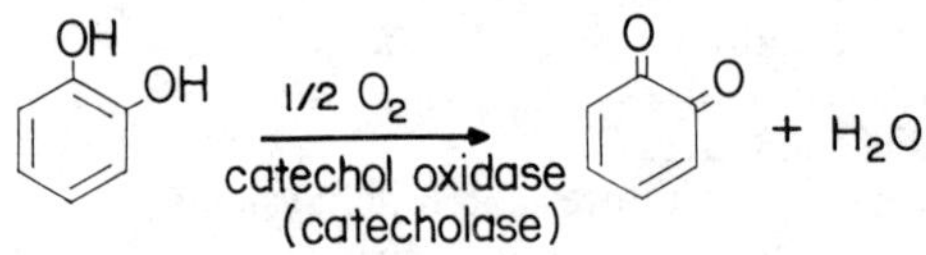

sidered. In this chapter we will restrict ourselves to polyphenol oxidases in fruits.

Since fruits are part of the diet of most human communities, it is not surprising that the browning of such fruits from bruising, crushing, extraction of juice, or simply during consumption, has been noted even by the most casual observer. Consequently, the browning reactions, which are usually catalysed by polyphenol oxidases, have been studied by many investigators. Virtually every fruit that has been studied with any care and precision has been shown to contain a polyphenol oxidase. The importance of enzyme in food processing has similarly drawn much attention. Reviews relating to this aspect of the polyphenol oxidases have periodically appeared (Joslyn and Ponting, 1951; Mathews and Parpia, 1976). Reviews on specific fruits or fruit products have also been written, for example in relation to coffee (Amorim and Amorim, 1977), tea (Sanderson and Coggan, 1977), and grapes and wine (Mayer, 1978). It is surprising how many polyphenol oxidases, mostly catechol oxidases, from fruits at the ripening stage have been described. A partial list of these is given in Table I, which also gives some of the properties of the enzymes described. It is immediately clear that the description of the enzymes studied is usually incomplete and essential data are often missing. But even more striking is the almost total absence of data on the enzymes during development of the fruit. Apparently most investigators have had access to the ripe fruit, but were unable or unwilling to follow fruit development from anthesis to ripe fruit. In a few cases such aspects have been considered in general (Dilley, 1970; van Buren, 1970) while Hobson (1967) described the course of development of polyphenol oxidase activity in tomatoes. Joshi and Shiralkar (1977) followed it in mango fruit. The data of Hobson (1967) show a rise in activity during fruit formation followed by a drop in the ripe fruit. Unfortunately the fruit was already sizeable when the measurements began. The data on mango indicate a peak of activity after 66 days in the pulp of the fruit, followed by a drop, while in the peel activity drops steadily from day 12 onwards. Unfortunately the work on mango did not take into account the presence of a laccase in the outer layers of the fruit (Joel *et al.*, 1978). These results are not in accord with a rise in enzyme activity reported by Subramanyam *et al.* (1975). In bananas the course of development of enzyme activity during ripening is apparently different in the peel and the pulp (Montgomery and Sgarbieri, 1975). A general rise has been reported by Patil and Magar (1971), but according to Montgomery activity drops in the flesh and rises in the peel. However only fully developed fruit was studied. Some of the data on mango, tomato and banana are shown in Table II.

A good deal of confusion exists about other fruits, to the extent that they have been investigated at all. For example, Flurkey and Jen (1978) show an initial peak of polyphenol oxidase activity in peach followed by a drop, but an earlier paper from the same group (Jen and Kahler, 1974) indicates that activity increases during ripening. An additional factor which is related to the browning reaction of fruits is the absence or presence of and the location of the phenolic substrate. The location of phenolics has been studied primarily in roots. They may be present in specialized cells, e.g. in banana roots (Beckman and Mueller, 1970; Barnell and Barnell, 1945) or they may be located in the vacuole of the same cell in which the enzyme is located (Roberts, 1952). Both the level of enzyme and of phenolic substrate appear to be involved in the browning reaction of ripe fruit, as has been shown for apples (Macheix, 1970; Weurman and Swain, 1955) and avocados (Golan, Kahn and Sadovski, 1977; Kahn, 1975). Inadequate attention has been paid to the different parts of the fruit and to the subcellular location of the enzymes, both in the ripe fruit and during fruit development. Generally most reports locate the enzyme either in chloroplasts or mitochondria or the enzyme is said to be soluble. The significance of this localization in processing of fruit juices has been discussed by Biederman (1956). In some cases, when different authors report on the enzyme in a given fruit, one isolates it from the particulate fraction while others report on a soluble enzyme. Thus the polyphenol oxidase described by Palmer (1964) from bananas was almost certainly a soluble enzyme but Padron *et al.* (1975) describe an apparently different enzyme also said to be soluble. The polyphenol oxidases described by Montgomery and Sgarbieri (1975), again different, is also said to be soluble. In avocado a soluble enzyme is present as described by Kahn (1977), but she also claims to have found the

TABLE I

Data on polyphenol oxidase reported in fruit

Fruit	K_m; substrate (mM)	pH optimum	Isoenzymes and Apparent mol. wt. (kilo dalton)	Subcellular location	Reference
Apple	8·5; 4-methyl catechol	7·3	+	Mitochondria	Harel *et al.* (1964)
	6·0; 4-methyl catechol	5·1	+(30–40, 60–70, 120–130)	Chloroplasts	Harel *et al.* (1965)
		(4·2 and 7·0)		Enzyme from peel only	Stelzig *et al.* (1972)
Avocado	0·9–7·0; 4-methyl catechol	4·8 or 5·5–6·5	+(5, 14, 28, 56, 112)	?	Dizik and Knapp (1970)
			+(35)	Microbodies	Knapp (1965)
					Kahn (1976, 1977)
					Sharon and Kahn (1979)
Apricot	8·3; 4-methyl catechol	5·0	(30, 60, 120)	Chloroplasts	Harel *et al.* (1966)
	17%; O_2	6·0–8·0		Soluble	Lerner *et al.* (1974)
Banana	0·6; dopamine	7·0		Soluble?	Palmer (1963), Palmer and Roberts (1967)
		6·0	+	Soluble	Padron *et al.* (1975)
		7·0	+	Soluble	Montgomery and Sgarbieri (1975)
Cacao		6·0		Soluble?	Quesnel and Jugmohansingh (1970)
					Chaplin (1978)
Cherry	10–50; catechin	4·2	+	?	Pifferi and Cultiera (1974)
	1·4–10; catechin, chlorogenic acid	6·5	+	?	Pifferi *et al.* (1974)
	50; catechol	7·0–7·8	+	?	Benjamin and Montgomery (1974)

Eggplant	4·0; catechol	5·2		?	Knapp (1965)
	0·3; caffeic acid				
	1·7; methyl catechol	6·8		?	Ramaswamy and Rege (1975)
Grape	8–9% O_2				Radler and Torokfalvy (1973)
	20% O_2	5·0	55–58	Chloroplasts	Harel and Mayer (1971)
	6·5; 4-methyl catechol	5·0	+	Chloroplasts	Dubernet and Ribereau-Gayon (1974)
			80	Chloroplasts	Kidron *et al.* (1977)
Mango	2·4; dopamine catechol	6·5–7·0	+	?	Joshi and Shiralkar (1977)
		5·4			Venkaiah and Patwardhan (1977)
				(Laccase also present)	(Joel *et al.* (1978))
Olive	5–10; chlorogenic acid and catechin	5·0	+ 200 and 42	Chloroplasts	Ben-Shalom *et al.* (1977a, b)
Peach	29; catechol	6·0	+	?	Jen and Kahler (1974)
	4·2–36; catechol	6·5–7·2			Wong *et al.* (1971)
	catechol oxidase:	5·1	+	Particulate	Mayer and Harel (1968)
					Harel and Mayer (1970)
	laccase: 12; quinol 11% O_2	6·0	+ (73·5)	Soluble	Lehman *et al.* (1974)
Pear	48; catechol	5·8–6·4	+	Soluble?	Tate *et al.* (1964)
	8; methyl catechol	4·5 and 5·0	+	Soluble?	Rivas and Whitaker (1973)
	5·8; methyl catechol				

The enzyme has also been reported in tomato (Hobson, 1967), melon, *Ribes nigra* vegetative parts (Trajovski, 1976), oranges (Bruemmer and Roe, 1970), green chilli (Luhadiya and Kilkarni, 1978), cranberry (Chan and Yang, 1971) as well as in coffee (Amorim and Amorim, 1977).

TABLE II
Formation of polyphenol oxidase in tomato, banana and mango

(A) *Tomato* (after Hobson, 1967)

Time from flower abscission (days)	(Fruit size and Colour)	Activity (units 10 g^{-1} fresh wt hr^{-1}) at 30° and pH 7·15. Substrate: catechol 0·15 M
10	½″	75
29	½″–1″	111
36	1–1½″	238
47	1½–2″	355
66	green	138
77	orange green	124
81	red	85

(B) *Mango* (after Joshi and Shiralkar, 1977)

Time after fertilization (Days)	Activity (units/mg protein) Pulp	Peel
12	v. little	26·2
40	19·2	11·5
54	—	9·5
66	26·0	—
85	v. little	v. little

(C) *Banana* (after Montgomery and Sgarbieri, 1975)

Day	Ripeness of fruit	Interior pulp	Activity[a] Exterior pulp	Peel
1	Green tip	46,300	17,800	1800
2	Slightly green	43,300	14,000	1500
3	Yellow	40,200	16,500	2100
5	Brown spot	34,900	12,900	3800
8	Overripe	29,800	13,300	4400

[a] (Units/ml extract, at pH 7·0, catechol substrate).

enzyme in a micro-body fraction (Sharon and Kahn, 1979). Since this is the first report of a polyphenol oxidase in such a fraction, this work needs confirmation by other workers.

From the foregoing it is clear that a great deal of confusion still exists. Much of this confusion is due to changes in enzyme property and location which occur during isolation and fractionation. Ripening, as will be seen in the following, is also accompanied by changes in enzyme properties. Many of the apparently contradictory reports can be accounted for by these changes as well as by the variability of the plant material examined. Furthermore, not all workers are sufficiently careful in defining the exact state of the fruit which they examined.

Some of the known characteristics of catechol oxidases and laccases in

TABLE III
Systematic distribution of polyphenol oxidases

	Catechol oxidase	Laccase
Bacteria	+	–
Algae	+	–
Fungi	++	+++
Bryophytes	++	–
Pteridophytes	++	–
Gymnospermae	++	+
Angiospermae	++++	Scattered

TABLE IV
Distribution of polyphenol oxidases within the plant

	Catechol oxidase	Laccase
Higher plants	Roots, stems Leaves, fruits	Limited Usually only secretory tissue
Fungi	Mycelium Fruiting bodies	Mycelium Fruiting bodies

TABLE V
Subcellular location of polyphenol oxidases

	Catechol oxidase	Laccase
Chloroplasts	+	–
Mitochondria	+	–
Microbodies	?	–
Soluble	+	+

plants and fungi will help to put the present discussion into perspective. These enzymes are ubiquitous in their distribution (Table III), and apparently appear in all parts of the plant or of the fungus (Table IV). The catechol oxidases may be particulate or soluble (Table V) while laccase, as far as is known, is always soluble (Table V). The basic molecular properties of the two enzymes are indicated in Table VI and ways of distinguishing between them by the use of suitable inhibitors are shown in Table VII.

A comparison of the data in Tables III–VII with the properties of the enzymes described for the fruit tissues shows that the fruit enzymes *per se* appear to have no characteristics which are unique. In general the fruit enzymes are typical catechol oxidases or laccases with low affinity (high K_m

TABLE VI
Some molecular properties of polyphenol oxidases

	Catechol oxidase	Laccase
Mol. wt.	35–120,000	60–350,000
Subunit composition	Probably 4 (2 heavy + 2 light or 4 equal)	Probably 2–4
Copper (form)	Cu^{+} Cu^{++} no ESR signal	Type I Cu (blue, ESR) (1) II Cu (not blue, ESR) (1) III Cu (not blue, no ESR) (2)
Copper (content)	One or two/polypeptide chain	4/polypeptide chain
Substrate specificity	Monophenol, diphenol, ortho-; not ascorbic acid	Monophenol, (only fungal) diphenol, ortho- or para-; ascorbic acid
Protein structure	Polypeptides	Glycopolypeptides
pH optimum	Variable Often no sharp peak	Variable Usually a sharp peak

TABLE VII
Inhibition of polyphenol oxidases

	Catechol oxidase	Laccase
Copper chelators	+	+
Carbon monoxide	+	−
Phenyl hydrazine	+	−
EDTA	±	±
Competitive inhibitors for substrate	+	+
Halides	+	+

values) for both phenolic substrate and oxygen. Their pH optimum lies usually between 4·5 and 7·0. None has an optimum at a very acid pH, indicating that they are present in a cell compartment separated from the rather acid cell vacuole of many fruits.

In the following sections we will discuss in detail the changes in level of enzyme activity, during development, in a few fruit tissues which have been studied in detail in our laboratory.

II. Catechol Oxidase in Apples

The development of enzyme activity during the ripening of apples has been studied in some detail. The basic information is given in Figs 1–3. From this it is clear that enzyme activity per apple has a peak about ten weeks after flowering when the apples are still quite small. This peak occurs for activities at pH values of 7·3 and 5·0. If activity is followed and expressed per unit fresh

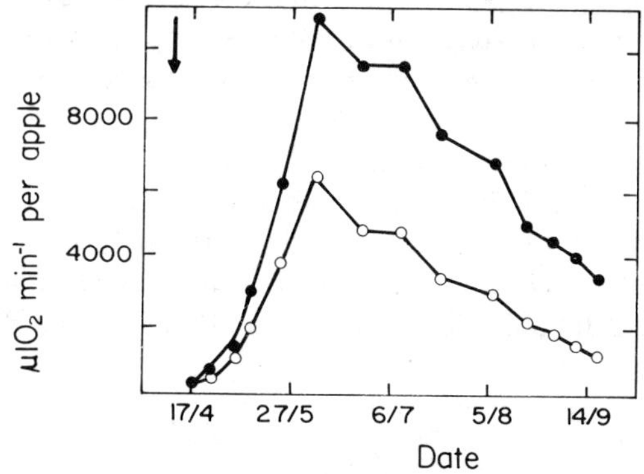

FIG. 1. Total catechol oxidase activity per apple during fruit development. Activity as $\mu l\ O_2$ in crude homogenate at pH 7·3 (○) and at pH 5·1 (●). Arrow—time of flowering. (From Harel *et al.*, 1966.)

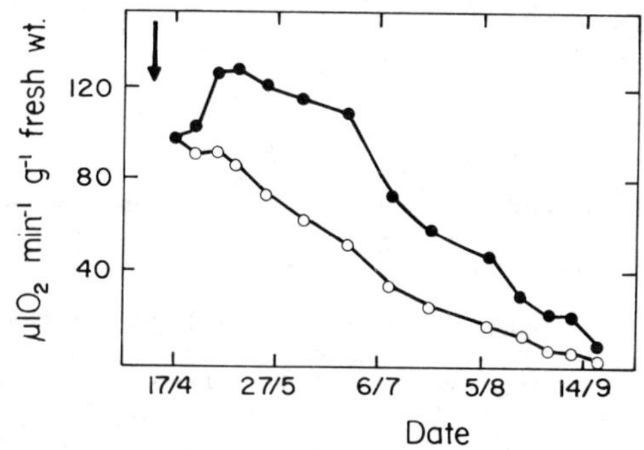

FIG. 2. Catechol oxidase activity in a particulate fraction from apples during fruit development. Arrow time of flowering. Activity at pH 7·3 (○) and at pII 5 1 (●). (From IIarcl *et al.*, 1966.)

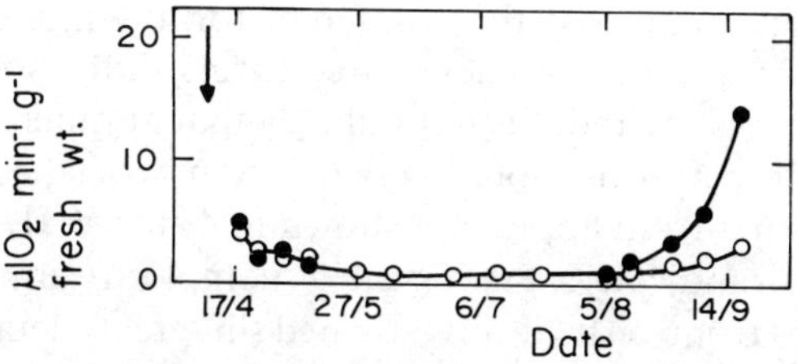

FIG. 3. Soluble catechol oxidase activity in apples during development. Symbols as in Fig. 2. (From Harel *et al.*, 1966.)

weight of the apple, the peak is less marked and occurs earlier. Only the enzyme from the particulate fractions shows a peak. A low level of soluble enzyme activity is also present in the very young fruit and in ripe apples. Soluble activity is virtually absent in young developing fruit. The significance of the difference in rate of development of the activities at two pH values lies in the fact that these two activities are associated with two distinct subcellular fractions. Activity at pH 7·3 is associated primarily with a mitochondrial fraction while that at pH 5·1 is due to activity in the chloroplasts (Fig. 4). If the data in Figs 1 and 2 are considered in this light there is some indication that the

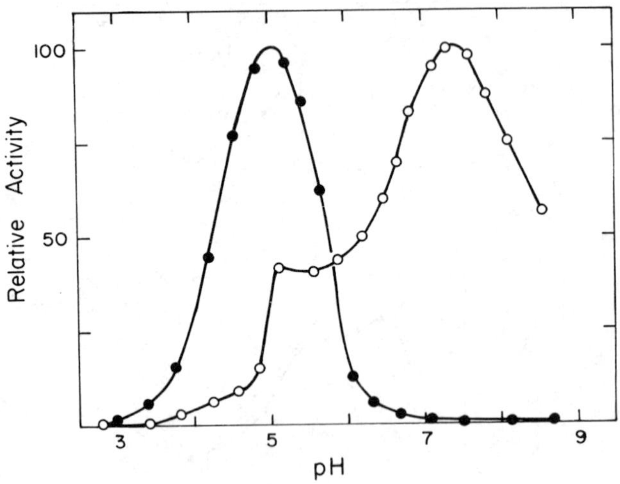

FIG. 4. Dependence of catechol oxidase activity extracted from particulate fractions of apples on pH. ● – Triton X-100 extract of chloroplasts; ○ – 1% Digitonin extract of mitochondrial fraction. (From Harel *et al.*, 1965.)

two organelles are not present in the same ratio at various stages of development of the fruit. Moreover, activity in the chloroplast is greater earlier on in development. In more mature apples the chloroplast begins to disintegrate. In the ripe apples, some of the enzyme activity appears to become soluble. It is reasonable to assume that this activity originates in the particulate fraction. It should also be recalled that the enzyme present in the chloroplasts occurs in at least three different forms—monomer, dimer and tetramer (Harel *et al.*, 1965). These three forms differ somewhat in their molecular characteristics and it is not unlikely that at different stages of fruit development different forms predominate. Some of these differences in chloroplasts isolated from apples are shown in Table VIII.

These data then indicate a great deal of complexity in the behaviour of catechol oxidase from a fruit investigated in great detail by numerous workers. Among papers on catechol oxidase in apples which may be mentioned are those of Walker (1964), Jones *et al.* (1965), Walker and Hulme

TABLE VIII
Differences in properties of monomer, dimer and tetramer of catechol oxidase extracted from apple chloroplasts

Fraction	pH optimum	K_m O_2	K_m 4-methylcatechol	Relative oxidation rate of chlorogenic acid
I (tetramer)	5·1	6·4%	—	62
II (dimer)	5·1	39·2%	15 mM	168
III (monomer)	5·1	15·4%	4·8 mM	77

(From data of Harel *et al.*, 1965 and Harel and Mayer, 1968).

(1966) and Stelzig *et al.* (1972), all of which fit into the overall picture presented here, in spite of some apparent discrepancies.

Most of the discrepancies in these and other reports can probably be accounted for if multiplicity, subcellular location and distribution within the fruit are taken into account.

III. CATECHOL OXIDASE AND LACCASE IN PEACHES

The basic data on the development of these two enzymes in peaches are given in Fig. 5. In peaches we are dealing with three distinct enzymes, a particulate and a soluble catechol oxidase and a soluble laccase. Each of these reaches a peak at a different stage in fruit development. The laccase appears late, the

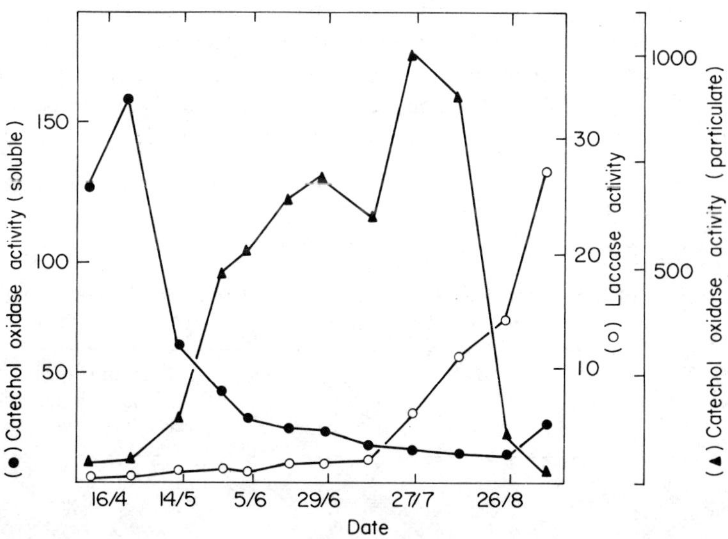

FIG. 5. Polyphenol oxidases during the development of peaches. Activity as 10^{-3} μl O_2 min^{-1} per fruit. Measurements began two weeks after fruit set. (From Harel *et al.*, 1970.)

soluble catechol oxidase is present at high levels during early development and then drops sharply and the particulate enzyme peaks towards the end of fruit development and then drops again very rapidly (Fig. 5). The three enzymes whose activity changes during development are quite distinct. The particulate catechol oxidase is located in a particulate fraction, probably chloro- or chromoplasts. As is the case for the enzyme in apples, it too can be resolved into three fractions which appear to correspond to the monomer, dimer and tetramer. The soluble enzyme detected in the young fruits has not been characterized in great detail, but it appears to be a typical catechol oxidase, as determined by substrate specificity and response to inhibitors (Harel *et al.*, 1970). The enzyme investigated in greatest detail in our laboratory is the laccase, which appears quite late in fruit development. A comparison of the properties of this enzyme and the particulate catechol oxidase from peaches is given in Table IX. Recently a report on catechol oxidase in peaches has appeared (Flurkey and Jen, 1978). Activity was followed in a $10,000 \times g$ supernatant after extraction with buffer containing 1 M KCl. Their data show a peak of activity per gram fruit after two weeks of growth, followed by a decline to a low level which persisted for seven to eight weeks. Activity in acetone powders of the peaches was much greater, apparently due to activation of the enzyme. In this case activity was highest in the first week and dropped to a very low value which persisted from week three onwards. It is difficult to compare precisely our data and that of Flurkey and Jen. It appears, however, that their observations with acetone powders are the result of the simultaneous measurement of a soluble and particulate catechol oxidase in the fruit. The overall pattern of change in enzyme level during fruit development in their work resembles our observations on the soluble enzyme. Flurkey and Jen (1978) also report several isoenzymes of the peach catechol oxidase, and apparent changes in form on storage of acetone powder in the cold. The results of the latter work in the peaches can be reconciled fairly well with the general picture of the behaviour of catechol oxidase as described above for apples.

TABLE IX

Comparison of some properties of the particulate catechol oxidase and soluble laccase from peaches

	pH optimum	No. of isoenzymes	Inhibition by 2,3-naphthalenediol 5 mM
Catechol oxidase (particulate)	5·0 and 7·3	3	70–80
Laccase	6·0	1	0

(From Mayer and Harel, 1968; Mayer *et al.*, 1964)

IV. Catechol Oxidase in Grapes

This enzyme has been studied by many workers, partly because of its significance in wine making (Mayer, 1978; Ivanov and Ivanova, 1968; Radler and Torokfalvy, 1973). The course of enzyme activity with development of the berry is shown in Fig. 6. Activity peaks immediately after fruit set and then drops steeply. Essentially the same results have been obtained by Dubernet (1974) although he found residual activity to be rather higher than that observed by us. The enzyme from grapes, with regard to its molecular properties, has been the subject of intense studies. The enzyme is present in a particulate fraction of the cell, the chloroplasts. Extracts of the chloroplasts show the presence of eight isoenzymes as determined by gel electrophoresis, which are at least in part interconvertible (Harel and Mayer, 1971; Harel *et al.*, 1973). The isoenzymes from grapes can be shown to consist in part of subunits of different molecular weight and in part of units of the same molecular weight but of different charge. The isoelectric point of the enzyme protein lies at about pH 4·0, but the precise point is different in different grape cultivars (Kidron *et al.*, 1978). The enzyme from grape shows an extremely interesting property, in that its activity is changed by exposure to acid pH (Fig. 7). In addition, its activity shows a sigmoidal dependence on the O_2 concentration (Fig. 8). These changes in activity are apparently associated with changes in the conformation of the enzyme. Such conformational changes have been demonstrated directly, by measuring the Stokes radius of the enzyme before and after its activation, the values obtained being about 3 nm at pH 7·0 and

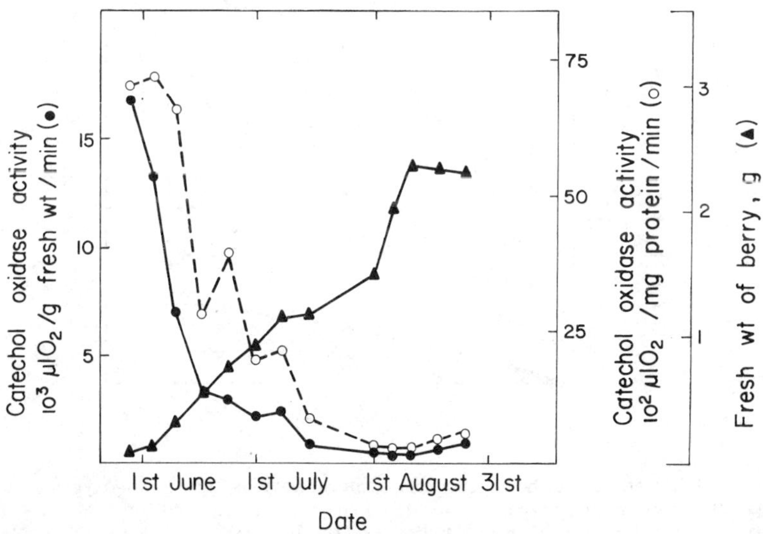

Fig. 6. Catechol oxidase activity during development of grape berry cv. Clairette. (From Kidron *et al.*, 1978.)

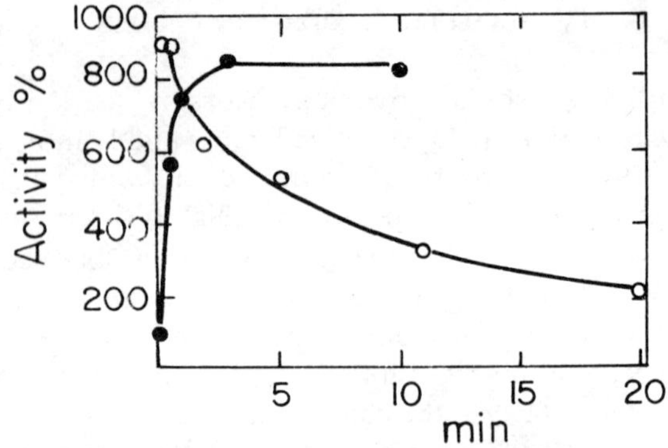

FIG. 7. Enhancement of catechol oxidase activity at pH 7·0 after pre-incubation at pH 5·0 and subsequent relaxation of activity. Enzyme activity as % of activity before pH shock. ● — activity at pH 7·0 after preincubation for various periods of time at pH 5·0; ○ — relaxation of activity during incubation at pH 7·0, following a 3 min shock at pH 5·0. (From Lerner *et al.*, 1972.)

8 nm after exposure to pH 5·0. Apparently, exposure of the enzyme to acid pH results in its unfolding, with a concomitant marked increase in activity. This unfolding is reversible (Lerner and Mayer, 1975). Further evidence for such conformational changes in the grape enzyme are provided by the study of the effect of phenylhydrazine on the enzyme. While phenylhydrazine specifically and powerfully inhibits the catecholase activity of the enzyme, it does not inhibit the oxidation of *p*-cresol, i.e. cresolase activity (Lerner *et al.*, 1971; 1974). These results must be considered within the framework of events

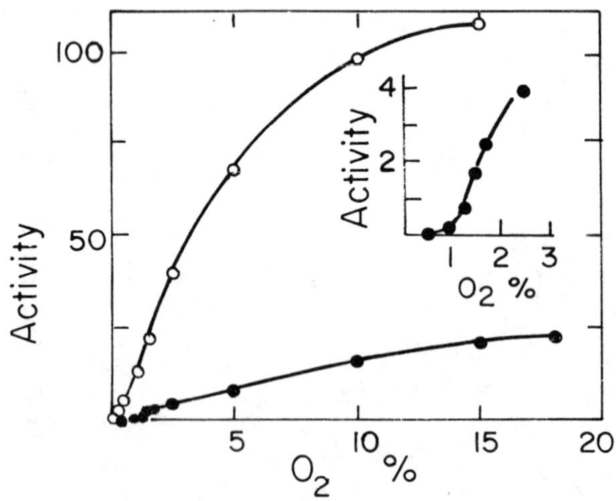

FIG. 8. The effect of urea on the dependence of catechol oxidase activity on oxygen concentration. ● — 0·3 ml untreated enzyme; ○ — 0·05 ml enzyme preincubated for 2 min in 4 M urea. Activity determined in 0·1 M phosphate buffer pH 7·0 with 5 mM 4-methylcatechol as substrate. Activity is expressed in scale units per min. Inset shows activity of untreated enzyme at low oxygen concentrations on an expanded scale. (From Lerner *et al.*, 1972.)

occurring in the intact tissue. The grape enzyme is located in the chloroplast, in which changes in pH and O_2 concentration occur, due to photosynthetic activity. Furthermore, it is well known that the ratio of cresolase to catecholase activity in catechol oxidase is modified by reducing and oxidizing agents (Lerner *et al.*, 1974; Mayer and Harel, 1979). The level of these also will change during the photosynthetic activity of the chloroplasts. Thus from the above discussion it is clear that given a finite amount of enzyme in a fruit, the expression of its activity could well be modified by the metabolic status of the tissue. This function is not usually taken into account in investigations of the polyphenol oxidase in fruit tissues. It is sometimes disputed that catechol oxidase is located in chloroplasts. Thus Sanderson (1964) claimed that the enzyme in tea reported by Li and Bonner (1974) to be chloroplast bound, is in fact a soluble enzyme. This fact is of course critical with regard to the arguments mentioned above. As we will show in the following, in some cases there can be no argument as to the subcellular location of catechol oxidase, e.g. in olives.

V. Catechol Oxidase in Olives

The olive is a striking example of a change in the subcellular location of catechol oxidase during ripening. The level of enzyme changes during development (Fig. 9). As we have already observed in other fruits there is a continuous drop in activity per unit fruit weight. A drop is also observed in total activity per fruit up to about mid-August. At this time there is a marked increase in fresh weight, while the activity per unit fresh weight declines slightly. As a result the curve of activity per gram fruit levels off at maturity. However, the change in activity was accompanied by a marked change in

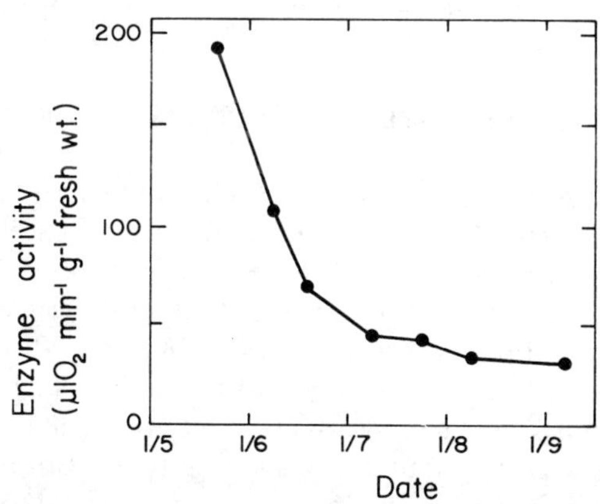

FIG. 9. Catechol oxidase activity of olives during fruit development. Activity towards 4-methylcatechol at pH 7·0. Activity was determined in a crude homogenate of the fruit, after depitting. Petal drop occurred about 15/5. (After Ben Shalom *et al.*, 1977b.)

enzyme locale. In the green olives activity was firmly bound to chloroplast membranes. It accompanied chlorophyll in all attempts at fractionation and purification. Furthermore, it could only be extracted from the chloroplast fraction following treatment of the fruit with acetone, i.e. after preparation of acetone powders (Ben Shalom *et al.*, 1977a). The enzyme present in the black olive was readily solubilized by buffer at pH 6·0 and was retained in a 40,000-g supernatant fraction, indicating its solubility. The pH optimum of the enzyme in the chloroplast fraction from green olives and that from black olives has a broad pH optimum between 5·5 and 7·5 in contrast to that observed for partially purified enzyme prepared from acetone powders. The substrate specificity of the two preparations was also very similar except for a somewhat higher rate of oxidation of catechol by the particulate enzyme. These data could indicate that the soluble enzyme of black olives originates in the chloroplasts which disintegrate during fruit ripening, which in its later stages is accompanied by a drop in the chlorophyll content (Ben Shalom, *et al.*, 1977b). It remained to be shown unequivocally that the enzyme is indeed located in the chloroplasts. This was accomplished by staining for the enzyme by utilizing the electron density of the oxidation products of one of its substrates—3,4-dihydroxyphenylalanine (Shomer *et al.*, 1979). This work clearly showed the location of the product of enzyme oxidation inside the thylakoids of the chloroplasts and hence it may be assumed that the enzyme too was located on the inner faces of the thylakoid membranes (Shomer *et al.*, 1979). In addition we also observed the products of oxidation within the mitochondria, strongly supporting the view that it is also present within these organelles and not merely bound to them. Olives represent an extreme case of change in the location of enzyme, which may be accompanied by a change in activity of the enzymes due to an activation process resulting from its liberation. Thus, liberation may simply facilitate access of substrate to the active site which could well be resticted within the chloroplast.

VI. Callus and Suspension Cultures as Model Systems

It is difficult to study the effect of various environmental factors on catechol oxidase in intact fruits. Controlled conditions cannot readily be imposed and experimentally it is inconvenient to try and modify enzyme content or properties in the fruit. In addition, it is tedious to follow enzyme activity in developing fruits and the procedure takes a long time. It would therefore be useful to have model systems in which to investigate such effects. Tissue or suspension cultures seemed a reasonable choice for such a model system. We chose apple fruits as the source for such cultures. Both callus and tissue cultures derived from apples contained catechol oxidase. As in the intact fruit, activity had two optima, at pH 4·9, and at 7·3. The enzyme from tissue cultures behaved electrophoretically in the same way as that extracted from apples. Furthermore, a considerable part of the enzyme activity was associated with a particulate fraction which sedimented at 1100 g. A very small amount was associated with a fraction sedimenting at 20,000 g. The percentage of activity present in the soluble fraction steadily increased to about 50% after 70 days while that in the 1100-g fraction dropped to one-third of its original value. A further feature observed in the callus cultures is illustrated in Fig. 10. It is quite obvious that when the initial explant had a low activity, activity rose, and fell again when the tissue aged, with a clear peak. When the explant had a high initial activity, activity first fell and then rose

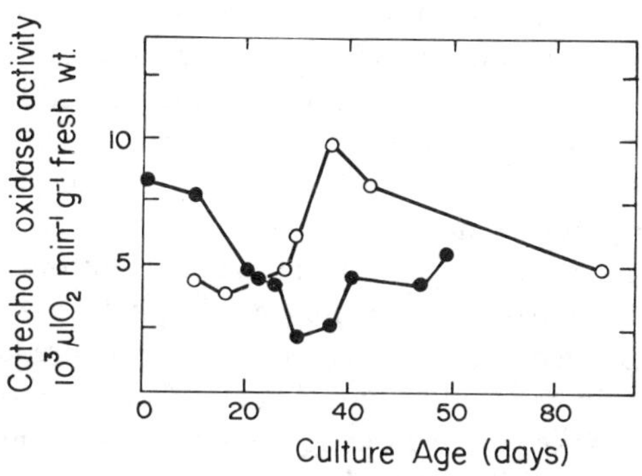

FIG. 10. Catechol oxidase activity in tissue cultures of apple fruit. Enzyme activity at pH 4·9. ○ — in a culture with initial low activity; ● — in a culture with initial high activity. (From results of Volk *et al.*, 1977.)

again, going through a trough. In experiments using suspension culture, we observed exactly the same phenomena, but the time span was shortened to about half that in the callus cultures. In most respects the catechol oxidase in suspension and callus cultures from apples resembled that in the developing apple fruit. The only major difference which was noted was that absolute activity per gram fresh weight was considerably higher in the cultures than in the fruit. Several reasons might account for this. In the developing apple fruit one is dealing with a tissue which grows primarily due to cell expansion while in the callus and suspension cultures the cells are both dividing and expanding. We found no correlation between catechol oxidase activity and growth of the cultures as measured by protein content or dry weight, and hence the number of divisions may be important. Another significant difference was that the cultures had a much lower content of phenolic compounds. Despite these differences it appears that a study of the behaviour of catechol oxidase in tissue cultures derived from fruit might be a useful model for events in the intact fruit. It also allows one to study the effect of various environmental factors which might affect enzyme level as well as to relate enzyme level to various endogenous and exogenous growth substances (Volk *et al.*, 1977, 1978).

VII. The Function of Catechol Oxidase in Fruits

It appears that all the fruits studied in detail have one thing in common. The level of catechol oxidase activity shows a high level or peak at a relatively early stage of development and thereafter its level drops. This is the case whether activity is expressed per unit weight, per fruit or per mg protein. The only exception to this is the peel of bananas (Table II) but in the pulp the generalization holds. A second general phenomenon is that as the fruit ages apparently more enzyme becomes soluble. The drop in activity during development on a whole fruit basis implies not only a cessation of enzyme synthesis, but also an actual disappearance of enzyme activity by inactivation or inhibition or due to turnover and perhaps decomposition. The increase in solubility and release from a particular fraction probably is due to ageing phenomena in the tissue. From this it would seem likely that the important role of the enzyme in the fruit is during its early development, while activity observed in the ripe fruit is simply residual. It should be remembered that the function of fruit is in principle to assist in the distribution of seeds. The seeds develop along with the fruit, and are mature long before the fruit senescences or is ripe for harvest. In order to protect the seeds from predation prior to their being mature it would be sensible that a high level of catechol oxidase together with its substrates makes the fruit less palatable. Undoubtedly many factors contribute to this, but oxidation of phenolics, tanning of protein and lowering of their digestibility might be important factors. As the seed matures and dries, the fruit ripens. At this stage, in the kind of fruit we have been considering, seed distribution is often achieved by consumption of the fruit and scattering of the residual seeds, or by passage of the seeds through the digestive tract of the animal consuming the fruit. At this stage there is no longer a need to protect against predation and hence perhaps no need for a high level of catechol oxidase. This is a very speculative suggestion but worthy of consideration. The same argument would apply with regard to prevention of infection by fungi and bacteria. This again would be of great importance prior to seed maturation and desiccation and much less so in the ripe fruit in which the seeds are ready for dispersal. However it is at least debatable whether catechol oxidase plays a vital role in the defence of plant tissues against infective agents. This problem has been discussed in detail recently (Mayer and Harel, 1979). Nevertheless, the enzyme might contribute to the defence and hence this could be at least part of its function in fruits.

An inevitable conclusion from the data brought and the conclusions drawn is that one should look for the physiological and biochemical functions of catechol oxidase in fruits, not when they are ready for eating but early on in development, when the activity is at a peak. The nuisance value, or significance of flavour formation, related to human consumption is surely a secondary factor which may have been overstressed in the past.

REFERENCES

Amorim, H. U. and Amorim, V. L. (1977). *In* "Enzymes in Food and Beverage Preserving", pp. 27–56. ACS Symposium Series 47.

Barnell, H. R. and Barnell, F. (1945). *Ann. Bot.* **9**, 77.

Beckman, C. H. and Mueller, W. C. (1970). *Phytopathol.* **60**, 79–82.

Benjamin, N. D. and Montgomery, M. W. (1974). *J. Fd Sci.* **38**, 799–806.

Ben-Shalom, N., Kahn, V., Harel, E. and Mayer, A. M. (1977a). *Phytochem.* **16**, 1153–1158.

Ben-Shalom, N., Kahn, V., Harel, E. and Mayer, A. M. (1977b). *J. Sci. Fd Agr.* **28**, 545–556.

Biederman, W. (1956). *Mitt Lebensm. Hyg.* **47**, 86–112.

Bruemmer, J. H. and Roe, B. (1970). *J. Fd Sci.* **35**, 116–119.

Chan, H. T. and Yang, H. Y. (1971). *J. Fd Sci.* **36**, 169–173.

Chaplin, M. F. (1978). *Phytochemistry* **17**, 1897–1899.

Dilley, D. R. (1970). *In* "The Biochemistry of Fruits and Their Products" (A. C. Hulme, Ed.), Vol. I, pp. 179–208. Academic Press, London and New York.

Dizik, N. S. and Knapp, F. W. (1970). *J. Fd Sci.* **35**, 282–285.

Dubernet, M. (1974). Ph.D. Thesis. University of Bordeaux II.

Dubernet, M. and Ribereau-Gayon, P. (1974). *Phytochemistry* **13**, 1085–1087.

Flurkey, W. H. and Jen, J. J. (1978). *J. Fd Sci.* **43**, 1826–1831.

Golan, A., Kahn, V. and Sadovski, A. Y. (1977). *J. agric. Fd Chem.* **25**, 1253–1260.

Harel, E. (1966). Ph.D. Thesis. Hebrew University, Jerusalem.

Harel, E. and Mayer, A. M. (1968). *Phytochemistry* **7**, 199–204.

Harel, E. and Mayer, A. M. (1970). *Phytochemistry* **9**, 2447–2452.

Harel, E. and Mayer, A. M. (1971). *Phytochemistry* **10**, 17–22.

Harel, E., Mayer, A. M. and Shain, Y. (1964). *Physiol. Plant.* **17**, 921–930.

Harel, E., Mayer, A. M. and Shain, Y. (1965). *Phytochemistry* **4**, 783–790.

Harel, E., Mayer, A. M. and Shain, Y. (1966). *J. Sci. Fd Agric.* **17**, 389–392.

Harel, E., Mayer, A. M. and Lerner, H. R. (1970). *J. Sci. Fd Agr.* **21**, 541–543.

Harel, E., Mayer, A. M. and Lehman, E. (1973). *Phytochemistry* **12**, 2649–2654.

Hobson, G. E. (1967). *J. Sci. Fd Agr.* **18**, 523–526.

Ivanov, T. and Ivanova, A. (1968). *Ann. Technol. Agric.* **17**, 333–340.

Jen, J. J. and Kahler, K. R. (1974). *Hort. Sci.* **9**, 590–591.

Joel, D., Marbach, I. and Mayer, A. M. (1978). *Phytochemistry* **17**, 796–797.

Jones, J. D., Hulme, A. C. and Wooltorton, L. S. C. (1965). *Phytochemistry* **4**, 659–676.

Joshi, P. R. and Shiralker, N. D. (1977). *J. Fd Sci. Tech.* **14**, 77–79.

Joslyn, M. and Ponting, J. (1951). *Adv. Fd Res.* **3**, 1–44.

Kahn, V. (1975). *J. Sci. Fd Agr.* **26**, 1319–1324.

Kahn, V. (1976). *Phytochemistry* **15**, 267–273.

Kahn, V. (1977). *J. Fd Sci.* **42**, 38–43.

Kidron, M., Harel, E. and Mayer, A. M. (1978). *Am. J. Enol. Vitic.* **29**, 30–35.

Knapp, F. W. (1965). *J. Fd Sci.* **30**, 930–936.

Lehman, E., Harel, E. and Mayer, A. M. (1974). *Phytochemistry* **13**, 1713–1717.

Lerner, H. R. (1974). Ph.D. Thesis, Hebrew University, Jerusalem.

Lerner, H. R. and Mayer, A. M. (1975). *Phytochemistry* **14**, 1955–1957.

Lerner, H. R., Harel, E., Lehman, E. and Mayer, A. M. (1971). *Phytochemistry* **10**, 2637–2640.

Lerner, H. R., Mayer, A. M. and Harel, E. (1972). *Phytochemistry* **11**, 2415–2421.

Lerner, H. R., Mayer, A. M. and Harel, E. (1974). *Phytochemistry* **13**, 397–401.

Li, L. P. and Bonner, J. (1947). *Bioch. J.* **41**, 105–110.

Luhadiya, A. P. and Kulkarni, R. R. (1978). *J. Fd Sci. Tech.* **15**, 214–215.

Macheix, J.-J. (1970). *Physiol. Veg.* **8**, 585–602.
Mathews, A. G. and Parpia, H. A. B. (1976). *Adv. Fd Res.* **19**, 75–145.
Mayer, A. M. (1978). *Ann. Technol. Agric.* **27**, 149–159.
Mayer, A. M. and Harel, E. (1968). *Phytochemirtry* **7**, 1253–1256.
Mayer, A. M. and Harel, E. (1979). *Phytochemistry* **18**, 193–215.
Mayer, A. M., Harel, E. and Shain, Y. (1964). *Phytochemistry* **3**, 447–451.
Montgomery, M. W. and Sgarbieri, U. C. (1975). *Phytochemistry* **14**, 1245–1249.
Padron, M. P., Lozano, J. and Gonzalez, G. A. (1975). *Phytochemistry* **14**, 1959–1963.
Palmer, J. K. (1963). *Plant Physiol.* **39**, 508–513.
Palmer, J. K. and Roberts, J. B. (1967). *Science* **157**, 200–201.
Patil, D. L. and Magar, N. G. (1971). *Sci. Cult.* **41**, 497–500.
Pifferi, P. G. and Cultiera, R. (1974). *J. Fd Sci.* **39**, 786–791.
Pifferi, P. G., Zamorani, A. and Mazzoco, F. (1974). *Agrochim.* **18**, 404–415.
Quesnel, V. C. and Jugmohansingh, K. (1970). *J. Sci. Fd Agr.* **21**, 537–541.
Radler, F. and Torokfalvy, E. (1973). *Z. Lebensm. Unters. Fursh.* **152**, 38–41.
Ramaswamy, S. and Rege, D. V. (1975). *Acta Alimon. Acad. Sci. Hung.* **4**, 355–365.
Rivas, N. J. and Whitaker, J. R. (1973). *Plant Physiol.* **52**, 501–507.
Roberts, E. A. H. (1952). *J. Sci. Fd Agr.* **3**, 193–198.
Sanderson, G. W. (1964). *Biochim. Biophys. Acta* **92**, 622–624.
Sanderson, G. W. and Coggan, P. (1977). *In* "Enzymes in Food and Beverage Preserving", pp. 12–28. ACS Symposium, Series 47.
Sharon, O. and Kahn, V. (1979). *Physiol. Plant.* **45**, 227–234.
Shomer, I., Ben-Shalom, N., Harel, E. and Mayer, A. M. (1979). *Ann. Bot.* **44**, 261–263.
Stelzig, D. A., Akhtor, S. and Ribeiro, S. (1972). *Phytochemistry* **11**, 535–539.
Subramanyam, H., Krishnamurthy, S. and Parpia, H. A. B. (1975). *Adv. Fd Res.* **21**, 224–307.
Tate, J. N., Luh, B. S. and York, G. K. (1964). *J. Fd Sci.* **29**, 829–836.
Trajovski, V. (1976). *Swedish J. Agri. Res.* **6**, 215–227.
van Buren, J. (1970). *In* "The Biochemistry of Fruits and Their Products" (A. C. Hulme, Ed.), Vol. I, pp. 269–304. Academic Press, London and New York.
Venkaiah, B., Patwardhan, M. U. (1977). *J. Fd Sci. Tech.* **14**, 173–175.
Volk, R., Harel, E., Mayer, A. M. and Gan-Zvi, E. (1977). *J. Expt. Bot.* **28**, 820 830.
Volk, R., Harel, E., Mayer, A. M. and Gan-Zvi, E. (1978). *J. Expt. Bot.* **29**, 1099–1109.
Walker, J. R. L. (1964). *Aust. J. Biol. Sci.* **17**, 360–377.
Walker, J. R. L. and Hulme, A. C. (1966). *Phytochemistry* **5**, 259–267.
Weurman, C. and Swain, T. (1955). *J. Sci. Fd Agr.* **6**, 186–192.
Wong, T. C., Luh, B. S. and Whitaker, J. R. (1971). *Plant Physiol.* **48**, 19–23.

CHAPTER 10

Molecular Properties of Plant Tyrosinases

D. A. ROBB

Department of Biochemistry, University of Strathclyde, The Todd Centre, 31 Taylor Street, Glasgow, Scotland

I. Introduction 181
II. Tyrosinase 182
A. Historical Perspective 182
B. Molecular Properties 183
C. Spectral Properties: Oxytyrosinase 185
D. Binuclear Copper 186
E. Ligands to the Copper 187
F. Inhibition by Anions 188
III. Concluding Remarks 190
References 191

I. Introduction

Enzymic browning of cut and damaged fruit and vegetables is a common occurrence affecting the quality of fresh produce and of processed foods. Peroxidase and enzymes of the polyphenol oxidase family are normally considered as likely causative agents. Mayer and Harel (1979) have recently reviewed the properties of the polyphenol oxidases classifying them as laccases (E.C. 1.10.3.2.) or catechol oxidases, and have shown that the former are of limited distribution. It appears that catechol oxidases can be further divided into two groups. There are enzymes which can catalyse the oxidation of *o*-diphenols, i.e. can convert DOPA to melanin, but which cannot oxidize tyrosine to melanin. Such enzymes have been isolated from pear (Rivas and Whitaker, 1973), tea (Gregory and Bendall, 1966), sugar cane (Combs *et al.*, 1974) and cherry (Benjamin and Montgomery, 1973) and classification as catechol oxidase (E.C. 1.10.3.1.) seems appropriate for this group. The second group as well as acting as catechol oxidases can also convert tyrosine to melanin, and the classical name for this group, tyrosinase, is retained in this article. Tyrosinases are classified as monophenol monooxygenases (E.C.

1.14.18.1.), and are the most widely distributed of the polyphenol oxidases. Since catechol can be oxidized by peroxidase, laccase, catechol oxidase and tyrosinase, care must be taken in characterization studies. Walker and McCallion (1980) have discussed ways of distinguishing laccase from catechol oxidase and tyrosinase in crude extracts, and peroxidase reactions can be distinguished by their inhibition on addition of catalase. The study of purified samples has allowed further distinctions to be made between laccases and tyrosinases. Laccases catalyse one electron transfers, converting catechol to its semiquinone; they are blue extracellular glycoproteins comprising a monomeric unit of two paramagnetic and two diamagnetic copper atoms per mole (Reinhammar, 1979). Tyrosinases, on the other hand, are straw-coloured and carry out a two electron transfer converting catechol directly to *o*-quinone (Mason *et al.*, 1961). In this contribution, recent studies on the structure of tyrosinases will be reviewed and evidence will be presented to show that they contain a minimum of two diamagnetic copper atoms per mole.

II. Tyrosinase

A. Historical Perspective

Serious study of the structure and mechanism of action of tyrosinase began when large-scale preparations, which are required for such work, were first made by Kubowitz in Warburg's laboratory, and by Keilin and Mann, both using vegetables as source material.

From studies of potato tyrosinase Kubowitz (1938) concluded that the enzyme contained two copper atoms at its active site and that there were similarities between it and haemocyanin. It was proposed that the oxidation of catechol ($C_6H_4(OH)_2$) by tyrosinase ($E\ Cu_2^{2+}$) would involve the copper shuttling between valence states, with oxygen binding to the cuprous state. A suitable representation is;

$$C_6H_4(OH)_2 + E\ Cu_2^{2+} \rightarrow C_6H_4O_2 + 2H^+ + E\ Cu_2^{+}$$
$$E\ Cu_2^{+} + O_2 + C_6H_4(OH)_2 \rightarrow C_6H_4O_2 + 2^{-}OH + E\ Cu_2^{2+}$$

Later studies questioned whether a change in valency occurred. Thus it was concluded that the metal was always in the cuprous state (Kertesz and Zito, 1962) and this view was reinforced by the discovery of a tyrosinase containing one copper per mole, and therefore unable to take part in a two electron transfer unless Cu(III) was implicated (Fling *et al.*, 1963). Thus tyrosinase became to be regarded as the classical example of a cuprous protein containing a basic sub-unit of molecular weight of 33,000 with one Cu; in *Neurospora* it was monomeric, in mushroom tetrameric. This view started to be questioned in two later reviews (Vanneste and Züberbühler, 1974; Coleman, 1974) and the evidence now available suggests that Kubowitz's ideas of forty-years' standing are essentially correct.

B. MOLECULAR PROPERTIES

Estimation of molecular weight of a protein is best achieved by a sedimentation equilibrium experiment (van Holde, 1975). Only mushroom tyrosinase has been adequately studied by this method, although some data has also been obtained for *Neurospora*, but interpretation is complicated due to aggregation of the protein. The molecular weight of some tyrosinases has also been deduced from (a) the combined use of sedimentation velocity and diffusion studies, and (b) a minimal molecular weight obtained by gel electrophoresis in the presence of sodium dodecyl sulphate or gel filtration in guanidine hydrochloride. All these methods require a relatively pure enzyme preparation which probably explains why so little data is available for plant tyrosinases. A representative set of data is given in Table I together with typical values for mammalian tyrosinase and for the basic unit of gastropod and arthropod haemocyanin. Also given in Table I are values for the copper content, and these are even more limited in availability. Values given for the amount of copper tend to increase with continued study and this is illustrated in the cases of the enzymes from *Streptomyces* and *Neurospora* where values of 0·2% (Lerch and Ettlinger, 1972; Fling *et al.*, 1963; Gutteridge and Robb, 1975) have recently been corrected to 0·4% and 0·3% respectively. Thus the figure for *Vitis* in Table I should be regarded as a minimum. As mentioned above, mushroom and *Neurospora* tyrosinases both show properties of aggregating systems as indeed do tyrosinases from other plants (Mayer and Harel, 1979). However, the mushroom enzyme is unique at present in containing dissimilar sub-units. Strothkamp *et al.* (1976) were the first to investigate this aspect and the work has lately been extended. It appears that the major form of mushroom tyrosinase is a tetramer containing two heavy sub-units ($2 \times \sim 45{,}000$) and two light sub-units ($2 \times \sim 13{,}000$). The situation is complicated by the occurrence of at least two types of heavy sub-unit, termed H^a and H^b (Robb, 1979) so that a range of tetrameric structures is possible, e.g. $H^a{}_2L_2$, $H^{ab}L_2$ and $H^b{}_2L_2$. These may also be partially distinguished by their substrate specificity. Some similarities in molecular size are apparent in Table I, e.g. between *Neurospora* and *Beta* tyrosinases which are also similar in size to the H sub-unit. More data for other plant tyrosinases is required before further comment can usefully be made. Differences should not be ignored either; thus tyrosinase is a more basic protein in *Neurospora* than mushroom and whereas fungal tyrosinases are cytoplasmic, the *Beta* enzyme is located in the chloroplast.

TABLE I
Molecular properties of some tyrosinases

Source	Molecular wt.	Method[a]	% Cu	No. of Cu atoms	pI	Reference
Vibrio tyrosinaticus	41,000	G				Pomerantz and Murthy (1974)
Streptomyces glaucescens	29,000	SE, G	0·2[b]	1[b]	6·95	Lerch and Ettlinger (1972)
Neurospora crassa	35,000–120,000	SE	0·195	1		Fling *et al.* (1963)
	31,000	G	0·21	1		Gutteridge and Robb (1975)
	42,000	G	0·3	2		Lerch (1976)
	46,000–170,000	SE			7·8	Robb and Gutteridge (1981)
Agaricus bisporus	128,000	SD	0·2	4		Kertesz and Zito (1962)
(commercial mushroom)	119,000	SE	0·2	4		Bouchilloux *et al.* (1963)
	110,000	SE			4·8	Robb and Gutteridge (1981)
	61,000	SE	0·2	2		Nakamura *et al.* (1966)
	32,000–118,000	SE				Duckworth and Coleman (1970)
Beta vulgaris (spinach beet)	40,000	G				Vaughan *et al.* (1975)
Vitis vinifera (grape)	80,000	SD	0·04			Kidron *et al.* (1977)
Mammalian (hamster)	55,000	G				Pomerantz and Murthy (1974)
	70,000					
Haemocyanin (i) gastropod	50,000[c]			2		Lontie and Witters (1973)
(ii) arthropod	73,000[c]			2		

[a] SE, sedimentation equilibrium; SD, sedimentation–diffusion; G, sub-unit wt. determined by gel electrophoresis or gel filtration.
[b] Recent determinations show two Cu atoms (Lerch, unpublished data).
[c] Minimal functional unit.

C. SPECTRAL PROPERTIES: OXYTYROSINASE

Warburg's laboratory, in its pioneering studies on tyrosinase, noted similarities between it and haemocyanin. This comparison is particularly apt when oxygenated forms are considered. Oxyhaemocyanin is a blue pigment with absorption maxima at 346 nm and 540 nm and is readily obtained from molluscs and arthropods. On storage, haemocyanin ages; it gradually loses its colour and is unable to bind oxygen. This process can be reversed by treatment with reducing agents, and hydrogen peroxide has proved particularly useful for this function (Lontie and Witters, 1973). Tyrosinase is markedly more difficult to obtain, requiring lengthy purification. If hydrogen peroxide or hydroxylammonium chloride is added, a species is obtained with a very similar spectrum to oxyhaemocyanin; it absorbs strongly at 345 nm and also at 600 nm. Mason and his group (Jolley *et al.*, 1974) have concluded from studies with the mushroom enzyme that this moderately stable complex is an oxygenated form of the enzyme and have proposed the term oxytyrosinase for it. Data is now available for the *Neurospora* enzyme also and the comparison to oxyhaemocyanin is striking (Table II). The amount of oxytyrosinase in purified mushroom tyrosinase is low (2–15%) but a significantly larger proportion (30%) exists in the purified *Neurospora* enzyme (Fig. 1).

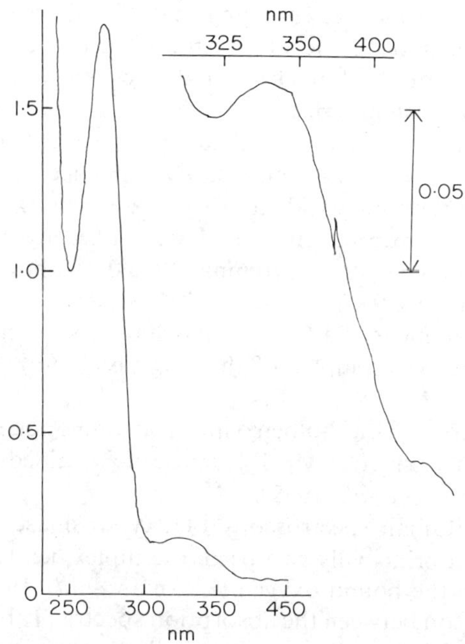

FIG. 1. Absorption spectrum of purified *Neurospora* tyrosinase showing a shoulder at 340 nm due to the presence of oxytyrosinase. Absorbance is plotted on the vertical axis.

TABLE II
Extinction coefficients for oxyhaemocyanin and oxytyrosinase

Protein	max	E_{max}/Cu	Reference
Gastropod	346	8000	Lontie and Witters (1973)
haemocyanin	580	540	
Agaricus	345	9000	Jolley *et al.* (1974)
tyrosinase	600	600	
Neurospora	345	9000	Lerch (1976)
tyrosinase	600	500	

D. BINUCLEAR COPPER

The presence of the metal in tyrosinase cannot be detected by magnetic susceptibility measurements (Makino *et al.*, 1974) or by electron paramagnetic resonance (Bouchilloux *et al.*, 1963). These properties are typical of univalent copper and thus the enzyme has been said to contain Cu(I). An alternative explanation requires the presence of two cupric atoms so close together that they become antiferromagnetically coupled. Any occurrence of enzymes containing only one copper per mole renders this alternative unlikely and two such cases existed unchallenged for several years (Fling *et al.*, 1963; Lerch and Ettlinger, 1972). Recent corrections have removed this obstacle (see Section II B) and agreement is emerging that fungal tyrosinases contain at least 2 mol of copper per mole of enzyme. One can also find additional evidence in support of the binuclear model, which taken together seems formidable. It may be summarized as:

(1) Isolated mushroom tyrosinase can accept one electron per copper atom present (Makino *et al.*, 1974), most easily explained as $Cu^{2+} + e \rightarrow Cu^{+}$.

(2) Binding of carbon monoxide and benzoate, analogues of oxygen and phenolic substrates respectively, are characterized by a binding ratio of Cu/CO=2·0, Cu/benzoate approximately 2·0 (Kubowitz, 1938; Duckworth and Coleman, 1970).

(3) Actual measurement of a Cu . . . Cu distance of 0·6 nm determined for the paramagnetic NO complex of the enzyme (Schoot-Uiterkamp *et al.*, 1974).

(4) Reconstitution of the holoenzyme from the apoenzyme and Cu(II) shows complicated kinetics which have been explained on the basis of a copper pair (Kertesz *et al.*, 1972).

(5) Resonance Raman spectroscopy of oxytyrosinase suggests that the oxygen is present principally as a peroxy complex, i.e. two electrons have been donated to the bound oxygen (Eickman *et al.*, 1979).

(6) The comparison between the absorption spectra (Table II) of oxytyrosinase and oxyhaemocyanin and the similar luminescence of their CO analogues (Kuiper *et al.*, 1980) suggests a common environment for the

copper and it is generally accepted that haemocyanin contains binuclear copper.

Acceptance of the proposal for a binuclear copper cluster at the active site allows the following model (Fig. 2) to relate the two forms of enzyme isolated (met-tyrosinase and oxytyrosinase).

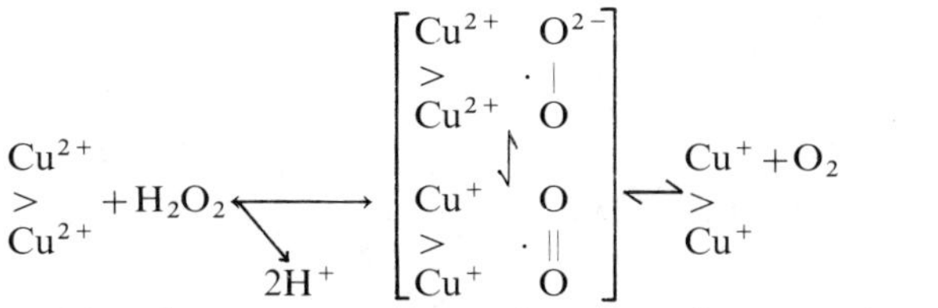

FIG. 2. Possible mechanism relating various forms of tyrosinase.

E. LIGANDS TO THE COPPER

In the absence of X-ray crystallographic analysis, ligands cannot be assigned with any degree of certainty. However the complete amino acid sequence is available for the *Neurospora* enzyme (Lerch, 1978) and this has helped in the quest for the residues constituting the active site. An exceptional feature of the primary structure is the presence of a unique thioether bridge formed by interaction of the cysteine and histidine residues 94 and 96. A search for a similar bridge in another tyrosinase, a laccase and a haemocyanin has been negative (K. Lerch, personal communication) and hence the importance of this feature is presently unclear. Two distinct approaches suggest that histidine is involved in copper binding. Photo-oxidation of *Neurospora* and mushroom tyrosinases in the presence of a photosensitizer causes a slow and irreversible loss of enzyme activity. This process is markedly enhanced by the addition of cyanide, a metal chelating agent, and displays a pH dependency characteristic of histidine (Gutteridge *et al.*, 1977). Also photo-oxidation of the *Neurospora* apoenzyme results in the loss of three histidine residues (residues 188, 193, 289) and the product can no longer be converted into the holoenzyme (Lerch, 1978).

Reaction inactivation has been a term of long standing, used to describe the poisoning of tyrosinase when oxidising *o*-diphenols, particularly catechol (Miller and Dawson, 1941). This phenomenon would now be termed suicide inactivation (Abeles and Maycock, 1976). The progress curve in Fig. 3 shows that inactivation is pronounced with purified *Neurospora* tyrosinase. It has been previously recorded that copper is lost during this process (Brooks and Dawson, 1966) and more recently one histidine residue (residue 306) has been reported to be destroyed also (Lerch, 1978). Interestingly, this recent study finds no evidence for adduct formation between the catechol moiety and the protein; such a product has been claimed for the mushroom enzyme (Wood and Ingraham, 1965).

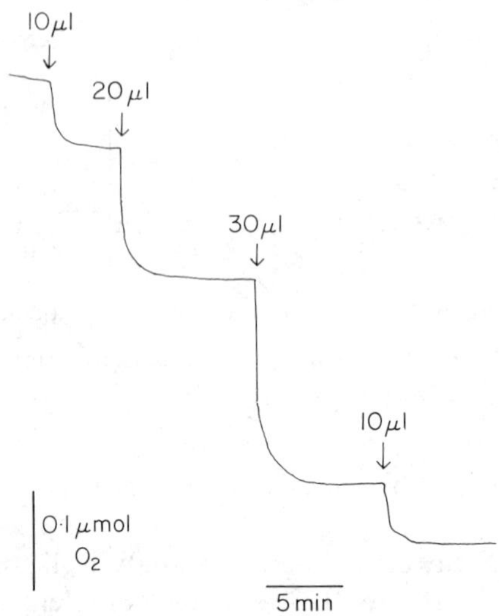

FIG. 3. Titration of *Neurospora* tyrosinase with catechol. An oxygen electrode was immersed in 2·5 ml 0·05 M phosphate buffer pH 6·5 containing 3 mg catechol. Enzyme was added at each arrow in the amounts shown. The oxygen concentration is given on the ordinate and time is shown on the abscissa.

F. INHIBITION BY ANIONS

Several early reports document that tyrosinase is inhibited by anions such as halide ions, azide and benzoate (Lerner, 1952; Krueger, 1955; Kuttner and Wagreich, 1953). Inhibition by these anions has been claimed to be both competitive and non-competitive in nature. It appears likely that these anions are reacting with the active site, and the pH dependency has been explained on the basis of the involvement of a side chain residue such as histidine (Krueger, 1955). Conversely it has been argued that the pH dependency reflects the ionisation of the inhibitor and that the undissociated acid is a much stronger inhibitor than the anion (Kuttner and Wagreich, 1953; Robb *et al.*, 1966). Data relevant to this question is shown in Fig. 4 where the apparent inhibition constant is plotted as a negative logarithm against the pH. Dixon's guidelines that slopes should have values of one or zero (Dixon and Webb, 1958) have been employed in drawing the lines. Inflexion points on such curves give the pK values of a group either on substrate or enzyme which is important in controlling the pH dependency. Also shown on the curve is the fact that over

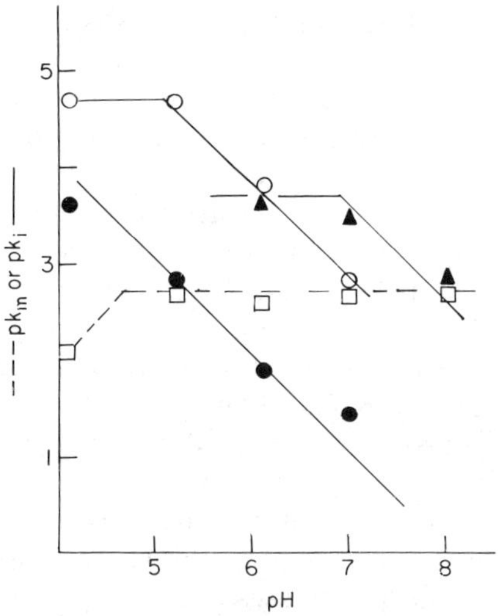

FIG. 4. Effect of pH on p*Km* for 4-tert-butylcatechol (□) and on p*Ki* for azide (○), 4-nitrocatechol (▲), and fluoride (●). The data was obtained with mushroom tyrosinase and is taken from Robb *et al.* (1966).

the pH range 5–8 the *Km* for butylcatechol is independent of pH. If a group on the enzyme is important, such as histidine, we would expect each inhibitor to show a similar p*K* near to neutrality. In fact for the inhibitors shown, fluoride shows no inflexion but azide and nitrocatechol show p*K*s of 5·1 and 6·9 respectively. These values are close to those for the ionization constants of hydrazoic acid and nitrocatechol suggesting that indeed the undissociated acids are strong inhibitors. It is unlikely that the acid would complex directly with the copper without dissociation of a proton. By analogy with other metalloenzymes, such as peroxidase and carbonic anhydrase whose structure has been extensively analysed, the possibility exists that a ligand is displaced on binding the anion (Pratt, 1975; Lindskog *et al.*, 1971). At present this aspect is under study and it is too early to reach any conclusions. One of several models under consideration postulates a hydroxyl group as a ligand at the active site. Binding of an undissociated acid and an anion would be represented as

$$E{-}Cu_2{}^{2+}OH^- + HA \rightarrow ECu_2{}^{2+}A^- + H_2O$$
$$E{-}Cu_2{}^{2+}OH^- + A^- \rightarrow ECu_2{}^{2+}A^- + {}^-OH$$

Such a proposal would explain why the binding of the undissociated acid is independent of pH and the binding of an anion is favoured by low pH (facilitates removal of ^-OH as H_2O).

III. Concluding Remarks

The role of tyrosinase in plant metabolism continues to interest enzymologists and physiologists alike. Changes in activity during growth and development of fruit is a subject discussed by Mayer (see Chapter 9, this volume); further aspects concern its biosynthetic role, protective functions (Mayer and Harel, 1979) and more recently a link with boron requirements has been proposed (Lewis 1980). Thus this article has ignored many studies in concentrating on the structural aspect, but this has been a particularly active area of late. Results obtained with two fungal tyrosinases should give fresh impetus to studies on higher plants on two accounts.

First, the discovery of oxytyrosinase and its similarity to oxyhaemocyanin suggests that a possible function as an oxygen carrier should be explored. Secondly, structural studies on plant tyrosinases are required to evaluate how far results obtained with fungi are representative. A comparison of the smaller *Beta vulgaris* and *Vicia faba* tyrosinases with the aggregating system found in *Solanum tuberosum* would be of great value for example. The discovery of dissimilar sub-units in mushroom tyrosinase should also promote further study. There is still reservation about the existence of these sub-units *in vivo*. It is possible that they are formed by limited proteolysis (nicking) during the isolation and purification procedures from a unit the size of mammalian tyrosinase. Proteolysis may also be a hazard in the purification of the *Neurospora* enzyme since two polypeptides of molecular weight 33,000 and 46,000 are routinely found in purified preparations from this source. Previous disagreements on the molecular size of the *Neurospora* enzyme may be partly due to this cause.

More information also needs to be sought on the disposition of the copper atoms. At present there is evidence implicating histidine as a ligand, and on the basis of the primary structure of *Neurospora* tyrosinase it appears that cysteine is not a ligand. Two possible models have already been characterized in other proteins. In one, superoxide dismutase, the two metal atoms, Cu and Zn are bridged by a histidine residue and are 0·6 nm apart. An artificial form of the enzyme, superoxide dimutase, in which Cu replaces the Zn has been prepared (Fee and Briggs, 1975) and although it shows antiferromagnetic coupling only one atom in the pair showed redox activity. It is not known whether this form possessed any tyrosinase-like activity. Alternatively the copper atoms may be complexed in a cage, in the manner determined for methemerythrin which contains a binuclear iron cluster (Stenkamp *et al.*, 1976). In this regard, confirmation of the 0·6 nm distance separating the copper atoms in mushroom tyrosinase is required. An attempt to make this measurement with the *Neurospora* enzyme failed (Deinum *et al.*, 1976), but XAFS spectroscopy (Garner, 1979) is an alternative technique which is capable of providing this information.

Thus if further work substantiates the binuclear model for tyrosinase one may anticipate useful comparisons between it and plant laccase and ascorbate oxidase which are also thought to contain a binuclear copper cluster (Reinhammar, 1979).

REFERENCES

Abeles, R. H. and Maycock, A. L. (1976). *Accounts Chem. Res.* **9**, 313–319.

Benjamin, N. D. and Montgomery, M. W. (1973). *J. Fd Sci.* **38**, 799–806.

Bouchilloux, S., McMahill, P. and Mason, H. S. (1963). *J. Biol. Chem.* **238**, 1699–1707.

Brooks, D. W. and Dawson, C. R. (1966). *In* "The Biochemistry of Copper" (J. Peisach, P. Aisen and W. E. Blumberg, Eds), pp. 343–355. Academic Press, New York and London.

Coleman, J. E. (1974). *In* "Food Related Enzymes" (J. R. Whitaker Ed.), pp. 267–304. American Chemical Society, Washington.

Combs, J., Baldry, C., Bucke, C. and Long, S. P. (1974). *Phytochemistry* **13**, 2703–2708.

Deinum, J., Lerch, K. and Reinhammar, B. (1976). *FEBS Letters* **69**, 161–164.

Dixon, M. and Webb, E. C. (1958). "Enzymes", 1st Ed., pp. 120–150. Longman, London.

Duckworth, H. and Coleman, J. E. (1970). *J. Biol. Chem.* **245**, 1613–1625.

Eickman, N. C., Soloman, E. I., Larrabee, J. A., Spiro, T. G. and Lerch, K. (1979). *J. Am. Chem. Soc.* **100**, 6529–6531.

Fee, J. and Briggs, R. G. (1975). *Biochim. Biophys. Acta* **400**, 439–450.

Fling, M., Horowitz, N. H. and Heinemann, S. F. (1963). *J. Biol. Chem.* **238**, 2045–2053.

Garner, C. D. (1979). *Nature* **277**, 89–90.

Gregory, R. P. F. and Bendall, D. S. (1966). *Biochem. J.* **101**, 569–581.

Gutteridge, S. and Robb, D. A. (1975). *Eur. J. Biochem.* **54**, 107–116.

Gutteridge, S., Dickson, G. and Robb, D. A. (1977). *Phytochemistry*, **16**, 517–519.

Jolley, R. L., Evans, L. H., Makino, N. and Mason, H. S. (1974). *J. Biol. Chem.* **249**, 335–345.

Kertesz, D. and Zito, R. (1962). *In* "Oxygenases" (O. Hayaishi Ed.), pp. 307–354. Academic Press, New York and London.

Kertesz, D., Rotilio, G., Brunori, M., Zito, R. and Antonini, E. (1972). *Biochem. Biophys. Res. Commun.* **49**, 1208–1215.

Kidron, M., Harel, E. and Mayer, A. M. (1977). *Phytochemistry* **16**, 1050–1051.

Krueger, R. C. (1955). *Arch. Biochem. Biophys.* **57**, 52–60.

Kubowitz, F. (1938). *Biochem. Zeit.* **299**, 32–57.

Kuiper, H. A., Lerch, K., Brunori, M. and Finazzi Agro, A. (1980). *FEBS Letters*, **111**, 232–234.

Kuttner, R. and Wagreich, H. (1953). *Arch. Biochem. Biophys.* **43**, 80–87.

Lerch, K. (1976). *FEBS Letters*, **69**, 157–160.

Lerch, K. (1978). *Proc. Natl. Acad. Sci. U.S.A.* **75**, 3635–3639.

Lerch, K. and Ettlinger, L. (1972). *Eur. J. Biochem.* **31**, 427–437.

Lerner, A. B. (1952). *Arch. Biochem. Biophys.* **36**, 473–481.

Lewis, D. H. (1980). *New Phytol.* **84**, 209–229.

Lindskog, S., Henderson, L. E., Kannan, K., Liljas, A., Nyman, P. and Strandberg, B. (1971). *In* "The Enzymes", 3rd Ed. (P. Boyer, Ed.), Vol. 5, pp. 587–665. Academic Press, New York and London.

Lontie, R. and Witters, R. (1973). *In* "Inorganic Biochemistry" (G. I. Eichhorn, Ed.), pp. 344–358. Elsevier, Amsterdam.

Makino, N., McMahill, P., Mason, H. S. and Moss, T. H. (1974). *J. Biol. Chem.* **249**, 6062–6066.

Mason, H. S., Spencer, E. and Yamazaki, I. (1961). *Biochem. Biophys. Res. Commun.* **4**, 236–238.
Mayer, A. M. and Harel, E. (1979). *Phytochemistry* **18**, 193–215.
Miller, W. H. and Dawson, C. R. (1941). *J. Am. Chem. Soc.* **63**, 3368–3374.
Nakamura, T., Sho, S. and Ogura, Y. (1966). *J. Biochem. (Tokyo)* **59**, 481–486.
Pomerantz, S. and Murthy, V. J. (1974). *Arch. Biochem. Biophys.* **160**, 73–82.
Pratt, J. M. (1975). *In* "Techniques and Topics in Bio-inorganic Chemistry" (C. A. McAuliffe, Ed.), pp. 109–204. MacMillan, London.
Reinhammar, B. (1979). *Adv. Inorg. Biochem.* **1**, 92–118.
Rivas, N. J. and Whitaker, J. R. (1973). *Plant Physiol.* **52**, 501–507.
Robb, D. A. (1979). *Biochem. Soc. Transac.* **7**, 131–132.
Robb, D. A. and Gutteridge, S. (1981). *Phytochemistry* **20**, 1481–1485.
Robb, D. A., Swain, T. and Mapson, L. W. (1966). *Phytochemistry* **5**, 665–675.
Schoot Uiterkamp, A. J. M., van der Deen, H., Berendsen, H. C. J. and Boas, J. F. (1974). *Biochim. Biophys. Acta* **372**, 407–425.
Stenkamp, R. E., Sieker, L. C. and Jensen, L. H. (1976). *Proc. Natl. Acad. Sci. U.S.A.* **73**, 349–351.
Strothkamp, K. G., Jolley, R. L. and Mason, H. S. (1976). *Biochem. Biophys. Res. Commun.* **70**, 519–524.
van Holde, K. E. (1975). *In* "The Proteins" 3rd Ed. (H. Neurath and R. L. Hill, Eds), pp. 226–291. Academic Press, New York and London.
Vanneste, W. H. and Züberbühler, A. (1974). *In* "Molecular Mechanisms of Oxygen Activation" (O. Hayaishi, Ed.), pp. 371–404. Academic Press, New York and London.
Vaughan, P. T. F., Eason, R., Paton, J. Y. and Ritchie, G. A. (1975). *Phytochemistry*, **14**, 2383–2386.
Walker, J. and McCallion, R. F. (1980). *Phytochemistry* **19**, 373–377.
Wood, B. J. B. and Ingraham, L. L. (1965). *Nature* **205**, 291–292.

CHAPTER 11

Changes in Phenolic Metabolism in Fruit and Vegetable Tissues Under Stress

M. J. C. RHODES, L. S. C. WOOLTORTON AND A. C. HILL

Agricultural Research Council, Food Research Institute, Norwich, England

I. Introduction 193
II. Phenolic Responses to Mechanical Damage in Brassica Roots . . . 193
III. Responses in Phenylpropanoid Metabolism to Low Temperature Treatments 208
References 218

I. INTRODUCTION

Many plant tissues respond to environmental stress with changes in the metabolism of phenolic compounds (Imaseki *et al.*, 1968; Rhodes and Wooltorton, 1978a). These changes include both the oxidation of pre-existing compounds and the increased synthesis of either monomeric or polymeric phenolic compounds. This chapter deals with the biosynthetic responses of plant tissues to two types of physiological stress, namely mechanical damage and low temperature stress. Two systems will be described: (1) the Brassica root which responds to mechanical damage by depositing a layer of polymeric lignin-like material on its cut surfaces, and (2) potato tubers and tomato fruit which respond to storage at low temperatures by increased metabolism of a monomeric phenol, chlorogenic acid. In both cases emphasis is placed on the enzymological aspects of these responses.

II. PHENOLIC RESPONSES TO MECHANICAL DAMAGE IN BRASSICA ROOTS

The "root" of *Brassica napus* (swede, Swedish turnip, rutabaga) in botanical terms consists of both root and hypocotyl tissue and during the first year of growth is composed largely of unlignified secondary parenchyma cells. When discs (10 mm diameter × 2 mm) of the tissue are cut and incubated, a layer of lignin-like material is deposited on the cut surfaces of the tissue. This increase in lignin-like material may be followed by staining whole discs with phloroglucinol/HCl. Freshly prepared discs stain very little with this reagent whilst aged discs, after a lag-phase of about four hours, show progressive increase in staining up to 24 h. More detailed examination of sections of the discs showed that freshly prepared discs stain with phloroglucinol/HCl only in the vascular elements, while with aged discs there was an increasing degree of staining restricted to the first and second layers of cells at the cut surface. Studies using UV microscopy at 280 nm showed that during ageing there was increasing absorption of UV light in the walls of cells at the cut surface. These results indicate that the material deposited was lignin. It was found that the deposition of lignin could be enhanced by incubating the excised discs in the presence of low levels of ethylene (0·1–10 p.p.m.).

Attempts were made to characterize the chemical nature of the lignin-like material in washed cell wall preparations from aged discs using oxidation under alkaline conditions with either nitrobenzene or cupric oxide. Lignin enriched fractions such as Klason lignin and thioglycollic acid lignin (Sarkanen and Ludwig, 1971) were also isolated from fresh and aged discs. Both the yield of the lignin fractions and of the oxidation products was considerably higher in aged compared with fresh discs (Rhodes and Wooltorton, 1973a). Ethylene treated material gave higher yields of lignin and its oxidation products than discs aged in air. Labelling experiments using ^{14}C-phenylalanine indicated that on ageing there was a marked change in the pattern of utilization of the amino acid with increasing incorporation into the lignin containing fraction at the expense of the protein fraction (Rhodes and Wooltorton, 1973a,b). The lignin-like material labelled by phenylalanine was resistant to mild alkaline hydrolysis and on oxidation yielded labelled lignin oxidation products. The yield and specific activity of the oxidation products, *p*-hydroxybenzaldehyde, vanillin and syringaldehyde, were higher in aged discs compared with freshly prepared discs, and among the aged discs ethylene treated material gave higher yields of more heavily labelled products than material aged in air (Rhodes and Wooltorton, 1973b). These findings taken together show that there is an increase in lignin production on ageing and that this increase is further stimulated by ethylene. The response in forming a surface layer of lignin, in response to mechanical damage, is also shown by a number of other plant tissues and these include the sweet potato (Uritani and Oba, 1978) and the radish (Asada and Matsumoto, 1969).

Figure 1 shows the likely route of biosynthesis of lignin from phenylalanine and illustrates the central role of the CoA thioesters of cinnamic acids in the formation of the three cinnamyl alcohols which are the immediate precursors of lignin polymerization. Since the discovery of the enzyme, phenylalanine ammonia lyase, P.A.L. (Koukol and Conn, 1961), most of the enzymes of the pathway have been characterized. The properties of the enzymes have been

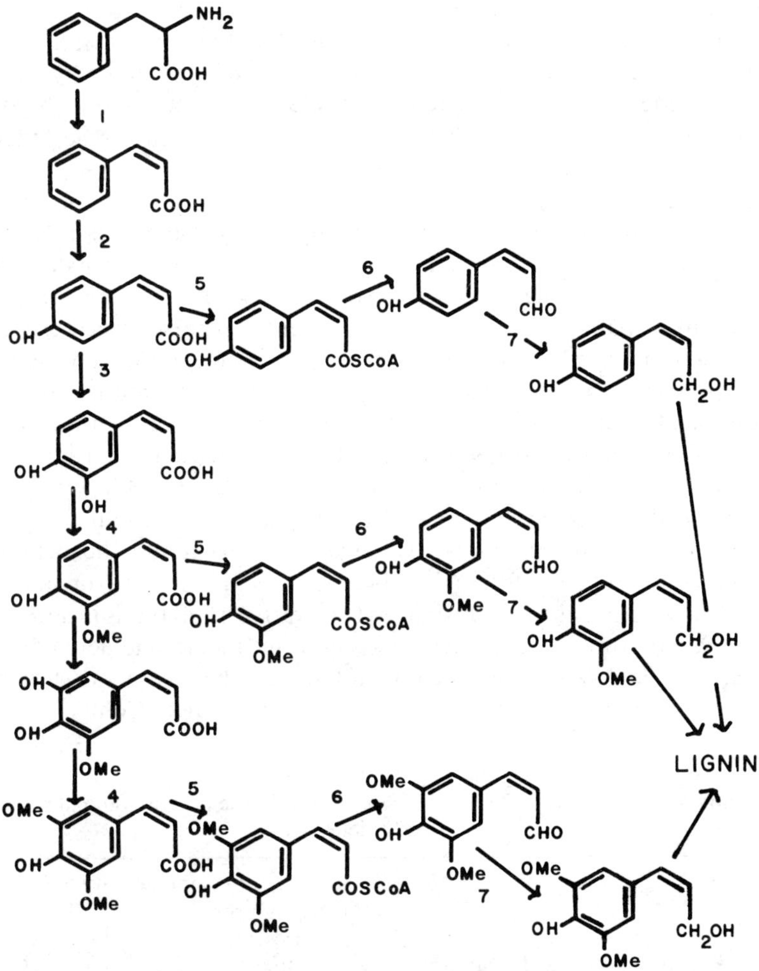

FIG. 1. The pathway of biosynthesis of lignin from L-phenylalanine. 1, P.A.L.; 2, C.A.4H.; 3, *p*-coumarate hydroxylase; 4, O.M.T.; 5, hydroxycinnamate CoA ligase; 6, hydroxycinnamoyl CoA reductase; 7, cinnamyl alcohol dehydrogenase. The cinnamic acid intermediates in this pathway are all trans-isomers.

the subject of a number of recent reviews (Grisebach, 1977; Gross, 1977; Hahlbrock and Grisebach, 1979; Rhodes and Wooltorton, 1978a) but in spite of considerable progress there are still gaps in our knowledge of the pathway. For instance, an enzyme to catalyse the hydroxylation in which ferulic acid is converted to 5-hydroxyferulic acid has yet to be described while there is still dispute on the nature of the enzymes catalysing the hydroxylation of *p*-coumaric acid to caffeic acid (Stafford, 1974; Mayer, Chapter 9, this volume). Vaughan and Butt (1969) and Stafford (1976) have proposed that a phenolase-like enzyme catalyses this hydroxylation in the presence of a suitable reducing agent such as ascorbic acid. Brassica tissue has been shown to actively metabolize *p*-coumaric acid to caffeic acid but lacks phenolase activity. Under the conditions described by Vaughan and Butt (1969) activity in hydroxylating *p*-coumaric acid was not detected with Brassica extracts prepared in a number of different ways. Thus, at least in Brassica tissue, there is uncertainty about this step in the pathway.

Table I shows changes in the activity of seven enzymes, involved in the pathway of lignin biosynthesis from phenylalanine, and shikimate dehydrogenase as a representative of the pathway by which phenylalanine is formed from simple sugars. The seven enzymes of the phenylpropanoid pathway are capable of catalysing all the steps of the sequence except the second and third ring hydroxylations (see Fig. 1). Some of the enzymes have a broad specificity for the substitution pattern of the aromatic ring of the cinnamic acids and thus can catalyse more than one step in the sequence. For instance, the *o*-methyltransferase (O.M.T.) of *Brassica* roots catalyses the methylation of both caffeic and 5-hydroxyferulic acids (Hill and Rhodes, 1975) while the hydroxycinnamate CoA liagase, hydroxycinnamoyl CoA reductase and cinnamyl alcohol dehydrogenases can catalyse the metabolism of both *p*-coumaroyl and feruloyl derivatives. The ligase and aromatic ADH have only low activities towards sinapoyl derivatives, and this is in accordance with the fact that *Brassica napus* lignin is poor in sinapic acid derived groups which would yield syringaldehyde on oxidation. Table I shows that there are major changes in activity of three of the enzymes studied. Comparing tissue

TABLE I
Changes in enzyme levels in *Brassica* discs aged in the presence and absence of ethylene

Enzyme	Initial discs		Enzyme activity pkat/mg protein (Relative changes) Discs aged in air		Discs aged in 6 p.p.m. C_2H_4	
(1) Shikimate Dehydrogenase	139	(1)	247	(1·8)	253	(1·8)
(2) P.A.L.	10	(1)	44	(4·4)	225	(22·5)
(3) C.A.4H[a]	0·9	(1)	11·4	(12·7)	20·5	(22·8)
(4) O.M.T.	12	(1)	27·2	(2·3)	44·3	(3·7)
(5) Hydroxycinnamate CoA ligase	6·6	(1)	59·4	(9·0)	148	(22·4)
(6) Feruloyl CoA reductase	22·5	(1)	—	—	62·3	(2·8)
(7) Coniferyl alcohol dehydrogenase	135	(1)	—	—	166	(1·2)
(8) Peroxidase[b]	$6{\cdot}1 \times 10^{-5}$	(1)	$6{\cdot}4 \times 10^{-5}$	(1·0)	$8{\cdot}7 \times 10^{-5}$	(1·5)

[a] Expressed per mg microsomal protein.
[b] Enzyme units/10 g fresh weight of tissue.

aged in ethylene with freshly prepared tissue, most of the enzymes show only small increases (1·2–3·7-fold) in activity, while P.A.L., Cinnamate 4-Hydroxylase (C.A.4H.) and the ligase show proportionately larger increases in activity (~20-fold). If discs aged in air and in 6 p.p.m. C_2H_4 are compared, C_2H_4 stimulated P.A.L. activity five-fold, had smaller effects on the ligase (2·5-fold) and C.A.4H. (1·8-fold) but had very little effect on the other enzymes studied. These results suggest that P.A.L., C.A.4H. and ligase play an important role in the control of the pathway and that the effect of ethylene is to enhance the increase in these activities.

P.A.L. catalyses the stereospecific deamination of L-phenylalanine to produce trans-cinnamic acid. The level of P.A.L. activity appears to play a major role in the control of phenylpropanoid metabolism in a number of plant tissues (Camm and Towers, 1973). However, this view has been challenged by Margna (1977) who proposed that in some tissues the availability of the substrate, phenylalanine, rather than the level of enzyme activity determined the rate of flow through the pathway. The level of P.A.L. activity in plants is sensitive to a number of environmental and hormonal factors (Camm and Towers, 1973). In cell suspension cultures of parsley, Hahlbrock and his colleagues have made an intensive study of the role of P.A.L. in the control of light induced flavonoid biosynthesis (Hahlbrock and Wellmann, 1973). They have concluded that light induces the *de novo* synthesis of P.A.L. protein and that this is mediated through a light dependent synthesis or activation of a P.A.L. specific messenger RNA species (Schroder *et al.*, 1979; Hahlbrock and Grisebach, 1979). In other systems *de novo* synthesis of P.A.L. also appears to be a major factor in the increase in activity of the pathway of phenylpropanoid biosynthesis in response to other stimuli (Duchesne *et al.*, 1977). There is evidence that *de novo* synthesis of P.A.L. in some systems is followed by the appearance of a P.A.L. inactivating system (Billett and Smith, 1978) and that activation–inactivation phenomena may regulate subsequent activity of the enzyme (Attridge and Smith, 1974).

In swede discs, P.A.L. activity during ageing, rose from low levels (0–10 pkat/mg protein) to over 200 pkat/mg protein in a period of 24 h. The time course showed a lag phase of one to two hours followed by a rapid rise for up to 16–24 h and then a steady level was established. Treatment with ethylene stimulated the rise in P.A.L. activity by 3–10 fold; a concentration of ethylene of 1·2 p.p.m. was required to give half maximal stimulation. Both propylene and acetylene could replace ethylene in this effect but higher concentrations were required to achieve the same degree of stimulation (Rhodes and Wooltorton, 1971). P.A.L. was mainly found in the soluble phase of aqueous extracts of swede tissue but a significant amount (up to 7% of the total) was associated with a membrane fraction which sedimented at 80,000 × g. The major portion of the membrane associated P.A.L. activity could not be removed by repeated washing of the membrane fraction. In sucrose density gradient studies the P.A.L. remained firmly bound to a particle with a buoyant density equivalent to about 1·18 which contained up to 90% of the C.A.4H. activity of the tissue.

C.A.4H. like P.A.L. showed a large change in activity during ageing and was stimulated by ethylene treatment but to a smaller degree than P.A.L. C.A.4H. catalyses the hydroxylation of trans cinnamic acid to produce trans *p*-coumaric acid (Russell, 1971; Potts *et al.*, 1974; Hill and Rhodes, 1975; Benveniste *et al.*, 1977; Rich and Lamb, 1977). The C.A.4H. reaction involves the simultaneous reduction of O_2 to H_2O and the oxidation of $NADPH_2$ coupled to the hydroxylation of cinnamic acid and is thus a mixed function oxidase. The enzyme is associated with a membrane system, and disruption of the membrane with ionic detergents or with chaotropic ions inhibits the enzyme activity. There is evidence that components of the lipid fraction of the membrane are essential for activity (Buche and Sanderman, 1973). C.A.4H. is inhibited by carbon monoxide and this inhibition is reversible by light with a maximum in the action spectrum for photoreversibility at 450 nm (Potts *et al.*, 1974). In a number of its properties the plant C.A.4H. resembles a group of animal and bacterial monoyxgenases which involve proteins bearing a haem

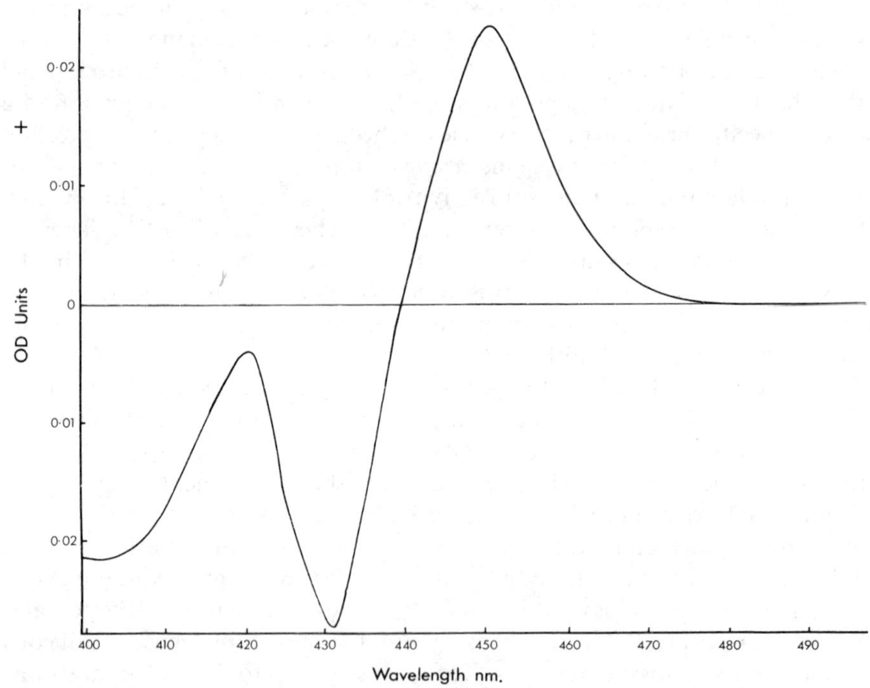

FIG. 2. The CO difference spectrum of swede microsomes. The difference spectrum between 400–500 nm was determined in an Amino Chance spectrophotometer. Equal amounts of microsomal preparation (2·5 mg protein/ml) were placed in the two cells and excess sodium dithionite was added to each. The base line was set to zero and then the sample cell was treated with CO and the difference spectrum recorded.

group and which form CO adducts with a maximum of absorbance at 450 nm. Such proteins are collectively called cytochrome P450s (Gunsalus *et al.*, 1975). Plant microsomes contain cytochrome P450 (Potts *et al.*, 1974; Yahiel *et al.*, 1974; Cotte-Martinon *et al.*, 1974; Hill, 1975; Rich *et al.*, 1975). The presence of cytochrome P450 in swede microsomes is illustrated in Fig. 2. A difference spectrum between dithionite reduced microsomes in the presence and absence of CO shows a positive peak at 450 nm and a second peak negative to the base line at 420 nm. Swede microsomes contain up to about 0·3 nmol cytochrome P450/mg protein.

The hydroxylation of cinnamic acid involves a characteristic proton shift from position 4 to position 3 of the aromatic ring during hydroxylation (Russell *et al.*, 1968) as is found in the analogous animal hydroxylation systems. C.A.4H. shows a high degree of specificity for its substrate, *t*-cinnamic acid;

TABLE II
Effect of $NADH_2$ on $NADPH_2$ supported C.A.4H. activity of swede microsomes

$NADPH_2$	$NADH_2$	C.A.4H. activity (pkat/mg protein)	
Concentration in the assay (μM)		Observed	Calculated from sum of individual rates with $NADH_2$ or $NADPH_2$
0	0	—	—
	50	0·14	—
	100	0·19	—
	500	1·10	—
	1000	2·20	—
50	0	18·9	18·9
	50	35·0	19·0
	100	34·2	19·1
	500	40·8	20·0
	1000	40·0	21·1
100	0	34·2	34·2
	50	50·6	34·3
	100	56·9	34·4
	500	59·2	35·3
	1000	58·3	36·4
500	0	61·4	61·4
	50	69·2	61·5
	100	73·9	61·6
	500	73·9	62·5
	1000	69·2	63·6
1000	0	72·5	72·5
	50	72·5	72·6
	100	76·4	72·7
	500	72·0	73·6
	1000	73·3	74·7

cis-cinnamic acid is a competitive inhibitor (Pflandler *et al.*, 1977). Other cinnamic acids, such as *p*-coumaric and ferulic acids, and phenylalanine are not substrates. The enzyme shows a high degree of specificity for the pyridine nucleotide cofactor; the *Km* for $NADPH_2$ being 46-fold lower than that for $NADH_2$ (Hill and Rhodes, 1975). There is an interesting synergism between the two nucleotides in some plant systems which is similar to that found in animal cytochrome P450 dependent systems. In swede microsomes (see Table II) low levels of NADH, which were themselves barely capable of supporting activity, more than doubled the rate of hydroxylation at low levels of $NADPH_2$ (50 μM). There was no evidence for the presence of a transhydrogenese in swede microsomes (Hill and Rhodes, 1975) and thus the apparent synergism cannot be explained in terms of interconversion of $NADH_2$ to $NADPH_2$. Similar synergistic interactions were observed for the C.A.4H. activity of microsomes of gherkin hypocotyls (Billett and Smith, 1978) and of Jerusalem artichoke tissue (Benveniste *et al.*, 1977) but in both cases low levels of $NADH_2$ were found to increase the rate of hydroxylation at both low and near saturating levels of $NADPH_2$. However, no synergism was observed between $NADH_2$ and $NADPH_2$ on the C.A.4H. activity of potato microsomes (Rich and Lamb, 1977).

Figure 3 shows that while 1 mM $NADH_2$ caused a slow reduction of cytochrome P450, addition of 0·1 mM $NADPH_2$ leads to an almost instantaneous reduction of the cytochrome. Other experiments showed that $NADPH_2$ reduced cytochrome P450 in the absence of $NADH_2$. Figure 4 shows the difference spectrum of $NADPH_2$ reduced microsomes in the presence and absence of CO. The spectrum shows a peak at 450 nm, a shoulder at 430 nm but no peak at 420 nm. Subsequent reduction with dithionite increased the height of the peak at 450 nm (indicating that $NADPH_2$ at 1 mM only reduced 60% of the total cytochrome) and the appearance of the peak at 420 nm. In animal systems the peak at 420 nm is thought to represent an enzymically inactive form of the enzyme, termed P420 (Gunsalus *et al.*, 1975). In plants, the nature of the material giving rise to the peak at 420 nm is uncertain although the presence of a species corresponding to the P420 of animal cells has been suggested in cauliflower microsomes (Rich and Bendall, 1975). Swede microsomes contain a substantial level of peroxidase which gives a CO difference spectrum with a peak at 420 nm. Washing the microsomes removes some of the peroxidase activity and a substantial portion of the material giving the 420 nm peak. Thus at least part of the absorbance at 420 nm may be due to peroxidase but underlying this may be another component corresponding to P420. This component, however, does not appear to be reducible by $NADPH_2$.

In animal cells a flavoprotein reductase is involved in the reduction of cytochrome P450 by $NADPH_2$. This component, which is measured by its ability to reduce cytochrome c and is thus referred to as $NADPH_2$ cytochrome c reductase, is present in plant microsomes. Table III shows that C.A.4H.,

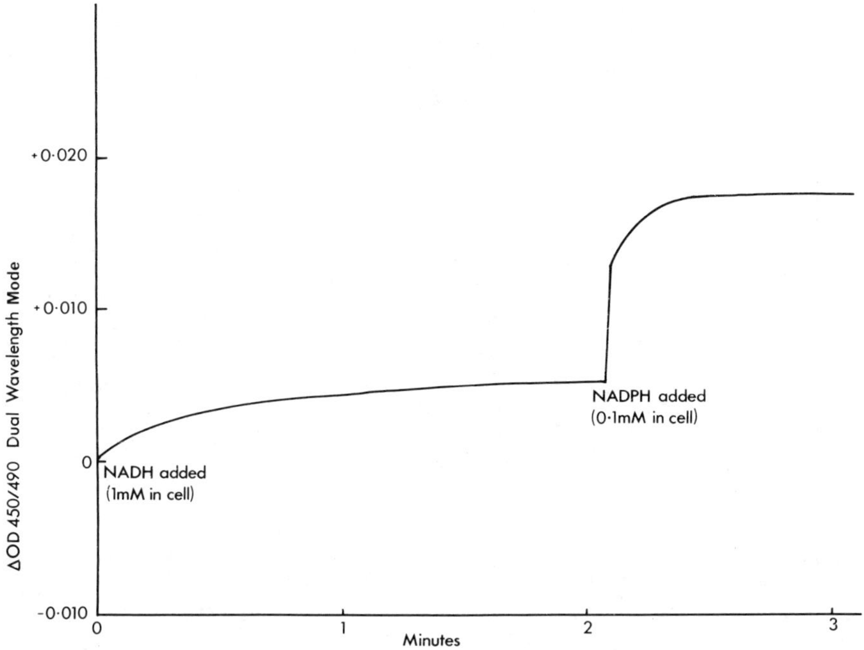

FIG. 3. Time course of reduction of cytochrome P450 in swede microsomes. The Amino Chance spectrophotometer was used in the dual wavelength mode using a single cell containing 2·6 mg/ml of microsomal suspension in the presence of CO. The cofactors were added as indicated above and their ability to reduce the cytochrome P450 and hence induce the formation of its CO adduct was followed.

Cytochrome P450 and the $NADPH_2$ cytochrome c reductase are all concentrated in the 80,000 × g fraction of swede homogenates, as is NADH-cytochrome c reductase which has been implicated in the synergism between $NADH_2$ and $NADPH_2$ in the hydroxylation reaction in animal cells by providing a $NADH_2$ dependent route for the reduction of cytochrome P450 (Gunsalus *et al.*, 1975). In contrast both the P420 component and peroxidase are found mainly in the supernatant fraction.

Figure 5 shows changes observed on ageing in the presence of ethylene of four microsomal components. The C.A.4H. activity rose from a low initial value following a short lag phase. The major increase occurred between 4 and 12 h and then the increase in activity slowed down. In contrast, the changes in the two cytochrome reductases and in the microsomal b cytochrome fraction were small with overall increases of the order of 30–80%. The level of cytochrome P450 showed a steady rise throughout the period of increasing C.A.4H. activity but the overall rise in cytochrome P450 is only about four-fold compared with 20–25-fold for C.A.4H. The C.A.4H. activity rose from 52 to 369 pkat/nmol cytochrome P450 during this period. Moreover, there was a

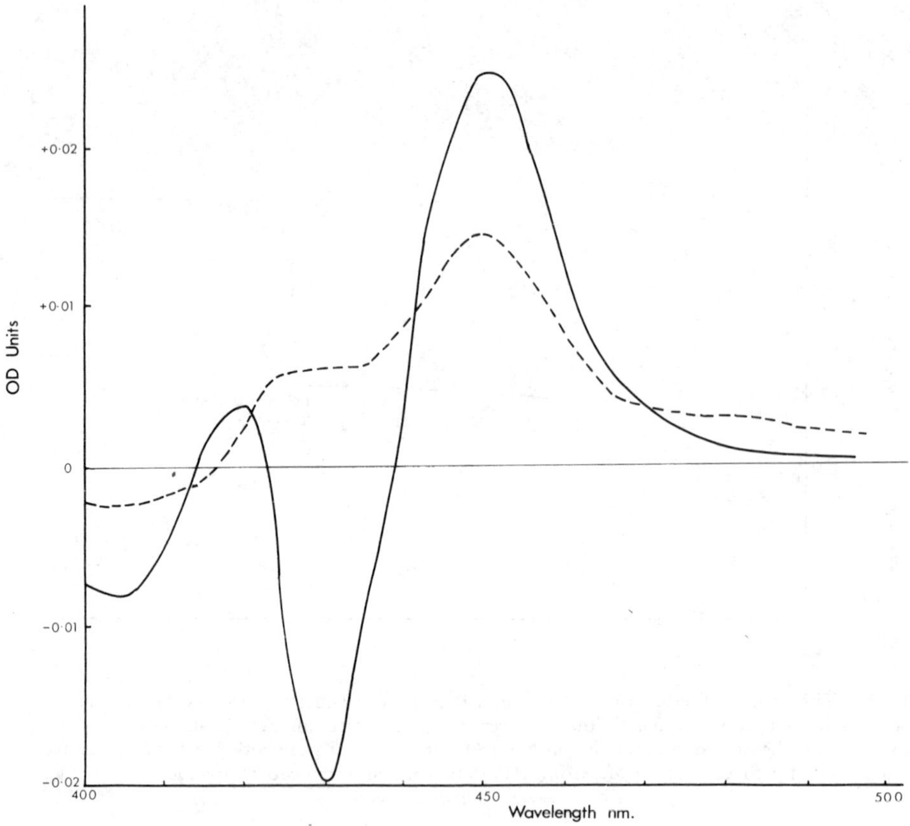

FIG. 4. CO difference spectrum of swede microsomes reduced with either 3 μmol $NADPH_2$ (– – – –) or excess dithionite (———).

substantial level of the cytochrome P450 in initial discs which had very low C.A.4H. activity. We have found some microsome preparation of long-stored roots which have easily measurable levels of cytochrome P450 but no detectable C.A.4H. activity. In plants, there is evidence that the flavoproteins and the b-type cytochrome are components of an electron tranport system linked to cinnamate hydroxylation (Rich and Lamb, 1977) but in the present experiment it appears unlikely that the level of these components play a substantial role in the control of the rise in C.A.4H. activity observed.

Table IV shows the effect of two inhibitors, cycloheximide and actinomycin D on changes in the various microsomal components during ageing. Both compounds inhibited the increase in C.A.4H. by more than 80%. Removing the inhibitor from the incubation medium had the effect of releasing the inhibition with cycloheximide but the effect with actinomycin D was only marginal. Similarly, with cytochrome P450, cycloheximide completely

TABLE III

Distribution of C.A.4H. and various other components in homogenates of aged swede root discs

Fraction	C.A.4H. pkat/mg protein	%	$NADPH_2$ Cytochrome c reductase nkat/mg protein	%	$NADH_2$ Cytochrome c reductase nkat/mg protein	%	Cytochrome P450 nmol/mg protein	%	P-420 nmol/mg protein	Peroxidase %	Peroxidase %	Protein mg/10 g fr. wt
10,000 × g Pellet	0·8	3	0	0	0·7	7	0	0	0·01	1	13	7
80,000 × g Pellet	30·6	97	1·5	93	8·9	87	0·1	100	0·11	4	5	3·0
80,000 × g Supernatant	0	0	0·1	7	0·6	6	0	0	0·50	95	82	15·0

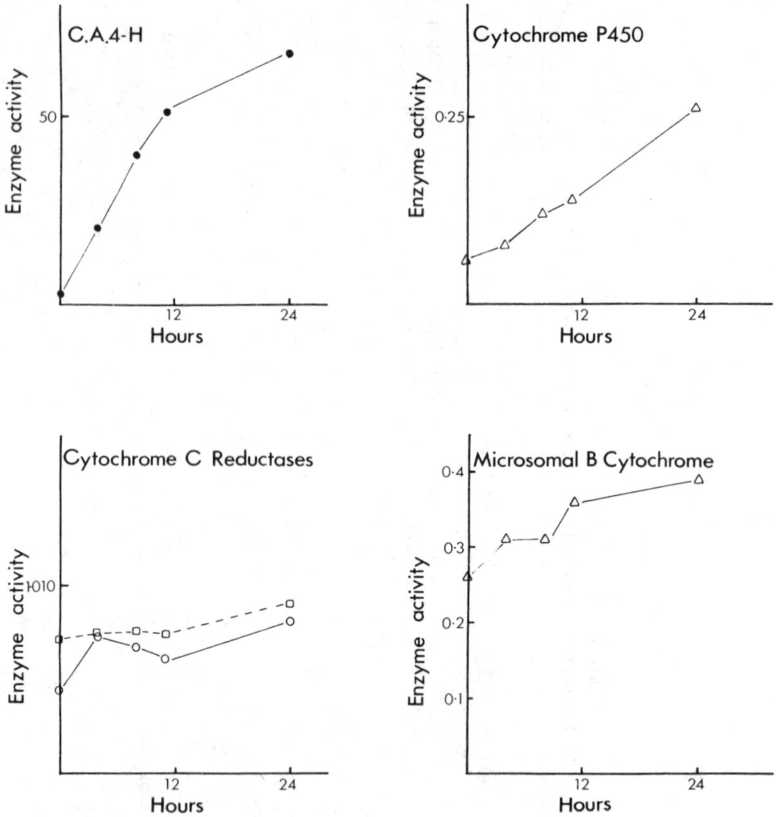

FIG. 5. Time course of changes in activity of four microsomal components during the ageing of swede root discs. $NADPH_2$—cytochrome c reductase (○ scale 0–1·0). $NADH_2$—cytochrome c reductase (□ scale 0–10).

inhibited the increase in the level of the cytochrome during ageing and this effect was reversed if the inhibitor was removed from the incubation medium. In contrast, actinomycin D, while it inhibited the increase in C.A.4H. by 81%, it inhibited the increase in cytochrome P450 by only 30%. Neither of the two inhibitors had much effect on changes in the levels of the two reductases and b-type cytochromes. This again suggests that these components do not play a controlling role in the overall rise in C.A.4H. activity.

The relationship between changes in cytochrome P450 level and the appearance of C.A.4H. activity is not a simple one. In cut potato tuber tissue (Rich and Lamb, 1977), the progress curves for increases in cytochrome P450 and C.A.4H. show a very similar pattern of a four–five-fold increase following a lag phase of eight hours. As both the reductases and cytochrome B_5 also showed a similar pattern it was concluded that this represented a non-specific response of membrane repair following damage during excision of the discs

TABLE IV
Effect of cycloheximide and actinomycin D on changes in microsomal components during ageing of swede root discs

	C.A.4H. (pkat/mg protein)	Cytochrome P450 (nmol/mg protein)	Cytochrome c reductases NADPH$_2$ (nkat/mg protein)	Cytochrome c reductases NADH$_2$ (nkat/mg protein)	B type cytochromes (relative level)
Expt I					
Initial discs	3	0·06	0·66	7·0	1
Discs aged 25 h	72	0·16	1·1	9·7	1·43
Discs aged 25 h + 1 μg/ml Cycloheximide (CH)	5	0·05	0·88	7·0	1·43
Discs aged 11 h + 1 μg/ml CH followed by 14 h in medium free of CH	31	0·12	1·3	5·6	1·71
Expt II					
Initial discs	2·5	0·06	0·05	5·6	0·1
Discs aged 24 h	65	0·13	1·1	11·6	1·78
Discs aged 24 h + 5 μg/ml Act. D	14	0·11	1·1	8·3	1·43
Discs aged 11 h + 5 μg/ml Act. D at 13 h to medium free of Act. D	17	0·11	1·1	8·3	2·0

and could not in itself be used as evidence for a functional relationship between C.A.4H. and the other individual membrane components. In the swede, the situation is more complex since three of the components do not show changes during the ageing period. There is no simple relationship between the levels of C.A.4H. activity and cytochrome P450 since they have very different time courses during ageing and different degrees of overall change. In animal cells, it has been shown that cytochrome P450s are heterogeneous and represent a group of related proteins bearing a characteristic haem group but which are functionally distinct. There is evidence that suggests that plant cytochrome P450s can catalyse a variety of hydroxylation and related reactions in diverse areas of metabolism. They have been implicated in gibberellin (Murphy and West, 1969), nerol and geraniol hydroxylations (Meehan and Coscia, 1973), and in fatty acid desaturation (Ben Abdelkader *et al.*, 1973) as well as in cinnamic acid hydroxylation. It seems likely that a range of different cytochrome P450 proteins are responsible for these reactions. The presence of cytochrome P450 in initial discs with low levels of C.A.4H. activity suggests that these discs may contain cytochrome P450s committed to other reactions. However, the cytochrome P450s induced during ageing may include proteins specific for cinnamic acid hydroxylation. Evidence that this may occur was obtained by study of the binding of cinnamic acid to microsomes of initial and aged discs. Schenkman *et al.* (1967) showed with animal tissues, that substrates for the cytochrome P450 based hydroxylation system caused characteristic spectral changes when added to microsomal preparations. They designated a Type I binding spectrum given by substrates for the reaction and a Type II spectrum given by some compounds, not substrates, for the hydroxylation system.

Figure 6 shows that cinnamic acid gives a Type I binding spectrum with microsomes from aged swede discs with a minimum at 428 nm and a peak at 386 nm. A Type I binding spectrum was not given by microsomes from freshly prepared discs and the spectral changes observed are consistent with the appearance of a specific species of cytochrome P450 linked to cinnamate hydroxylation during the ageing of swede root discs.

The third enzyme which responds to cutting and ageing of swede root discs by giving a large increase in activity is hydroxycinnamate CoA ligase. The swede ligase, which was described in detail (Rhodes and Wooltorton, 1973c), is similar in its general properties to the enzymes found in Forsythia (Gross

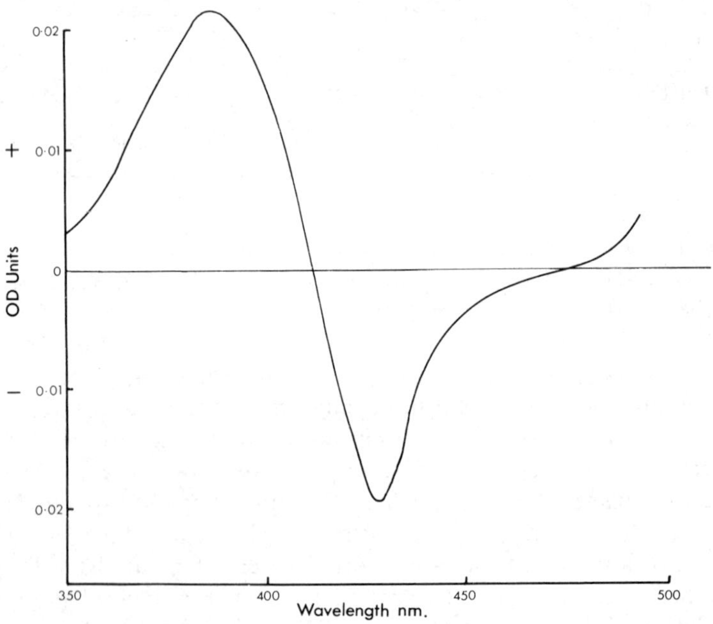

FIG. 6. The spectral changes caused by the addition of cinnamic acid to a microsomal suspension prepared from swede root discs aged for 24 h. The concentration of microsomal protein was 3·3 mg/ml and the final concentration of cinnamic acid was 5 μmol/ml.

and Zenk, 1974) and parsley (Knobloch and Hahlbrock, 1977). Table V shows that the enzyme from swede showed the highest activity and lowest *Km* values with *p*-coumaric acid but was also fairly active towards caffeic, ferulic and isoferulic acids. The enzyme shows only low activity towards sinapic acid and is inactive towards other methoxylated cinnamic acids. In this it clearly resembles the ligase II of Knobloch and Hahlbrock (1975). These authors also describe a ligase I which shows high activity towards both sinapic acid and other methoxylated cinnamic acids. This isoenzyme appears to be absent from swede tissue and this may be related to the low level of syringyl residues in wound lignin in swede root discs. As shown in Table I the swede ligase increased by more than 22-fold during ageing and was stimulated about two-fold by ethylene treatment.

There are a number of systems in which there is an apparent co-ordinated increase in the activities of the three enzymes, P.A.L., C.A.4H., and hydroxycinnamate CoA ligase. A co-ordinated increase in enzyme activities has been found in the light induced flavonoid synthesis in parsley (Ebel and Hahlbrock, 1977), and soyabean cultures (Ebel *et al.*, 1974) and in the light induced synthesis of chlorogenic acid in potato discs (Lamb, 1977). In the case of P.A.L., changes in enzyme activity have been related either to *de novo* synthesis (Duchesne *et al.*, 1977) or to activation phenomena (Smith, 1972). In the case of the parsley enzyme there is good evidence that *de novo* synthesis occurs and this is linked to the appearance of new species of *m*-RNA specific for P.A.L. (Schroder *et al.*, 1979; Hahlbrock and Grisebach, 1979). In the case of the other two enzymes there is little evidence in the literature on the mechanism of the observed increase in activity. Evidence from inhibitor experiments (Rhodes *et al.*, 1976) suggests that the increases in these two enzymes may also involve *de novo* synthesis but in view of the lack of specificity of the inhibitors, these findings are not conclusive. In the case of lignin biosynthesis in swede discs, it appears that the steps catalysed by P.A.L. and C.A.4H. and ligase are of prime importance in the control of the overall rate of the pathway.

TABLE V
Substrate specificity of swede root *p*-coumarate CoA ligase

Substrate	*Vm* (M)	*Km* relative to *p*-coumarate = 100
p-Coumarate	$3{\cdot}5 \times 10^{-6}$	100
Caffeate	$2{\cdot}5 \times 10^{-5}$	68
Sinapate	—	2
Isoferulate	$9{\cdot}8 \times 10^{-5}$	68
Ferulate	$13{\cdot}3 \times 10^{-5}$	72

The following acids were not activated: cinnamic acid, *m*-, *o*-, and *p*-methoxycinnamic acid, 3, 4 and 2,4-dimethoxycinnamic acid, 3,4,5-trimethoxycinnamic acid.

III. Responses in Phenylpropanoid Metabolism to Low Temperature Treatments

It has been shown that exposure of some plant tissues to low non-freezing temperatures can stimulate phenolic metabolism. Lieberman *et al.* (1959) showed that low temperature storage of sweet potatoes led to a rise in the chlorogenic acid content of the tissue. Low temperature treatments also stimulate the formation of hydroxylated cinnamic acids in gherkin seedlings under certain conditions (Englesma, 1970) and of anthocyanins in apple skins (Creasy, 1968) and strawberry leaves (Creasy, 1974). In a number of these examples the low temperature effect involves a cold-induced stimulation of P.A.L. activity. We have found that the metabolism of chlorogenic acid is stimulated in a number of plant tissues, including tomato, potato and sweet potato, during cold treatments and that the temperature requirements for this stimulation have very precise limits.

Figure 7 shows changes in the chlorogenic acid content of potato tubers during storage at 0, 2 or 10 °C. At 0 °C there was a rapid increase in chlorogenic acid to a peak after 50 days and then a tendency to decline. At 2 °C there was slower rise to a peak at about 75 days while at 10 °C there was a slow decline in the chlorogenic acid levels over the 150 days of storage. Figure

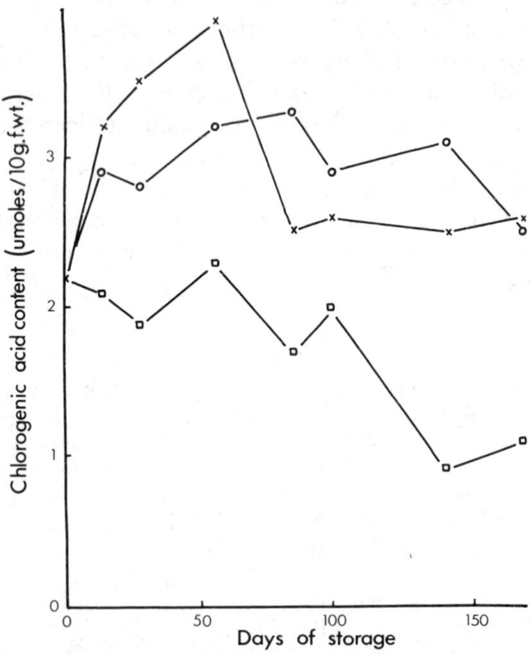

Fig. 7. Changes in the chlorogenic acid content of potato tubers stored at 0°C (× – ×), 2°C (○ – ○), and 10°C (□ – □).

8 shows changes in the specific activity of chlorogenic acid extracted from plugs of potato tuber tissue which had been incubated with ^{14}C-cinnamic acid. The plugs were prepared from tubers which had previously been stored for up to 56 days at either 0 or 10 °C. The specific activity of the chlorogenic acid fell during storage at 10 °C while at 0 °C there was a four to five-fold increase in the labelling of chlorogenic acid to a peak after 21–28 days. These results suggest that in tubers stored at below 2 °C there is a low temperature induced rise in the rate of chlorogenic acid biosynthesis.

Figure 9 shows changes in the activity of three enzymes of chlorogenic acid biosynthesis in potato tubers stored at either 0 °C or 10 °C. P.A.L. in potatoes stored at 10 °C was not detectable under our conditions of assay. However, at 0 °C there was a transitory rise in activity to a peak of about 0·5 nkat/20g fr. wt after about 8–10 days followed by a rapid fall back to very low levels. The level of *p*-coumarate CoA ligase was low (0·05 nkat/20g fr. wt) in tubers stored at 10 °C and remained so throughout the period of storage. However, there was a major rise in activity in tubers stored at 0 °C to a peak after about ten days. A

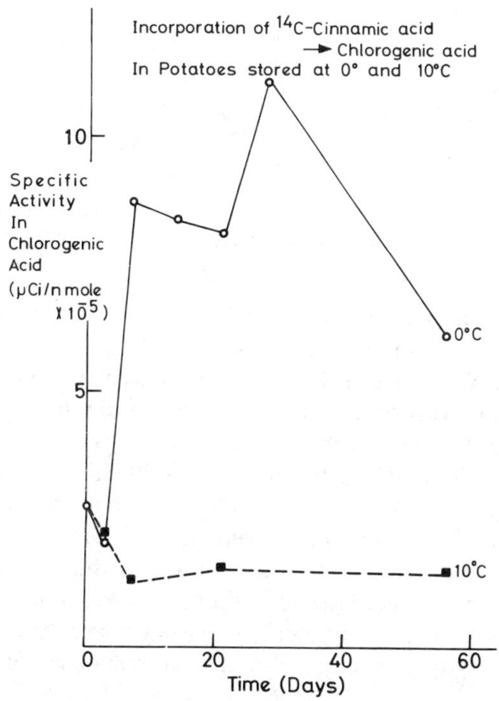

FIG. 8. Changes in the specific activity of chlorogenic acid in potato tissue taken from tubers which had previously been stored at 0 or 10°C for up to 56 days. At the various time intervals tubers were taken from storage, discs (10 mm × 2 mm diameter) were prepared and incubated for one hour in the presence of 1 μCi ^{14} C-cinnamic acid. After incubation the discs were extracted with methanol and the specific gravity of the chlorogenic acid was determined.

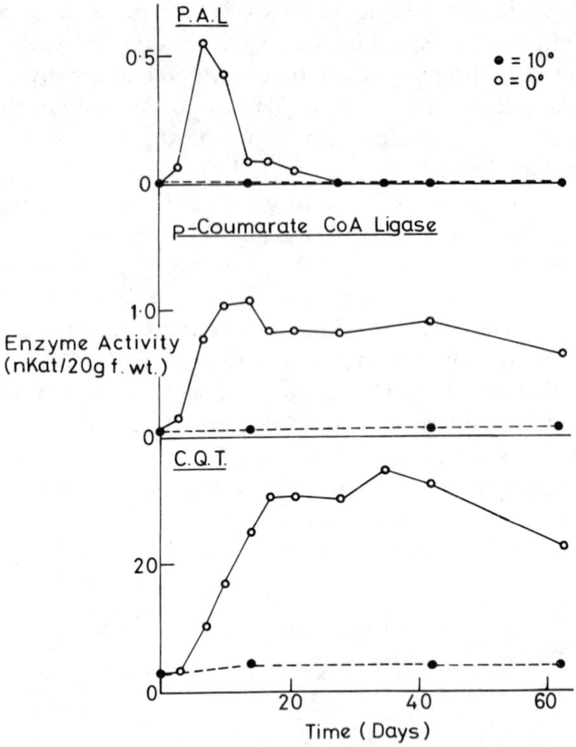

FIG. 9. Changes in the activity of three enzymes extracted from potato tubers which had been stored at either 0°C (continuous line) or 10°C (discontinuous line) for up to 62 days.

similar pattern of changes was observed in the enzyme, hydroxycinnamoyl CoA: quinate hydroxycinnamoyl transferase, C.Q.T. (Stockigt and Zenk, 1974; Rhodes and Wooltorton, 1976), which is involved in the final stages of chlorogenic acid biosynthesis. The level was low and fairly constant in tubers stored at 10 °C but there was a major rise in activity (approximately ten fold) to a peak of about 30 nkat/20g fr. wt after 35–40 days at 0 °C. As discussed elsewhere (Rhodes and Wooltorton, 1978b) the pattern of changes in the activities of these three enzymes in the cold induced stimulation of chlorogenic acid biosynthesis is different from that observed during the light induced formation of chlorogenic acid in excised potato discs. In particular, the large rise in C.Q.T. appears to be characteristic of low temperature induced changes and this led us to study in greater detail the responses of this activity to temperature.

Figure 10 shows changes in C.Q.T. activity in potatoes stored at 0, 2, 5 and 10 °C for up to 170 days. It can be seen that while there were no changes in C.Q.T. activity in tubers stored at 5 or 10 °C, large rises (4–13-fold) occurred during storage at 0 and 2 °C; the larger increase occurring at the lower

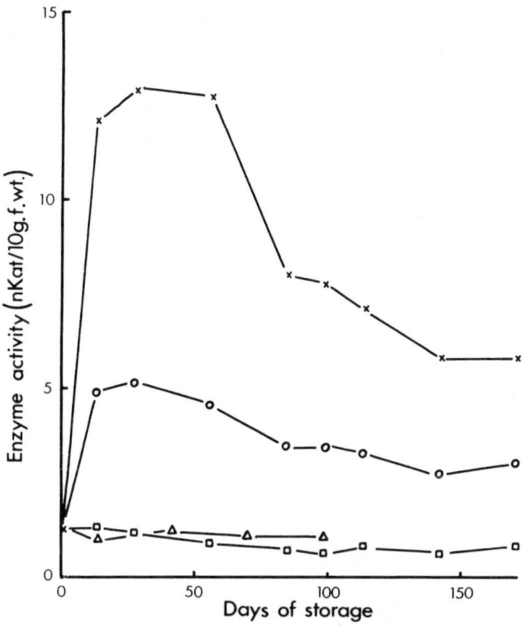

FIG. 10. Changes in the activity of C.Q.T. in potatoes which had been stored at either 0°C (× – ×), 2°C (○ – ○), 5°C (□ – □), or 10°C (△ – △), for up to 170 days.

temperature. These results suggest that there is a threshold temperature for the induction of the rise in C.Q.T. in potatoes between 2–5 °C. Figures 11 and 12 show similar data for the tomato fruit which also shows a low temperature stimulation of C.Q.T. activity. Both P.A.L. and C.Q.T. showed a significant rise in activity during storage at 2 °C, but not 20 °C, for tomatoes harvested at the "breaker" stage (Fig. 11) while both C.A.4H. and ligase tended to fall during storage at 2 °C. If fruit which had been stored at 2 °C was returned to 20 °C there was a rapid fall in both P.A.L. and C.Q.T. activity. Figure 12 shows changes in C.Q.T. activity in fruit stored at a range of temperatures and suggests that there is a threshold of about 10–12 °C below which there is stimulation of C.Q.T. activity. In tomato fruits there are major physiological changes occurring in the fruit during storage but it has been shown that the low temperature induced rise in C.Q.T. occurs at all stages of ripening and is thus independent of the ripening process (Rhodes and Wooltorton, 1977). Similar but less complete studies of changes in C.Q.T. activity have been made during storage of sweet potato roots and here a threshold for cold induced rise in C.Q.T. was shown to occur between 7·5 and 15 °C (Rhodes and Wooltorton, 1978b).

In potatoes, the degree of response in C.Q.T. activity on transfer to 0 °C is to some extent affected by the prior treatment of the tubers. In all the previous experiments with potatoes described, the tubers were stored at 10 °C from

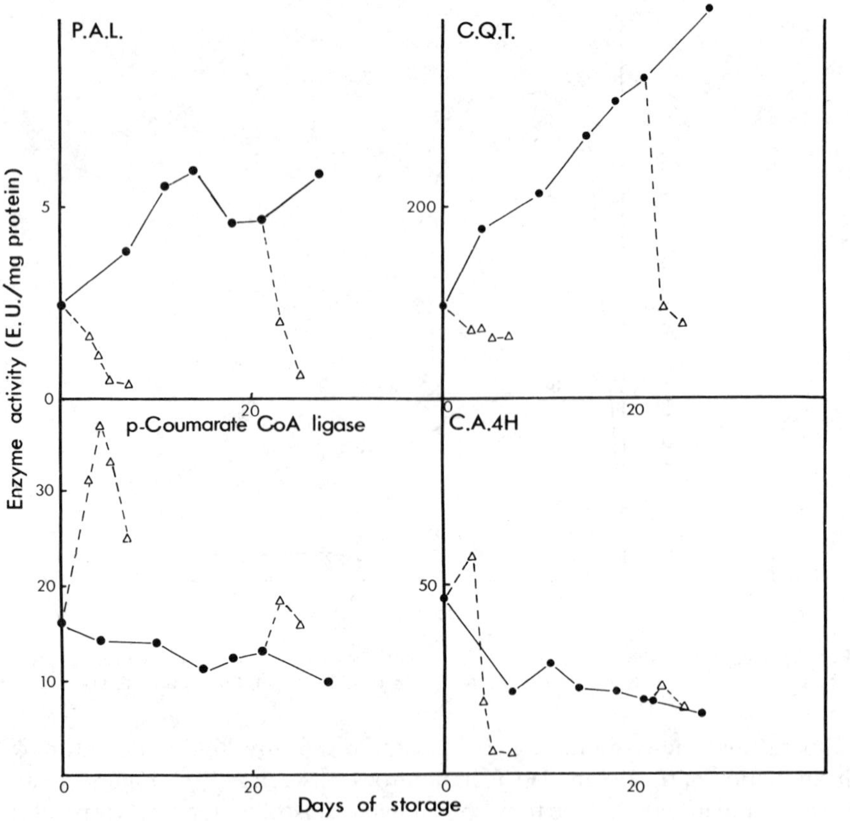

FIG. 11. Changes in the activity of four enzymes in tomato fruits picked at the breaker stage and subsequently stored for up to 30 days at either 20°C (△·—·△), or at 2°C (●—●). A sample of fruit which had been at 2°C for 21 days was subsequently returned to 20°C.

harvest and this material (see Fig. 13), when transferred to 0 °C, gave large rises in C.Q.T. activity though the time course of these changes altered during the season. However, if this material after a short period at 10 °C was stored at 5 °C, a procedure which has no effect in inducing a rise in C.Q.T. activity, the ability to respond to transfer to 0 °C was severely reduced as the storage season progressed (see Fig. 13). Thus, the response in C.Q.T. appears not only to be a response to the absolute temperature to which the tubers are exposed but also is conditioned by the extent of the drop in temperature. The precise temperature requirements for the cold induced changes in C.Q.T. activity and the fact that the changes in activity are reversible when the temperature is raised pose interesting problems in the role of low temperature in regulating enzyme levels. The fact that fully ripe as well as immature tomatoes responded to give the rise in activity at 2 °C, initially led us to suggest that the rise in

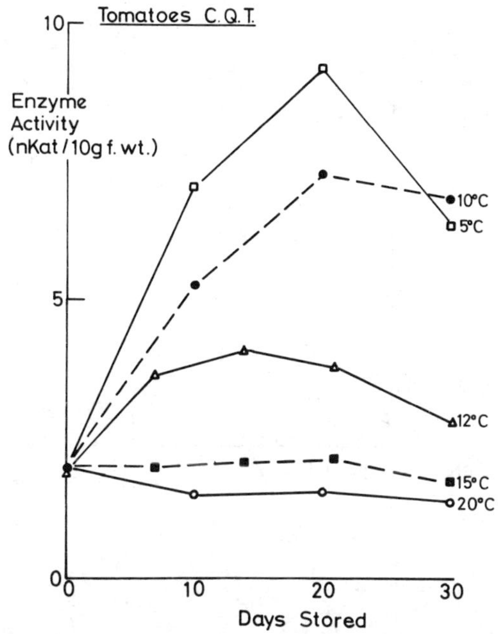

FIG. 12. Changes in the activity of C.Q.T. in "breaker" tomatoes stored at five different temperatures for up to 30 days.

activity might represent the activation of a pre-existing inactive enzyme since fully ripe tomatoes have a very low capacity for protein synthesis (De Swardt *et al.*, 1973). Attempts to activate extracts of unchilled tomatoes by treatment with hydrogen bond breaking reagents, mild detergents and proteolytic enzymes all proved negative. Experiments with inhibitors of protein synthesis are difficult with a bulky organ such as the tomato but some success was obtained using cycloheximide and puromycin infiltrated into the fruit by way of the peduncle. Fruit was picked at the breaker stage with a length of the peduncle still attached. The peduncle was broken at the abscission zone and the fruit was placed facing downwards with the cut tip under a solution of the inhibitor or water held in a small glass well. Fruits were held at 15 °C for 24 h and then removed from the inhibitor and transferred to 2 °C. Table VI shows that concentrations of cycloheximide of between 10^{-2}–10^{-3}M almost completely prevent the low temperature induced rise in C.Q.T. While puromycin at 10^{-2}–10^{-3}M caused a 50% inhibition of the increase in activity. Estimates of the amounts of the liquids taken up during infiltration suggest that about 0·1 ml solution enters the fruit and this suggests the average concentration in the inhibitor in the tissue would be of the order of 10^{-6}–10^{-7}M. The specificity of inhibitors such as cycloheximide and puromycin is open to question since they can also have secondary effects on

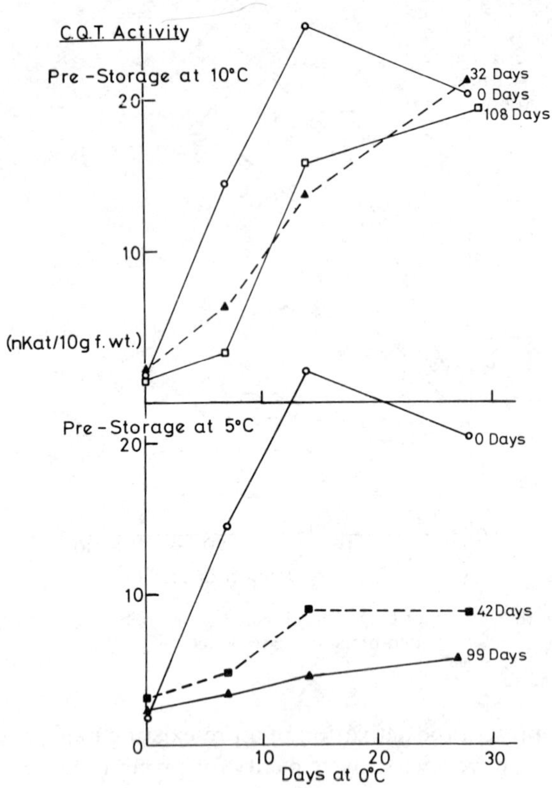

FIG. 13. The effect of previous storage history on the response in C.Q.T. activity on transfer of potato tubers to 0°C. Potato tubers after curing (at 10°C) were either stored at 5 or 10°C and under these conditions the activity of C.Q.T. remains constant. The effect of transferring these tubers to 0°C was studied after various periods of storage at 5 or 10°C (as indicated on the figure).

processes such as respiration and ion uptake (Ellis and MacDonald, 1970). Thus, it is not possible on such data alone to conclude that the rise in C.Q.T. is due to *de novo* protein synthesis. In a number of cases, increases in the activity of enzymes of phenylpropanoid metabolism (particularly P.A.L.) have been shown to be either insensitive to cycloheximide or stimulated by the presence of the inhibitor (Smith, 1972) and it has been shown that this is due to the activation of an inactive form of the enzyme (Attridge and Smith, 1974; Schopfer, 1977). Thus although it is not possible to conclude on the basis of our experiments that the rise in C.Q.T. is due to *de novo* synthesis, they do seem to make a mechanism based on the activation of an inactive enzyme less likely.

TABLE VI
Effect of cycloheximide (CH) and puromycin on changes in C.Q.T. activity in tomatoes stored at 2°C

	C.Q.T. activity (nkat/10 g fr. wt) Treatments				
	Control	H_2O	10^{-2}M CH	10^{-3}M CH	10^{-4}M CH
Expt 1					
Initial fruit	2·4	—	—	—	—
Fruit stored at 2°C for 13 days	8·0	9·0	2·0	3·6	6·6
Fruit stored at 2°C for 19 days	—	9·7	1·8	2·9	6·0
	Control	H_2O	10^{-2}M Puromycin	10^{-3}M Puromycin	
Expt 2					
Initial fruit	2·4	—	—	—	
Fruit stored at 2°C for 20 days	—	8·2	5·6	6·7	
Fruit stored at 2°C for 20 days	—	8·3	6·5	6·0	

Futher attempts were made to investigate the low temperature induced rise in C.Q.T. using plugs of tomato wall and potato tuber tissue incubated in the cold. Problems due to wounding effects on phenolic metabolism were minimized by the use of relatively large plugs of tissue (10 mm diameter plugs of the tomato wall and 10 mm diameter × 10 mm thick plugs of potato tissue). Incubation of tomato plugs led to a large rise in C.Q.T. over 10 days of incubation at 5 °C but not at 15 °C. Table VII shows that cycloheximide at a concentration between 10^{-5}–10^{-6}M completely prevented the rise in C.Q.T. while puromycin and the inhibitor of nucleic acid synthesis, actinomycin D inhibited the effect to a lesser extent. Incubation of potato plugs at 0 °C, but not at 5 °C, led to a large rise in C.Q.T. activity during incubation. Table VIII shows that the rise at 0 °C was sensitive to a range of inhibitors of protein synthesis including cycloheximide and puromycin. These experiments with excised tissue tend to confirm the work with whole tomatoes and show that the cold induced rise in C.Q.T. is very sensitive to inhibitors such as cycloheximide.

The C.Q.T. of cold treated potatoes has been purified to near homogeneity by a method involving affinity chromatography on blue Sepharose with specific desorption with CoA. The purified fraction of C.Q.T. from chilled tubers was resolved into three isoenzymic forms. Extracts from unchilled tubers contained two of these isoenzymic forms but at lower concentrations than the chilled tissue. However, the major fraction of C.Q.T. which eluted first from the column in extracts of chilled tubers was absent from unchilled tubers (Rhodes *et al.* 1979).

Lamb (1977) has shown in discs of potato tubers aged at higher temperatures that treatment with cinnamic acid could induce a rise in C.Q.T. activity, and thus the possible involvement of pathway intermediates in

TABLE VII
Effect of inhibitors of protein synthesis on changes in C.Q.T. activity during ageing of tomato plugs at 5°C

	C.Q.T. activity (nkat/10 g fr. wt)
Expt 1	
Initial discs	1·44
Aged discs—5°C	
7 days Control	3·32
10^{-5}M CH	1·36
10^{-6}M CH	1·48
10 days Control	2·42
10^{-5}M CH	1·45
10^{-6}M CH	1·48
Expt 2	
Initial discs	1·05
Aged discs—5°C	
7 days Control	3·60
0·5 mM Puromycin	2·31
40 μg/ml Actinomycin D	1·59

controlling changes in C.Q.T. activity was examined in both tomato discs and potato plugs aged in the cold. In both systems (see Table IX) a range of cinnamic acids were shown to reduce the rise in C.Q.T. in the cold. The most effective of these intermediates was cinnamic acid and at 2 mM this acid almost completely prevented the rise in C.Q.T. Since it has been shown that cinnamic acid is an inhibitor of C.Q.T. (Rhodes *et al.*, 1979) care was taken

TABLE VIII
Effect of inhibitors on changes in C.Q.T. level in potato plugs incubated at 0°C

	C.Q.T. activity (nkat/10 g fr. wt)
Initial discs	0·3
14 day aged plugs	
Control	19·9
1 mM Fluoro-β-phenylalanine	8·6
0·5 mM Puromycin	6·4
32 μM Actinomycin D	9·8
10^{-4}M Cycloheximide	0·33
10^{-5} Cycloheximide	0·3
10^{-6} Cycloheximide	0·5
10^{-7} Cycloheximide	10·2

TABLE IX
Effect of cinnamic acid derivatives on changes in C.Q.T. activity in tomato discs and potato plugs aged in the cold

	nkat/10 g tissue
(1) Tomato discs	
Freshly prepared	1·74
Aged at 5°C for 7 days	
Control	4·95
Cinnamic acid 0·5 mM	3·76
1 mM	2·84
2 mM	1·70
p-Coumaric acid 1 mM	3·40
Caffeic acid 1 mM	3·86
Ferulic acid 1 mM	4·51
Sinapic acid 1 mM	3·48
	nkat/20 g tissue
(2) Potato plugs	
Freshly prepared	0·48
Aged at 0°C for 14 days	
Control	16·1
Cinnamic acid 0·5 mM	10·1
1 mM	3·9
2 mM	1·5
p-Coumaric acid 1 mM	8·6
Caffeic acid 1 mM	10·4
Ferulic acid 1 mM	8·5
Sinapic acid 1 mM	15·8

to remove any residual cinnamic acid from the tissue extracts by chromatography on Sephadex G25 prior to assay for C.Q.T. The effect of cinnamic acid on the rise in C.Q.T. was reversible if the discs were returned to medium free of cinnamic acid. In a further experiment on the potato plugs the effects of cinnamic, *p*-coumaric and ferulic acids on the time course of changes in activity of three enzymes were studied (see Table X). All three acids caused a major inhibition of increase in P.A.L. but had smaller effects on both *p*-coumarate CoA ligase and C.Q.T. In all cases, cinnamic acid was the most effective inhibitor of P.A.L., ligase and C.Q.T. but the rise in C.Q.T. was relatively insensitive to the inhibitors. Indeed, in short periods of exposure the three acids tended to stimulate C.Q.T. activity, although in longer periods of exposure they were inhibitory. These results suggest that cinnamic acid intermediates block the increase in activity of P.A.L. but their accumulation may, at least in the short-term, stimulate C.Q.T. activity. The basic questions

TABLE X

Effect of cinnamic acid derivatives on time course of changes in P.A.L., *p*-Coumarate CoA ligase and C.Q.T. activity in potato plugs incubated at 0°C

	P.A.L.	Ligase (nkat/20 g fr. wt)	C.Q.T.
Initial discs	0	0·48	0·72
Aged 3 days at 0°C			
Control	0·047	0·27	1·25
+1 mm Cinnamic acid	0	0·10	2·77
+1 mm *p*-Coumaric acid			1·54
+1 mm Ferulic acid	0	0·20	2·04
Aged 7 days at 0°C			
Control	0·85	0·92	10·50
+1 mm Cinnamic acid	0	0·08	6·90
+1 mm *p*-Coumaric acid	0·10	0·52	6·5
+1 mm Ferulic acid	0·06	0·37	15·1
Aged 10 days at 0°C			
Control	2·60	2·12	40·8
+1 mm Cinnamic acid	0	0·09	12·0
+1 mm *p*-Coumaric acid	0·24	0·81	19·2
+1 mm Ferulic acid	0·07	0·39	22·8

on the mechanism of the cold induced rise in C.Q.T. still remain unanswered but it may well be that synthesis rather than the activation of an inactive form is involved.

REFERENCES

Asada, Y. and Matsumoto, I. (1969). *Phytopath. Soc. Japan* **35**, 160–167.

Attridge, T. H. and Smith, H. (1974). *Biochim. Biophys. Acta* **343**, 452.

Ben Abdelkader, A., Cherif, A., Demandre, C. and Mazliak, P. (1973). *Europ. J. Biochem.* **32**, 155.

Benveniste, I., Salaun, J-P. and Durst, F. (1977). *Phytochemistry* **16**, 69.

Billett, E. E. and Smith, H. (1978). *Phytochemistry* **17**, 1511.

Billett, E. E., Smith, D., Sanderman, H. J. and Grisebach, H. (1977). *Archi. Biochim. Biophys.* **178**, 1511.

Billett, E. E., Wallace, W. and Smith, H. (1978). *Biochim. Biophys. Acta* **524**, 219.

Buche, T. and Sanderman, H. J. (1973). *Arch. Biochem. Biophys.* **158**, 445.

Camm, E. L. and Towers, G. H. N. (1973). *Phytochemistry* **12**, 961.

Cotte-Martinon, M. G., Yahiel, V. and Ducet, G. (1974). *Phytochemistry* **13**, 2085.

Creasy, L. L. (1968). *Proc. Am. Soc. Hort. Sci.* **93,** 716.

Creasy, L. L. (1974). *Phytochemistry* **13,** 1391.

De Swardt, G. H., Swanepoel, J. H. and Duvenage, A. J. (1973). *Z. Pflanzephysiol.*, **Bd 70**, S358.

Duchesne, M., Fritig, B. and Hirth, L. (1977). *Biochim. Biophys. Acta* **485**, 465.
Ebel, J. and Hahlbrock, K. (1977). *Europ. J. Biochem.* **75**, 201.
Ebel, J., Schaller-Hekeler, B., Knobloch, K. H., Wellmann, E., Grisebach, H. and Hahlbrock, K. (1974). *Biochim. Biophys. Acta* **362**, 417.
Ellis, R. J. and MacDonald, I. R. (1970). *Plant Physiol.* **46**, 227.
Englesma, G. (1970). *Planta* **91**, 246.
Grisebach, H. (1977). *Naturwissenschaften* **64**, 619.
Gross, G. G. (1977). *Recent Adv. Phytochem.* **11**, 141.
Gross, G. G. and Zenk, M. H. (1974). *Europ. J. Biochem.* **42**, 453.
Gunsalus, I. C., Pederson, T. C. and Seigar, S. G. (1975). *Ann. Rev. Biochem.* **44**, 377.
Hahlbrock, K. (1976). *Europ. J. Biochem.* **63**, 137.
Hahlbrock, K. and Grisebach, H. (1979). *Ann. Rev. Plant Physiol.* **30**, 105.
Hahlbrock, K. and Wellmann, E. (1973). *Biochim. Biophys. Acta* **304**, 702.
Hill, A. C. (1975). PhD. Thesis, University of East Anglia.
Hill, A. C. and Rhodes, M. J. C. (1975). *Phytochemistry* **14**, 2387.
Imaseki, H., Asahi, T. and Uritani, I. (1968). *In* "Biochemical Regulation in Diseased Plants and Injury" (T. Hirai, Ed.), pp. 189–201. Kyritsu Printing Co., Tokyo.
Knobloch, K. H. and Hahlbrock, K. (1975). *Europ. J. Biochem.* **52**, 311.
Knobloch, K. H. and Hahlbrock, K. (1977). *Arch. Biochem. Biophys.* **184**, 237.
Koukol, J. and Conn, E. E. (1961). *J. Biol. Chem.* **236**, 2692.
Lamb, C. J. (1977). *FEBS Letters* **75**, 37.
Lieberman, M., Craft, C. A. and Wilcox, M. S. (1959). *Proc. Am. Soc. Hort. Sci.* **74**, 642.
Margna, U. (1977). *Phytochemistry* **16**, 419.
Meehan, T. D. and Coscia, C. J. (1973). *Biochem. Biophys. Res. Comm.* **53**, 1043.
Murphy, P. J. and West, C. A. (1969). *Arch. Biochem. Biophys.* **133**, 395.
Pflander, R., Scheel, D., Sanderman, H. Jr and Grisebach, H. (1977). *Arch. Biochem. Biophys.* **178**, 315.
Potts, J. R. M., Weklych, R. and Conn, E. E. (1974). *J. Biol. Chem.* **254**, 57.
Rhodes, M. J. C. and Wooltorton, L. S. C. (1971). *Phytochemistry* **10**, 1989.
Rhodes, M. J. C. and Wooltorton, L. S. C. (1973a). *Phytochemistry* **12**, 107.
Rhodes, M. J. C. and Wooltorton, L. S. C. (1973b). *Qual. Plant. Mat. Veg.* **23**, 145.
Rhodes, M. J. C. and Wooltorton, L. S. C. (1973c). *Phytochemistry* **12**, 2381.
Rhodes, M. J. C. and Wooltorton, L. S. C. (1976). *Phytochemistry* **15**, 947.
Rhodes, M. J. C. and Wooltorton, L. S. C. (1977). *Phytochemistry* **16**, 655.
Rhodes, M. J. C. and Wooltorton, L. S. C. (1978a). *In* "Biochemistry of Wounded Plant Tissue" (G. Kahl, Ed.), pp. 243–286. W. de Gruyter, Berlin.
Rhodes, M. J. C. and Wooltorton, L. S. C. (1978b). *Phytochemistry* **17**, 1225.
Rhodes, M. J. C., Hill, A. C. and Wooltorton, L. S. C. (1976). *Phytochemistry* **17**, 1225.
Rhodes, M. J. C., Wooltorton, L. S. C. and Lourenco, E. J. (1979). *Phytochemistry* **18**, 1125.
Rich, P. R. and Bendall, D. S. (1975). *Europ. J. Biochem.* **55**, 333.
Rich, P. R. and Lamb, C. J. (1977). *Europ. J. Biochem.* **72**, 353.
Rich, P. R., Cammack, R. and Bendall, D. S. (1975). *Europ. J. Biochem.* **59**, 281.
Russell, D. W. (1971). *J. Biol. Chem.* **246**, 3870.
Russell, D. W., Conn, E. E., Sulter, A. and Grisebach, H. (1968). *Biochim. Biophys. Acta* **70**, 210.
Sarkanen, K. V. and Ludwig, G. H. (1971). *In* "Lignins, Occurrence, Formation, Structure and Reactions". Wiley-Interscience, New York.
Schenkman, J. B., Remmer, H. and Estabrook, R. W. (1967). *Mol. Pharmacol.* **3**, 113.
Schopfer, P. (1977). *Ann. Rev. Plant Physiol.* **28**, 223.
Schroder, J., Kreuzaler, F., Schafer, E. and Hahlbrock, K. (1979). *J. Biol. Chem.* **254**, 57.

Smith, H. (1972). in *Phytochem. Soc. Symposia Series No. 9*, 303.
Stafford, H. (1974). *Ann. Rev. Plant Physiol.* **25**, 459.
Stafford, H. (1976). *Plant Physiol.* **57**, 320.
Stockigt, J. and Zenk, M. H. (1974). *FEBS Letters* **42**, 131.
Uritani, I. and Oba, K. (1978). *In* "Biochemistry of Wounded Plant Tissue" (G. Kahl, Ed.), pp. 287–308. W. de Gruyter, Berlin.
Vaughan, P. F. T. and Butt, V. (1969). *Biochem. J.* **113**, 109.
Yahiel, V., Cotte-Martinon, M. G. and Ducet, G. (1974). *Phytochemistry* **13**, 1649.

CHAPTER 12

Anthocyanins in Fruits and Vegetables

C. F. TIMBERLAKE

Long Ashton Research Station, Bristol, England

I. Introduction 221
II. Chemistry and Structural Transformations 222
III. Stabilization and Colour Augmentation 225
IV. Biosynthesis 229
V. Vegetable Anthocyanins 233
A. Anthocyanins of the Cruciferae 233
B. Anthocyanins of the Umbelliferae 234
C. Anthocyanins of Other Vegetables 235
VI. Fruit Anthocyanins 236
A. Anthocyanins of the Rosaceae 236
B. Anthocyanins of the Saxifragaceae 238
C. Anthocyanins of the Vitaceae 238
D. Anthocyanins of the Moraceae 241
E. Anthocyanins of the Ericaceae 242
F. Anthocyanins of Other Fruits 242
References 244

I. Introduction

The anthocyanins are the largest group of water-soluble pigments in plants. Much is already known of their chemistry and properties but there are many features, particularly those relating to the physico–chemical basis of flower, fruit and vegetable colour, which are not yet fully understood. In order to provide a background for discussion of anthocyanins in fruits and vegetables, recent work on structural transformations of the anthocyanins *in vitro* is described first and is followed by an account of the mechanisms by which anthocyanin pigments are thought to be stabilized *in vivo*. In this and in the subsequent very condensed summary of their biosynthesis, emphasis is given to flower pigments on which most work is being carried out. But their discussion is a necessary prelude to observations on fruit and vegetable pigments and helps to put the latter in proper perspective.

The physiology, function, metabolism, etc. of anthocyanins as part of the flavonoid group have been reviewed recently (Harborne, 1976a; McClure, 1975; Barz and Hösel, 1975). The final negative answer to the controversial contribution of photosynthesis to high intensity response of anthocyanin accumulation seems to have been given (Tselas *et al.*, 1979). It may be "unfortunate that the vast majority of research on photocontrols of flavonoids has been done on anthocyanins which are easy for the researcher to quantitate but more complex for the plant to synthesise" (McClure, 1975). It is certainly easy to obtain values for the absorbance of coloured acid extracts of plant materials which contain anthocyanins. However these measurements, usually termed "total pigment" figures, give little information about the individual components of what are usually mixtures of anthocyanins. The major difficulty in quantifying the components of mixtures of anthocyanins is that the compounds can occur in many forms, some of them colourless. In this chapter, it has been possible to mention only a few of the most recent findings on physiological aspects since an attempt has been made to introduce different facets of the anthocyanins where they impinge on the fruit or vegetable under discussion. Likewise, the identifications and distribution of the anthocyanins are not exhaustive, but have had to be limited to certain recent findings, selected for their interest.

II. Chemistry and Structural Transformations

The common anthocyanidins and their glycosides (anthocyanins) are depicted in Fig. 1. Others have been fully documented elsewhere (Harborne, 1967; Timberlake and Bridle, 1975, 1980). Anthocyanin sugars include (a) monosides—glucose, galactose, rhamnose and arabinose, (b) biosides—rutinose, sophorose, gentiobiose, sambubiose, lathyrose, laminariobiose and robinobiose; and (c) triosides—linear and branched chain. The sugars at position 3 can be acylated, usually with phenolic acids such as *p* coumaric, caffeic, ferulic or sinapic acids and occasionally, malonic, acetic and *p*-hydroxybenzoic acids.

An intriguing feature of the anthocyanins is the way that their colours vary with time and with pH. The anthocyanins are red in acid solutions but the colour decreases as the pH is raised. Neutral and alkaline solutions are violet or blue when freshly made, but fade slowly. The reasons for such changes have been studied extensively but it is only very recently that the precise natures, kinetics and equilibria of the chemical structures which are involved have been demonstrated unequivocally (Brouillard and Dubois, 1977; Brouillard and Delaporte, 1977; Brouillard and Delaporte, 1978; Brouillard *et al.*, 1978). The four structures which can exist are illustrated in Fig. 2 for malvidin 3-

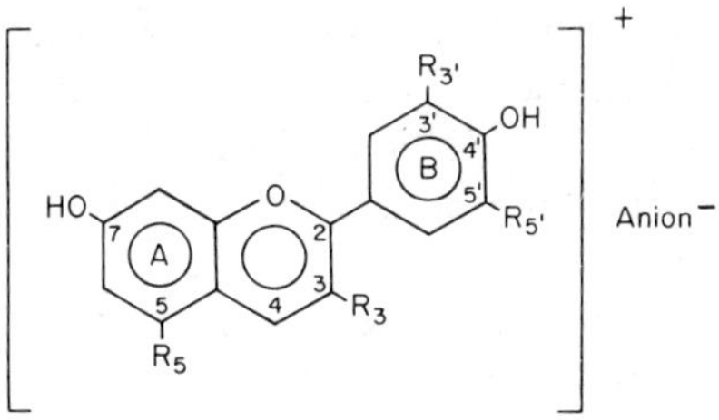

Anthocyanidins : R_3=OH , R_5=OH

		$R_{3'}$	$R_{5'}$
Pelargonidin	(Pg)	H	H
Cyanidin	(Cy)	OH	H
Peonidin	(Pn)	OCH_3	H
Delphinidin	(Dp)	OH	OH
Petunidin	(Pt)	OCH_3	OH
Malvidin	(Mv)	OCH_3	OCH_3

Anthocyanins: Pg,Cy,Pn,Dp,Pt,Mv with
R_3=O−sugar or O−acylated sugar
R_5=OH or O−glucose

FIG. 1. The common anthocyanidins and anthocyanins.

glucoside. Formerly it was assumed that the quinoidal-base (A) arose by dehydration of the carbinol pseudo-base (B); hence it was usually called an anhydro-base. This term should now be discontinued following the finding that the quinoidal base (A) is produced directly, by proton transfer, from the flavylium cation (AH^+). The latter also hydrates to a carbinol pseudo-base (B) which itself exists in tautomeric equilibrium with small amounts of chalcone (C). Changes in the relative proportions of these structures affect

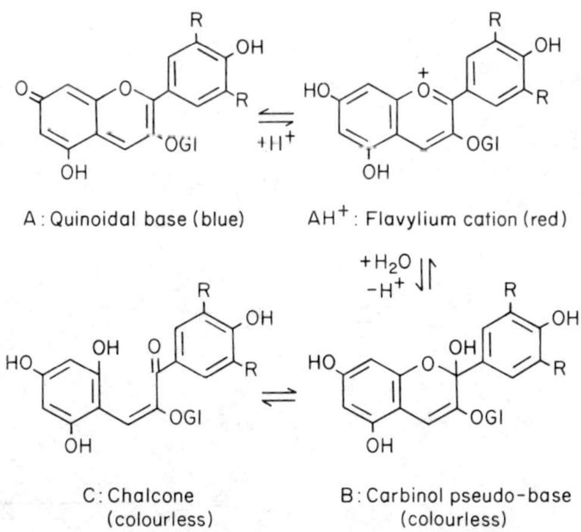

FIG. 2. Anthocyanin structural transformations with pH for malvidin 3-glucoside (25°C; 0·2 M ionic strength) (Brouillard and Delaporte, 1977).

colour because the quinoidal base (A) is violet–blue, the flavylium cation (AH^+) is red, the carbinol pseudo-base (B) is colourless and the chalcone is practically colourless (λ_{max} 340–350 nm, Brouillard *et al.*, 1979). The rates of the various equilibria are all measurable and different, the slowest being that between the carbinol pseudo-base and chalcone. This may not be of great significance at room temperature when the proportion of chalcone is small, but it assumes greater importance at elevated temperatures when all forms tend to be converted to the chalcone because of the endothermic nature of the $A \rightarrow AH^+ \rightarrow B \rightarrow C$ sequence of reactions. The colour of the anthocyanin will fade, not because of pigment degradation (although this can occur, depending upon the conditions) but simply on account of the changing equilibria. On cooling, sufficient time should be allowed for the chalcone to reform the coloured structures; the colour will then slowly return. Although colour changes induced in ripening fruit (apples, grapes) by environmental temperature variations are likely to be due to changes in the underlying biochemical enzyme activities, the effect of temperature on the equilibria between the different pigment forms should perhaps not be completely discounted.

The distributions of the various structures of malvidin 3-glucoside and malvidin 3, 5-diglucoside are represented in Fig. 3, compiled from the results of Brouillard (loc. cit.). Because of the reduced concentration of the red

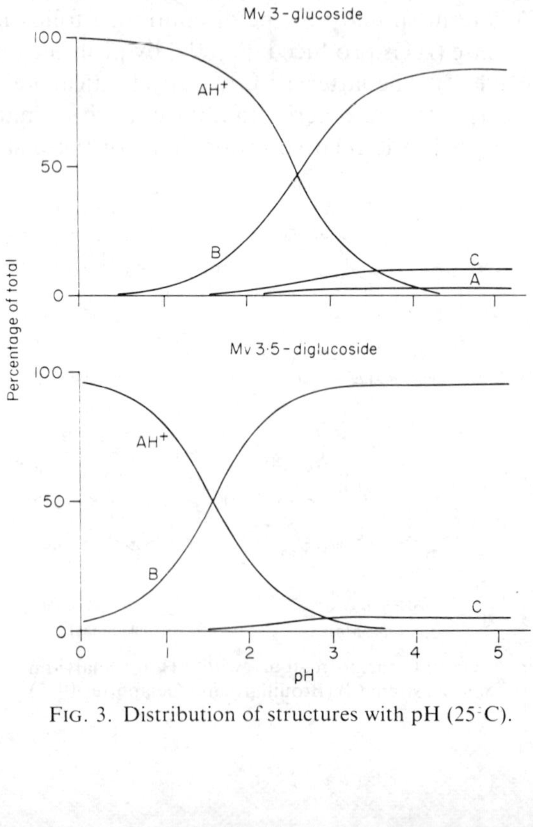

FIG. 3. Distribution of structures with pH (25°C).

flavylium cation the diglucoside is less coloured that the monoglucoside at pH 1–4. This is a consequence of the *pK* value of the hydration reaction $AH^+ + H_2O \rightleftharpoons B + H_3O^+$ which is approximately one unit lower for 3,5-diglucosides than 3-glucosides (Timberlake and Bridle, 1967). With increasing pH, the colours of both glycosides decrease. At pH 5, malvidin 3-glucoside exhibits only a faint purple colour, due to very small proportions of the quinoidal base (A) which persists. The colour of malvidin 3,5-diglucoside at equilibrium at pH 5 is even less; the proportions of quinoidal-base (A) are negligible (Fig. 3). Thus solutions of anthocyanin 3,5-diglycosides can be colourless at pH 4–5·5 at the concentrations at which they are usually handled *in vitro* (Timberlake and Bridle, 1967) and provided that no other organic compounds (apart from buffer) are present. Thus an explanation must be sought for the many vivid colours of fruits, flowers, leaves, etc., the pH values of which are such that little colour would be expected.

III. Stabilization and Colour Augmentation

In nature, colour in much plant material is made possible by operation of physico–chemical phenomena which stabilize and augment the coloured anthocyanin structures. A property peculiar to anthocyanins in the pH region 3–4 is that colour increases more than linearly with increase in pigment concentration (Scheffeldt and Hrazdina, 1978). At the high *in vivo* concentrations in flower petals (as great as 10^{-2}M; Asen *et al.*, 1972) the cationic form (AH^+) is stabilized and its colour becomes much more evident. For lack of a better explanation, this effect has been attributed to "self-association" of the anthocyanin molecules; but its true cause remains obscure, since the careful work of Brouillard (loc. cit. and personal communication) has failed to reveal anthocyanin species additional to the four already described.

In natural products the anthocyanins do not occur in isolation but in conjunction with other organic components, which may act as co-pigments and with metals, with which they may form complexes. Apparently, co-pigmentation has a bathochromic effect on the peak wavelength of the flavylium cation, but does not increase its absorbance. Quinoidal bases are stabilized more prominently, with increased absorbance as well as blue shifts. Typical *in vitro* results reproduced from the work of Asen *et al.* (1972) showed that *C*-glucosylflavones such as swertisin are effective co-pigments (Table I). The mechanism is believed to be inter-molecular hydrogen bonding between anthocyanin quinoidal groups and the 7 and 4′-phenolic groups of the flavonoid co-pigment.* The phenolic group in the 5 position of flavones (likewise the 3-OH group in flavonols) may be less involved, since it is already hydrogen bonded to the 4-carbonyl group (Simpson and Garden, 1952a,b).

* Sweeny, J. G., Wilkinson, M. M. and Iacokucci, G. A. ((1981). *J. Agric. Food Chem.* **29**, 563) now support the mechanism proposed by Goto *et al.* (1979).

TABLE I
Colour augmentation by co-pigmentation (after Asen *et al.*, 1972)

Cyanidin 3,5-diglucoside (0·002 M), pH 3·32 Alone λ_{max} = 508 nm, A^{1mm}_{max} = 0·500		
Co-pigment (0·006 M)	$\Delta\lambda_{max}$	A_{max} increased by
Flavan 3-ols ((+)-catechin)	6	0·8 times
Caffeic acid	7	0·6 times
Flavonols		
quercetin	19	2·2 times
rutin	20	2·3 times
Aurisidin	32	3·3 times
C-glycosyl flavones (swertisin)	33	4·7 times

Swertisin

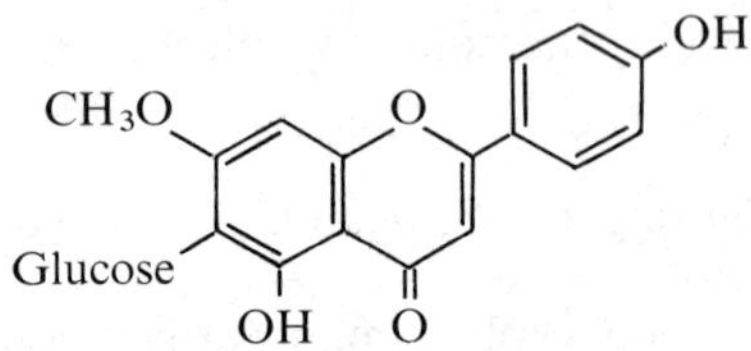

Colour augmentation by co-pigmentation increases with increasing anthocyanin concentration and increasing ratio of co-pigment to anthocyanin. However, it is limited in very strong pigment solutions by "self association" of the anthocyanin cationic forms which effectively reduces the amounts of anthocyanin available for co-pigmentation (Scheffeldt and Hrazdina, 1978).

Recently, a co-pigmentation constant (K_c) has been defined as the equivalent of co-pigment causing an absorbance increase of one-half the maximum obtained with that co-pigment (Hoshino *et al.*, 1980). The K_c values of delphinidin 3-glucoside and delphinidin 3,5-diglucoside were very similar at pH 6·0 indicating that the glucose moiety at position 5 has little effect on the strength of co-pigmentation. Likewise at pH 3·2, in the presence of co-pigment rutin, the absorbance of malvidin 3,5-diglucoside was similar (a little lower) to that of malvidin 3-glucoside (Scheffeldt and Hrazdina, 1978). However, the *relative* increase in absorbance of the diglucoside caused by co-pigmentation was much greater than that of the monoglucoside. This might be explained by the much lower concentration of quinoidal base in solutions of the diglucoside than the monoglucoside (Fig. 3). Thus, the ratio of co-pigment to quinoidal base was much greater and the relative effect of co-pigmentation on the diglucoside was more pronounced.

Hoshino *et al.* (1980) also studied the effect of an acyl group in the anthocyanin on co-pigmentation at pH 6·0 by comparing the behaviour of malvidin 3-*p*-coumaroylglucoside-5-glucoside with that of malvidin 3,5-

glucoside using swertisin 4′-glucoside (flavocommelin) as co-pigment. The acyl moiety (*p*-coumaric acid) was found to increase the strength of the co-pigment complex and reduced its dissociation even in dilute solution. However in terms of its relative effect on colour, acylation of the anthocyanin had a more variable effect at pH 2–5 (Williams and Hrazdina, 1979). At pH 3·2, the effect of *p*-coumaric acid was to halve the percentage increase in colour (from 10 times to 5 times) of malvidin 3,5-diglucoside, but to augment the relative increase in colour (from 1·45 times to 5 times) of malvidin 3-glucoside, in the presence of rutin as co-pigment.

Metal chelation is a further factor capable of stabilizing anthocyanin quinoidal bases. Thus iron and/or aluminium appears to be essential for blueness in cornflowers (Bayer *et al.*, 1966; Saito and Hayashi, 1965; Asen and Jurd, 1967). Red cornflowers are red presumably because the pelargonidin glycosides which they contain are incapable of combining with metals and not because of deficiency of co-pigments. The blue pigment commelinin from petals of *Commelina communis* contains magnesium as a natural component (Hayashi and Takeda, 1970) but it seems from recent work that magnesium may not be essential for blueness. Thus Goto *et al.* (1979) propose that commelinin is a 1:1 complex in which the anthocyanin awobanin (delphinidin 3-*p*-coumaroylglucoside-5-glucoside) and flavocommelin (swertisin 4′-glucoside) are stacked together by hydrophobic interactions between the aromatic rings, with a covering by the sugar moieties which endows hydrophilic nature and water solubility. Commelinin, identical with the natural product, has been synthesized *in vitro* from its component parts (Takeda and Hayashi, 1977), as have similar blue metallic pigments with the magnesium replaced by other divalent metals (Takeda, 1977). Goto *et al.* (1979) have suggested that these complexes may be salts since commelinin has a negative charge. Takeda has also studied the effect of acylation (Takeda *et al.*, 1980 and personal communication). Cyanidin 3-*p*-coumaroylglucoside-5-glucoside (shisonin) in admixture with flavocommelin formed stable metallic complexes analogous to those involving awobanin. Complexes did not form when the anthocyanins were delphinidin 3,5-diglucoside (de-acylated awobanin) and cyanidin 3,5-diglucoside (de-acylated shisonin), indicating the importance of the *p*-coumaroyl moiety for stable complex formation. An intriguing additional finding was that the length and nature of the sugar-chain are also critical, since delphinidin 3-*p*-coumaroylrhamnoglucoside-5-glucoside (violanin) did not form stable metal complexes under the same conditions as did awobanin (Takeda *et al.*, 1980).

Lastly, there are some blue flower pigments which possess *in vitro* stability in isolation, without the need for co-pigmentation or metal complexing (Table II). Their remarkable stabilities have been attributed to some sort of intra-molecular interaction or bonding between the acylating groups and the phenolic groups of the B ring.* A comparative study under identical conditions

* But see footnote to Table II.

TABLE II
Stable blue flower pigments

Platycodon grandiflorum	Dp 3-(dicaffeoyl)rutinoside-5-glucoside	(Saito *et al.*, 1972)
Cineraria (*Senecio cruentus* D.C.)	Dp 3,7,3′-triglucoside + 2 mol caffeic acid	(Yoshitama and Hayashi, 1974; Yoshitama *et al.*, 1975)
Tradescantia reflexa	Dp 3,7,3′-triglucoside + 3 mol caffeic acid. Cy pigment (purple)	(Yoshitama, 1978)
Lobelia erinus	Dp 3-rutinoside-5,3′,5′-triglucoside + *p*-coumaric, caffeic and ferulic acids	(Yoshitama, 1977)
Ipomoea tricolour	Pn 3-(dicaffeoylsophoroside)-5-glucoside	(Asen *et al.*, 1977)

The exceptional colour stability of a further pigment from *Zebrina* has been explained (Brouillard (1981). *Phytochemistry* **20**, 143).

of these pigments has not yet been made, so it is not possible to assess the full implications of these compounds in terms of any order of stability dependent upon the numbers or natures of acylating groups or of the B ring substitution pattern. While most of the pigments are based on delphinidin, stability is shown even by the peonidin derivative from *Ipomea tricolour*, which lacks the *o*-hydroxyl grouping in the B ring. Indeed, the stability of this anthocyanin is such that it has warranted patenting as a food colour additive; its colours vary from vivid purple–red to deep blue within pH range 2–8 (Asen *et al.*, 1979).

It is noticeable that all of the pigments in Table II are di-glycosides, with glucose moieties in positions 5 or 7 as well as glycosylation in position 3. It seems possible that this may not be without significance and that the additional glucose molecules may play a part in stabilization of the quinoidal bases, perhaps in a manner similar to that suggested by Goto *et al.* (1979) for commelinin.

Acylated 3,7-diglycosides have not yet been identified in fruits and vegetables, but many acylated 3,5-diglycosides are known (Table III). Several of these pigments are violet or blue. It is appropriate to enquire whether their hues are a result of co-pigmentation or whether the anthocyanins themselves are intrinsically stable in the manner of the blue flower pigments just discussed. This is a matter for speculation at the moment but it emphasizes the need for further work in this area using mild, neutral extraction methods which enable the pigments to be isolated in the same states as they occur in the plants.

IV. Biosynthesis

Current opinion, based on work with labelled precursors and with flower mutants, is that dihydroflavonols (flavononols) which may be formed from flavanones by means of microsomal mixed-function oxidases, are the true intermediates in anthocyanin biosynthesis (Grisebach, 1979; Fig. 4). Glycosylation of anthocyanidins, such as cyanidin at position 3 is catalysed by enzymes using UDP-D-glucose as glycosyl donor, isolated from young red cabbage seedlings and cell cultures of *Haplopappus gracilis*. Similar 3-*O*-glycosyltransferases have been demonstrated recently in corollas of *Petunia hydrida* (Kho *et al.*, 1978) and petals of the Red Campion *Silene dioica* (Kamsteeg *et al.*, 1978a). The latter also contain enzymes catalysing (a) glucosylation of the 5-hydroxyl group of cyanidin-3-rhamnosylglucoside, and controlled by a single dominant gene M (Kamsteeg *et al.*, 1978b); (b) transfer of rhamnose from UDP-L-rhamnose to the glucose of cyanidin 3-glucoside and controlled by a single dominant gene N (Kamsteeg *et al.*, 1979a). Gene A controls anthocyanin formation in the petals; gene Ac governs the binding of cinnamic acid derivatives to the sugar at position 3; gene P determines whether the acyl moiety is *p*-coumaric acid or caffeic acid since it hydroxylates *p*-

TABLE III
Acylated 3,5-diglucosides in fruits and vegetables

Source	Anthocyanins	Reference
Aubergine	Dp 3-(*p*-coumaroylglucoside)-5-glucoside	Watanabe *et al*. (1966)
(*Solanum melongena*)	Dp 3-(*p*-coumaroylrutinoside)-5-glucoside	Tigchelaar *et al*. (1968)
	Dp 3-(caffeoyldiglucoside)-5-glucoside	Casoli and Dall'aglio (1969)
		Pifferi and Zamorani (1969)
Huckleberry	Pt 3-(*p*-coumaroylrutinoside)-5-glucoside	Francis and Harborne (1966)
(*Solanum intrusum*)	Pt 3-(di-*p*-coumaroylrutinoside)-5-glucoside	Saito *et al*. (1965)
	Mv 3-(*p*-coumaroylrutinoside)-5-glucoside	
Potato	Pg, Pn, Pt and Mv 3-(*p*-coumaroylrutinoside)-5-glucoside	Harborne (1960)
(*Solanum tuberosum*)	Pn 3-(feruloylrutinoside)-5-glucoside	Howard *et al*. (1970)
		Sachse (1973)
Sweet potato	Cy 3-(caffeoylsophoroside)-5-glucoside	Harborne (1967)
(*Ipomoea batatas*)	Cy and Pn 3-(dicaffeoyldiglucoside)-5-glucoside	Imbert *et al*. (1966)
Red cabbage	Cy 3-(*p*-coumaroylsophoroside)-5-glucoside	Hrazdina *et al*. (1977)
(*Brassica oleracea*)	Cy 3-(di-*p*-coumaroylsophoroside)-5-glucoside	
	Cy 3-(feruloylsophoroside)-5-glucoside	
	Cy 3-(diferuloylsophoroside)-5-glucoside	
	Cy 3-(sinapoylsophoroside)-5-glucoside	
	Cy 3-(disinapoylsophoroside)-5-glucoside	
	Cy 3-(malonylsophoroside)-5-glucoside	
Radish	Pg and Cy 3-(*p*-coumaroylsophoroside)-5-glucoside	Harborne (1964)
(*Raphinus sativus*)	Pg and Cy 3-(feruloylsophoroside)-5-glucoside	Fuleki (1969)
Yam		
(*Dioscorea tryphida*)	Mv 3-(feruloyldiglucoside)-5-glucoside	Carreno-Diaz and Grau (1977)
(*Dioscorea batatas*)	Mv 3-(*p*-coumaroylgentiobioside)-5-glucoside	Tsukui *et al*. (1977)
Grapes (cv. Ives)	Dp, Pt and Mv 3-(*p*-coumaroylglucoside)-5-glucoside	Williams *et al*. (1978)
(*Vitis* hybrid)		
Mung bean	Dp 3-(*p*-coumaroylglucoside)-5-glucoside	Proctor (1970)
(*Phaseolus aureus*)		

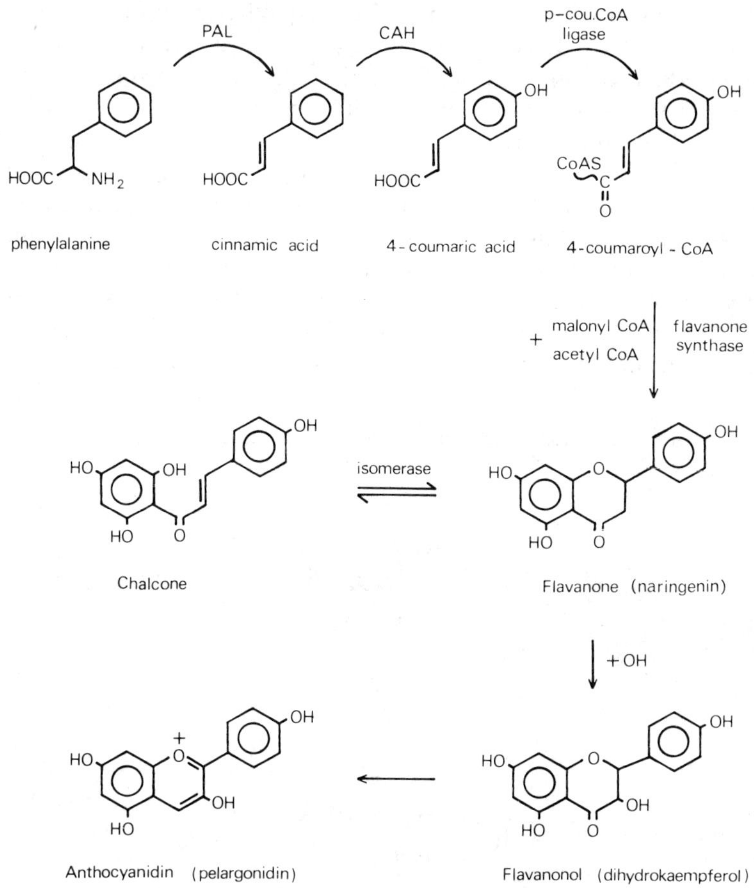

FIG. 4. Biosynthesis of anthocyadins.

coumaroyl CoA to caffeoyl CoA subsequently used as a precursor for synthesis of the anthocyanidin skeleton (Kamsteeg *et al.*, 1979b). In contrast, Forkmann (1980) concluded that the 3′-hydroxyl group in *Matthiola incana* anthocyanins does not arise by specific incorporation of caffeic acid during synthesis of the flavonoid skeleton, but rather by hydroxylation at the dihydroflavonol stage.

If as it seems, glycosylation is the last step in anthocyanin biosynthesis, a mechanism must be invoked to account for the apparent *in vivo* stability of the notoriously unstable anthocyanidins *in vitro*. One has yet to be formulated.*

Two further points need mentioning, viz. (a) that *O*-methylation probably

* Because of anthocyanidin instability, Brouillard suggests that the precursor is a 3-deoxyanthocyanidin in its chalcone form (Brouillard, R. (in press)). *In* "Anthocyanins as Food Colours" (P. Markakis, Ed.).

occurs at the C 15 stage by transfer of the methyl group of *S*-adenosyl-L-methionine, and (b) that the mechanism of conversion of flavanonols to anthocyanidins remains a matter for conjecture. The cyclic process proposed by Jurd (1969) has been considered feasible by Fritsch and Grisebach (1975).

An alternative more direct route to the anthocyanidins has been proposed by Jacques *et al.* (1977; Fig. 5) leading to the formation of a flav-3-en-3-ol. The latter is the precursor of the procyanidins, but under appropriate metabolic circumstances it may also be oxidatively metabolized to anthocyanidin and account for pigmentation occurring during the autumn coloration of leaves, the ripening of fruit and the transient "spring" flush. The recent report of photochemical deoxygenation of dihydroflavonols to flavanones may be significant in this connection (van der Westhuizen *et al.*, 1980).

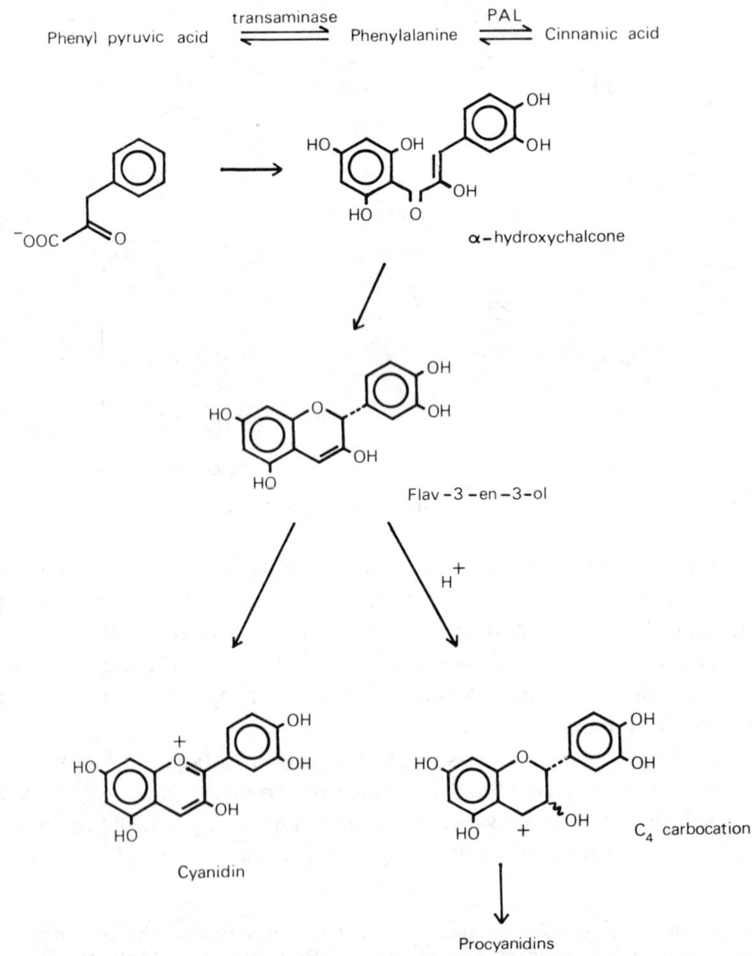

FIG. 5. Mechanism proposed by Jacques, Opie, Porter and Haslam (1977).

V. Vegetable Anthocyanins

A. Anthocyanins of the Cruciferae

Red cabbage (Brassica oleracea). Red cabbage anthocyanins are particularly suited for biosynthetic studies since their composition is known in considerable detail (Table III). Photo-control of their synthesis was studied by Hathout Bassim and Pecket (1975). Recently Hrazdina and Creasy (1979) isolated the enzyme flavanone synthase and studied changes in its activity together with that of phenylalanine ammonia lyase (PAL) and anthocyanin production in young seedlings (Fig. 6). Pigment was rapidly produced on illumination and increased up to six days. PAL activity increased rapidly to a maximum after eight hours but then declined to dark levels. (The stimulatory

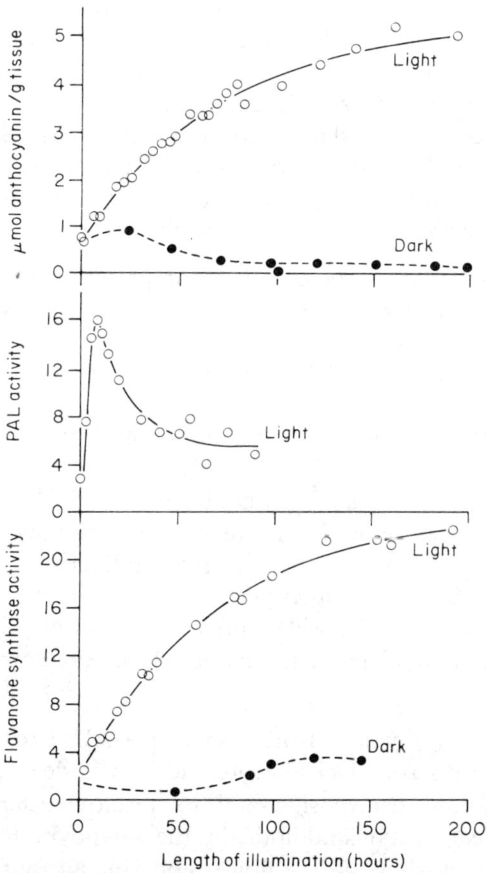

FIG. 6. Red cabbage seedlings (*Brassica oleracea*) (after Hrazdina and Creasy, 1979).

effect of light on PAL activity is a general phenomenon and is usually only transient in character. Substrate supply may be a limiting factor (Margna, 1977).) During this period, flavanone synthase activity paralleled pigment concentration. Similar changes occurred in the cotyledons and hypocotyls separately; the former were the main site of anthocyanin biosynthesis. The properties of the flavanone synthase enzyme were similar to those described previously for the same enzyme isolated from parsley cell culture (Wellmann *et al.*, 1976). Only *p*-coumaroyl CoA (neither caffeoyl nor feruloyl CoA) was utilized as substrate and naringenin was the first flavonoid formed.

A later step in the mechanism of biosynthesis of the red cabbage pigment was briefly referred to in the previous section, i.e. the first time *in vitro* isolation and purification of a glucosyl transferase catalysing the glucosylation of cyanidin at the 3-position (Saleh *et al.*, 1976). The enzyme did not glycosylate cyanidin 3-glucoside, cyanidin 3-sophoroside nor cyanidin 3,5-diglucoside. It is likely that other glucosyltransferases are involved in the pathway of glucosylation, which was indicated to be in the following sequence: cyanidin → cyanidin 3-glucoside → cyanidin 3-sophoroside → cyanidin 3-sophoroside-5-glucoside.

Red cabbage has a very sophisticated acylation pattern. Cyanidin 3-sophoroside-5-glucoside is acylated with one and two moles of *p*-coumaric, but thereafter the methoxylated derivatives, i.e. ferulic and the rare sinapic acid are preferred to caffeic acid; malonic acid is also present. Altogether an intriguing but somewhat daunting task for elucidation.

Radish (Raphanus sativus). The glycosylation pattern is the same as for red cabbage. Pelargonidin derivatives are also present but acylation is not as extensive (Table III).

B. ANTHOCYANINS OF THE UMBELLIFERAE

Pigmentation in this family is mostly confined to vegetable tissue, and little previous work has been done. Harborne (1976b) has now shown that only cyanidin glycosides are present which are characterized by a new linear triglycoside (cyanidin xylosylglucosylgalactoside) and the rare lathyroside (xylosylgalactoside). The triglycoside is often acylated with *p*-coumaric, ferulic and sinapic acids but not caffeic acid. In this respect it resembles the pigments of red cabbage.

Indian black carrot (Daucus carota *s.sp* sativa). Unlike the more usual orange–red form, this root lacks carotenoid and is deep purple–black. It contains the cyanidin triglycoside and its feruloyl ester, and cyanidin lathyroside. The leaf contains additionally the sinapoyl ester of cyanidin 3-xylosylglucosylgalactoside and the flower contains all four pigments.

Another sub-species restricted to maritime habitats in Europe, i.e.

maritimus, lacks xylose; its pigments consist of the feruloyl and sinapoyl derivatives of cyanidin 3-glucosylgalactoside.

Callus cells of the more usual *D. carota* have been used in biosynthetic work. The phenyl propanoid pathways vary according to medium composition. Without gibberellic acid, the cells accumulated cyanidin, but in its presence [^{14}C] phenylalanine was converted into chlorogenic acid. The hydroxy cinnamate CoA ligases extracted from the two cell populations had different substrate specificities (Heinzmann *et al.*, 1977).

Celery (Apium graveolens). The Old World celery contains the three acylated triglycosides. In contrast the New World *Apium australe* lacks the *p*-coumaroyl ester of the triglycoside.

Fennel (Foeniculum vulgare) contains the feruloyl and sinapoyl triglycosides.

Samphire (Crithium maritum). This herb, a native on sea cliffs on the western and southern coasts of the British Isles, contains only the sinapoyl ester of the triglycoside.

The Umbelliferae are very simple at the aglycone level (cyanidin); perhaps to be expected since the anthocyanins do not function here to attract pollinators. In contrast they are very complex at the glycosidic level which may reflect their specialized morphology (Harborne, 1976b).

C. ANTHOCYANINS OF OTHER VEGETABLES

1. Leguminosae

Pea (Pisum sativum). Two new anthocyanin pigments, viz. delphinidin 3-sophoroside-5-glucoside and delphinidin 3-sambubioside-5-glucoside, were isolated from purple pods together with the corresponding cyanidin derivatives; the flowers contained additional pigments. The biochemical genetics were clarified (Statham *et al.*, 1972). The activity of gene Cr in replacing 3-rhamnoside by 3-sambubioside and 3-sophoroside in flower-colour mutants was investigated by Statham and Crowden (1974). Red-stemmed pea seedlings were shorter than green-stemmed seedlings, but the anthocyanins were not a direct cause of this difference (Nozzolillo, 1979).

2. Compositae

Globe artichoke (Cynara scolymus). Globe artichoke contains cyanidin 3-caffeoylglucoside, 3-caffeoylsophoroside and 3-dicaffeoylsophoroside (Pifferi and Vaccari, 1978). Cyanidin is characteristic of the Compositae, but this appears to be the first report of acylation.

3. *Liliaceae*

Asparagus (Asparagus officinalis). Asparagus shoot contains largely the unusual cyanidin 3-rhamnosylglucosylglucoside, with smaller amounts of cyanidin 3-rutinoside and peonidin derivatives (Wann and Thompson, 1965; Francis, 1967).

Onion (Allium cepa). Bulbs of the Spanish red onion contain cyanidin 3-glucoside and another rare derivative, cyanidin 3-laminario-bioside, of which they are a good source. Laminariobiose is $\beta 1 \rightarrow 3$ glucosylglucoside compared with sophorose $\beta 1 \rightarrow 2$, cellobiose $\beta 1 \rightarrow 4$ and gentiobiose $\beta 1 \rightarrow 6$ (Du *et al.*, 1974).

4. *Araceae*

Taro (Colocasia esculenta). Taro is a root crop of secondary importance in many parts of the wet tropics. The pigmented edible corm contains pelargonidin 3-glucoside, cyanidin 3-rhamnoside and cyanidin 3-glucoside (Chan *et al.*, 1977). The findings accord with the only other identifications made in the Araceae, viz. cyanidin 3-glucoside in *Pistia stratiotes* (Zennie and McClure, 1977) and pelargonidin and cyanidin 3-rutinosides in *Anthurium andreanum* (Iwata *et al.*, 1979).

VI. Fruit Anthocyanins

A. Anthocyanins of the Rosaceae

1. Malus *Anthocyanins*

The major pigment of apples (*Malus pumila*) was identified as cyanidin 3-galactoside by Sando as long ago as 1937. Paper chromatography has since revealed minor components, the structures of which are more controversial. Cyanidin 3-arabinoside and cyanidin 7-arabinoside were identified by Sun and Francis (1968) in American Red Delicious fruit. The latter anthocyanin was not found in some English fruits; in these the minor components consisted of cyanidin 3-glucoside, 3-arabinoside and the hitherto undescribed 3-xyloside, together with derivatives presumed acylated with non-ultraviolet absorbing compounds (Timberlake and Bridle, 1971). The latter authors found that the apple pigments were difficult to separate chromatographically from associated procyanidins; which might account for the unusually high ratio of E_{275}/E_{max} for cyanidin 7-arabinoside described by Sun and Francis (1968).

Anthocyanin synthesis in the skin of apple fruit has been studied extensively and is influenced by light (Proctor and Creasy, 1971; Proctor, 1974; Bishop and Klein, 1975; Greene and Lord, 1975) and by temperature, growth regulators, protein synthesis inhibitors, fruit maturity, wounding and ethylene production (Chalmers and Faragher, 1977a,b). The effect of high nitrogen nutrition in reducing anthocyanin skin colour in dessert fruit was confirmed with cider fruit by Lea and Beech (1978). Such an effect, also prominent in other fruits could be direct, i.e. due to biochemical diversion of the carbohydrate pool to protein rather than phenolic synthesis, or indirect, i.e. due to increased vegetative growth and decreased light intensity.

Faragher and Chalmers (1977) followed pigment production and PAL activity induced by exposure to white light and concluded that while PAL activity may control the rate of anthocyanin synthesis in the whole fruit, other critical enzymes are involved in its synthesis in the skin. This conclusion appears well founded for as has been seen with red cabbage, flavanone synthase, which acts as a later stage in the biosynthetic sequence was better related to pigment production.

Tan (1979) studied the effect of light, low temperature and antibiotics on anthocyanin synthesis and the activities of PAL and its inactivating system PAL-IS in skin of whole apples and in skin discs. An intriguing use of an anthocyanin progeny marker to determine the value of hive pollen dispensers in apple orchards was described by Williams *et al.* (1979).

2. Rubus *Anthocyanins*

The anthocyanins in *Rubus* were identified and characterized by Harborne (1967); the following is a summary of more recent work which has augmented his initial findings (Misic, 1973; Barritt and Torre, 1973, 1975a,b; Torre and Barritt, 1977).

Cyanidin is the major aglycone. Blackberries contain cyanidin 3-glucoside, sometimes accompanied by small amounts of cyanidin 3-rutinoside. Fruit anthocyanin content varies from 82–326 mg/100 g. In red raspberries (*R. idaeus*) the characteristic component is cyanidin 3-sophoroside, with lesser amounts of cyanidin 3-glucosylrutinoside, 3-glucoside and 3-rutinoside. The formation of the rhamnose containing pigments is under control of a major gene R. Some cultivars, e.g. Newburgh and its derivatives Malling Exploit, Malling Promise and Willamette contain only cyanidin 3-sophoroside and 3-glucoside. Total pigment is 23–59 mg/100 g fruit. The two red raspberry/blackberry hybrids, Boysenberry and Loganberry, have pigments typical of red raspberry but amounts more typical of blackberries to which they are more morphologically similar.

Black raspberries (*R. occidentalis*) are characterized by two xylose containing pigments cyanidin 3-xylosylrutinoside and cyanidin 3-sambubioside, and also contain cyanidin 3-rutinoside and cyanidin 3-glucoside. Total pigment is very high—214–428 mg/100 g. The trisaccharides are branched chain, the glucose moiety of cyanidin 3-glucosylrutinoside (in *R. idaeus*) being replaced by xylose to give cyanidin 3-xylosylrutinoside (in *R. occidentalis*) (Harborne, 1967).

Pelargonidin is rare in the genus but can occur in some species. The red raspberry cultivar September contains all four pelargonidin derivatives corresponding to those of cyanidin and major amounts are found in five other species, viz. *R. odoratus*, *R. parviflorus*, *R. crataegifolius*, *R. illecebrosus* and *R. morifolius*.

3. Prunus *Anthocyanins*

Sweet cherries (*Prunus avium*) contain cyanidin 3-glucoside and 3-rutinoside (Harborne, 1967), but small amounts of the corresponding peonidin derivatives may also be present.

Sour cherries (*Prunus cerasus*) contain cyanidin 3-glucosylrutinoside, 3-sophoroside, 3-rutinoside and 3-glucoside (Harborne, 1967). The cultivar Montmorency contains additional minor components cyanidin 3-xylosylrutinoside (Shrikhande and Francis, 1973) and peonidin 3-rutinoside and 3-glucoside (Dekazos, 1970). Apparently, cyanidin 3-gentiobioside (von Elbe *et al.*, 1969 and earlier reports) has been reported incorrectly (Du *et al.*, 1975b). Ornamental cherries (*Prunus sargentii*) contain mainly cyanidin 3-glucoside and rutinoside but with traces of pelargonidin pigments and aliphatic acylated anthocyanins (Du *et al.*, 1975b).

The effects of growth regulators have been studied on sweet cherries (Ryugo, 1966; Drake *et al.*, 1978) and sour cherries (Dekazos and Worley, 1970). PAL was found to control phenolic synthesis in sweet cherries on the tree, but during ripening increased PAL activity was not directly related to pigment synthesis (Melin *et al.*, 1977).

B. ANTHOCYANINS OF THE SAXIFRAGACEAE

1. Ribes *Anthocyanins*

Cyanidin triglycosides are present also in *Ribes* and their distribution has been fully documented (Harborne, 1967). Also studied has been the inheritance of four cyanidin 3-glycosides in red currants (Oydvin, 1974) and the inheritance of fruit colour in red and black currants and gooseberries (Keep and Knight, 1970).

C. ANTHOCYANINS OF THE VITACEAE

1. Grapes

It is appropriate that grapes should be discussed since they are the largest commercial cultivated fruit crop. The distribution of anthocyanins in the fruit is complex and varies according to the species. Investigations have been facilitated by the use of high-pressure liquid chromatography and the separations which are possible are illustrated in Fig. 7 for *Vitis vinifera* (cultivar Cabernet Sauvignon), reproduced from the work of Nagel and Wulf (1979). All five common anthocyanidins are present and they occur in three groups (i) as 3-glucosides, (ii) as 3-acetylglucosides, and (iii) as 3-*p*-coumaroylglucosides (also as the 3-caffeoylglucoside of malvidin). They are based mainly on malvidin, with lesser amounts of delphinidin, petunidin and peonidin, and traces only of cyanidin. Their chromatographic separation depends upon the substitution pattern of the B ring. Retention times decrease with increasing hydroxylation (delphinidin < cyanidin). Petunidin derivatives are retained for shorter times than those of peonidin, while malvidin derivatives have the longest retention times.

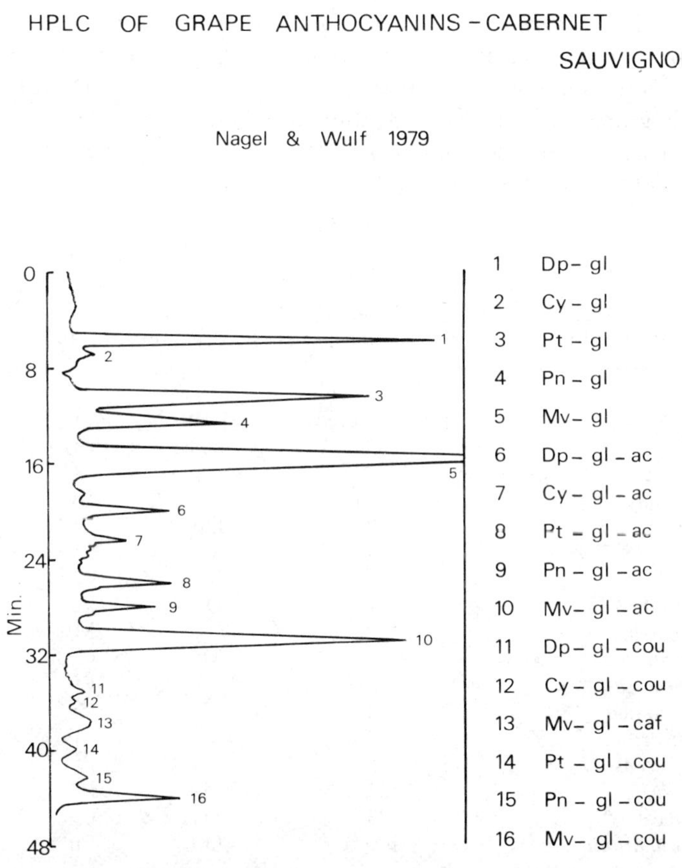

FIG. 7. HPLC of grape anthocyanins—Cabernet Sauvignon (after Nagel and Wulf, 1979).

Of interest are the anthocyanin 3-glucose acetates. Anthocyanins acylated with acetic acid were first discovered in skins of *Vitis cinerea* as recently as 1970 (Anderson *et al.*, 1970a,b). They are difficult to characterize because their absorption spectra are identical with their corresponding 3-glucosides. Their ease ot separation by HPLC illustrates the usefulness of this technique and invites speculation as to the extent that similar acetic acid acylated pigments might occur in other plant material. So far they have been found only in the grape and indeed if this is their only source, the reason why they should be confined to *Vitis* (and maybe only certain species) is not immediately apparent.

There are many other species of *Vitis* and most of these are characterized by the presence of varying amounts of 3,5-diglucosides. Table IV summarizes their paper chromatographic investigation by Ribereau-Gayon (1974). Malvidin is not always the major aglycone and can be displaced by delphinidin or cyanidin. Where cyanidin is the major component (*V. aestivalis*, *V. coriacea* and *V. lincecumii*), the orders appear reversed and little or no malvidin is present. Of these American species, *V. labrusca* is the most important for fruit production—a typical cultivar being the Concord grape—and is characterized by a distinctive flavour. Another American native is *V. rotundifolia* (muscadine grapes) with a unique and intense fruity aroma. Others, notably *V. riparia* (the river grape) and *V. rupestris* (the sand grape) are useful as breeding material or rootsocks. They confer vine resistance, particularly

TABLE IV
Anthocyanin distribution in *Vitis* species (Ribereau-Gayon, 1974)

	3,5-diglucosides (inc. acylated) (%)	Anthocyanidin (most)	Anthocyanidin (least)
vinifera	0	Mv	Cy
monticola	0	Dp	Cy
labrusca	4	Mv	Cy
berlandii	5	Mv	Cy
rubra	7	Dp	Pn
aestivalis	9	Cy	Mv
cordifolia	11	Mv	Cy
coriacea	12	Cy	Pt
arizonica	22	Mv	Cy
lincecumii	22	Cy	Mv
amurensis	55	Mv	Pt
riparia[a]	63	Mv	Pn
rupestris[a]	83	Mv	Cy
rotundifolia	100	Dp	Pn

[a] Species most used in hybridization

against the root louse *Phylloxera*, but their distinct "wild" flavours are usually carried over into crosses with *V. vinifera*.

In Europe the detection of 3,5-diglucosides in wines is taken as evidence of adulteration of the quality *vinifera* wines with those made from hybrid grapes (and supposedly inferior). However, since the diglucoside trait in *V. vinifera* is recessive, there is an inherent limit in the method. Thus, a cross between *V. riparia* and *V. vinifera* yields an F_1 population of hybrids, all with diglucoside character, but if one of these F_1 hybrids is crossed with *V. vinifera* one-half of the subsequent F_2 hybrids will not possess 3,5-diglucosides (Ribereau-Gayon, 1974).

Anthocyanins of some of these other species (e.g. *V. labrusca*) have also been separated by HPLC (Williams *et al.*, 1978; Hrazdina, 1979). The 3,5-diglucosides being the most polar were the first eluted, followed by the 3-glucosides. Then followed the diglucosides acylated with *p*-coumaric acid; the similarly acylated monoglucosides were retained most. Pigments acylated with acetic acid were absent. It is not yet known the extent of occurrence of the anthocyanin glucose acetates amongst the various *Vitis* species and HPLC should thus be of considerable aid to further genetic studies.

D. ANTHOCYANINS OF THE MORACEAE

1. Mulberry

The common or black mulberry (*Morus nigra*) contains cyanidin 3-glucoside but purple fruits of the Asian white mulberry *Morus alba* have been reported as containing cyanidin 3-rutinoside as its main component by Ishikura (1975), but cyanidin 3-glucoside, pelargonidin 3-glucoside and petunidin 3-rutinoside by Maki and Inamoto (1972).

2. Fig (Ficus carica)

The Robinsons (1932) reported cyanidin 3-glucoside in fig, confirmed by Duro and Condorelli (1977), but recent work (Puech *et al.*, 1975) showed that cyanidin 3-rutinoside was the major component (75%) in the intense blue–black Mission fig. Thus, in Moraceae the 3-rutinosides may be more prominent than previously supposed. Application of Ethephon to the Mission fig promoted the rate of anthocyanin accumulation but not the total amount of pigment synthesized (Puech *et al.*, 1976).

E. ANTHOCYANINS OF THE ERICACEAE

1. Cranberry (Vaccinium macrocarpon) *and low bush blueberry* (V. angustifolium)

There is little information over that reported earlier (Harborne, 1967)

2. Highbush blueberry (V. corymbosum)

The major anthocyanins are malvidin 3-galactoside, delphinidin 3-galactoside, delphinidin 3-arabinoside, petunidin 3-galactoside, petunidin 3-arabinoside and malvidin 3-arabinoside. Also present are peonidin 3-arabinoside and peonidin 3-galactoside with small amounts of delphinidin, petunidin, malvidin, peonidin and cyanidin 3-glucosides and cyanidin 3-galactoside (Ballinger *et al.*, 1970, 1972). The complexity of the anthocyanin pattern parallels that of *V. angustifolium*. The anthocyanins were also characterized during fruit ripening in cv. Wolcott. In unripe red fruit cyanidin 3-glucoside and 3-galactoside comprised 40% of the total pigment; all combinations of the five aglycones with glucose, galactose or arabinose, i.e. 15 pigments were detected. In ripe and overripe fruit, all were again found except peonidin 3-galactoside; malvidin 3-glucoside and 3-galactoside were now the main components (60%). The relative proportions of glucosides, galactosides and arabinosides were fairly constant during ripening (Makus and Ballinger, 1973).

3. Vaccinium *species*

Anthocyanins of ripe fruits of 12 species of the sub-genus *Cyanococcus* (true blueberries) and one species of the sub-genus *Polycodium* (*V. staminium*) have been examined by Ballinger *et al.* (1979). The *Cyanococcus* species contained five aglycones (in decreasing quantities—delphinidin, cyanidin, malvidin, petunidin and peonidin) and two sugars (galactose > arabinose); seven species also yielded glucose. The sole representative of *Polycodium* was more similar to that of cranberries (*V. macrocarpon*) in containing cyanidin (with a trace peonidin) and glucose. Thus, anthocyanin differences appear associated with sub-generic classification in *Vaccinium*.

F. ANTHOCYANINS OF OTHER FRUITS

1. Guttiferae

Mangosteen (Garcinia mangostana). The rind of this purplish–brown East Indian tropical fruit contains cyanidin 3-sophoroside and cyanidin 3-glucoside (Du and Francis, 1977); the first identifications in this family.

2. Anacardiaceae

Mango (Mangifera indica). The mango fruit contains a single anthocyanin, the rare peonidin 3-galactoside (Proctor and Creasy, 1969). Fruit colour was enhanced in response to a pre-harvest antitranspirant spray (Barmore *et al.*, 1974).

3. Sapindaceae

Litchi (Litchi chinensis). The fruit contains four pigments—cyanidin 3-glucoside and 3-galactoside, pelargonidin 3-glucoside and 3,5-diglucoside—the first report in this family. Colour production was enhanced by light and the cyanidin glycosides appeared before the pelargonidin derivatives during ripening (Prasad and Jha, 1978).

4. Punicaceae

Pomegranate (Punica granatum). Delphinidin 3,5-diglucoside was reported as the main fruit juice component (Harborne, 1967) but Du *et al.* (1975a) found cyanidin 3-glucoside the major component of the seed coat and fruit peel, accompanied by five other pigments.

5. Sapotaceae

Miracle fruit (Synsepalum dulcificum). The deep red oval berries contain largely cyanidin 3-galactoside and 3-glucoside, with smaller amounts of cyanidin 3-arabinoside, delphinidin 3-galactoside and delphinidin 3-arabinoside (Buckmire and Francis, 1976). The berries also contain an intensely sweet compound, which offers potential as a non-nutritive sweetener—hence its descriptive name.

6. Oleaceae

Olive (Olea europaea). Manzanillo olives contain mainly cyanidin 3-glucoside, 3-rutinoside and 3-glycosylrutinosides (reported linear and branched chain) and also acylated with caffeic acid (Luh and Mahecha, 1972; Maestro and Vazquez, 1976). The only other anthocyanins identified in the Oleaceae are in fruits of *Ligustrum* species which likewise contain cyanidin 3-rutinoside and cyanidin 3-glucoside, amongst others (Medrano *et al.*, 1977; Ishikura and Sugahara, 1979).

7. Solanaceae

Tamarillo (Cyphomandra betacea). The red strains of the Tamarillo or tree tomato contain pelargonidin, cyanidin and delphinidin 3-rutinosides and 3-

glucosides. The main skin component is cyanidin 3-rutinoside and the purple pigment in the jelly surrounding the seeds is largely delphinidin 3-rutinoside (Wrolstad and Heatherbell, 1974).

REFERENCES

Anderson, D. W., Gueffroy, D. E., Webb, A. D. and Kepner, R. E. (1970a). *Phytochemistry* **9**, 1579.

Anderson, D. W., Julian, E. A., Kepner, R. E. and Webb, A. D. (1970b). *Phytochemistry* **9**, 1569.

Asen, S. and Jurd, L. (1967). *Phytochemistry* **6**, 577.

Asen, S., Stewart, R. N. and Norris, K. H. (1972). *Phytochemistry* **11**, 1139.

Asen, S., Stewart, R. N. and Norris, K. H. (1977). *Phytochemistry* **16**, 1118.

Asen, S., Stewart, R. N. and Norris, K. H. (1979). U.S. Patent No. 4,172,902, October 30, 1979.

Ballinger, W. E., Maness, E. P. and Kushman, L. J. (1970). *J. Amer. Soc. Hort. Sci.* **95**, 283.

Ballinger, W. E., Maness, E. P., Galletta, G. J. and Kushman, L. J. (1972). *J. Amer. Soc. Hort. Sci.* **97**, 381.

Ballinger, W. E., Galletta, G. J. and Maness, E. P. (1979). *J. Amer. Soc. Hort. Sci.* **104**, 554.

Barmore, C. R., Spalding, D. H. and McMillan, R. T. (1974). *J. Hort. Sci.* **9**, 114.

Barritt, B. H. and Torre, L. C. (1973). *J. Chromatography* **75**, 151.

Barritt, B. H. and Torre, L. C. (1975a). *J. Amer. Soc. Hort. Sci.* **100**, 98.

Barritt, B. H. and Torre, L. C. (1975b). *J. hort. Sci.* **10**, 526.

Barz, W. and Hösel, W. (1975). *In* "The Flavonoids" (J. B. Harborne, T. J. Mabry and H. Mabry, Eds), pp. 916–969. Academic Press, London and New York.

Bayer, E., Egeter, H., Fink, A., Nether, K. and Wegmann, K. (1966). *Angew. Chem. Internatl Ed.* **5**, 791.

Bishop, R. C. and Klein, R. M. (1975). *J. hort. Sci.* **10**, 126.

Brouillard, R. and Delaporte, B. (1977). *J. Am. Chem. Soc.* **99**, 8461.

Brouillard, R. and Delaporte, B. (1978). "Protons and Ions Involved in Fast Dynamic Phenomena", pp. 403–412. Elsevier, Amsterdam.

Brouillard, R. and Dubois, J. E. (1977). *J. Am. Chem. Soc.* **99**, 1359.

Brouillard, R., Delaporte, B. and Dubois, J. E. (1978). *J. Am. Chem. Soc.* **100**, 6202.

Brouillard, R., Delaporte, B., El Hage Chahine, J. M. and Dubois, J. E. (1979). *J. de Chimie Physique* **76**, 273.

Buckmire, R. M. and Francis, F. J. (1976). *J. Food Sci.* **41**, 1363.

Carreno-Diaz, R. and Grau, N. (1977). *J. Food Sci.* **42**, 615.

Casoli, U. and Dall'aglio, G. (1969). *Ind. Conserve* **44**, 18, via *Chem. Abstracts* **71**, 69379.

Chalmers, D. J. and Faragher, J. D. (1977a). *Aust. J. Plant Physiol.* **4**, 111.

Chalmers, D. J. and Faragher, J. D. (1977b). *Aust. J. Plant Physiol.* **4**, 123.

Chan, H. T., Kao-Jao, T. H. C. and Nakayama, T. O. M. (1977). *J. Food Sci.* **42**, 19.

Dekazos, E. D. (1970). *J. Food Sci.* **35**, 237.

Dekazos, E. D. and Worley, J. F. (1970). *J. Amer. Soc. Hort. Sci.* **95**, 703.

Drake, S. R., Proebsting, E. L. and Nelson, J. W. (1978). *J. Food Sci.* **43**, 1695.

Du, C. T. and Francis, F. J. (1977). *J. Food Sci.* **42**, 1667.

Du, C. T., Wang, P. L. and Francis, F. J. (1974). *J. Food Sci.* **39**, 1265.

Du, C. T., Wang, P. L. and Francis, F. J. (1975a). *J. Food Sci.* **40**, 417.

Du, C. T., Wang, P. L. and Francis, F. J. (1975b). *J. Food Sci.* **40**, 1142.
Duro, F. and Condorelli, P. (1977). *Quad. Merceol* **16**, 37, via *Chem. Abstr.* **87**, 164201.
Faragher, J. D. and Chalmers, D. J. (1977). *Aust. J. Plant Physiol.* **4**, 133.
Forkmann, G. (1980). *Planta* **148**, 157.
Francis, F. J. (1967). *J. Food Sci.* **32**, 430.
Francis, F. J. and Harborne, J. B. (1966). *J. Food Sci.* **31**, 524.
Fritsch, H. and Grisebach, H. (1975). *Phytochemistry* **14**, 2437.
Fuleki, T. (1969). *J. Food Sci.* **34**, 365.
Goto, T., Hoshino, T. and Takase, S. (1979). Tetrahedron Letters No. 31, 2905.
Greene, D. W. and Lord, W. J. (1975). *J. hort. Sci.* **10**, 254.
Grisebach, H. (1979). *In* "Biochemistry of Plant Phenolics". Recent Advances in Phytochemistry, Vol. 12, pp. 221–248. Plenum Press, New York and London.
Harborne, J. B. (1960). *Biochem J.* **74**, 262.
Harborne, J. B. (1964). *Phytochemistry* **3**, 151.
Harborne, J. B. (1967). "Comparative Biochemistry of the Flavonoids". Academic Press, London and New York.
Harborne, J. B. (1976a). *In* "Chemistry and Biochemistry of Plant Pigments", Vol. 1 (T. W. Goodwin, Ed.), pp. 736–774. Academic Press, London and New York.
Harborne, J. B. (1976b). *Biochem, Syst. Ecol.* **4**, 31.
Hathout Bassim, T. A. and Pecket, R. C. (1975). *Phytochemistry* **14**, 731.
Hayashi, K. and Takeda, K. (1970). *Proc. Japan. Acad.* **46**, 535.
Heinzmann, U., Seitz, U. and Seitz, U. (1977). *Planta* **135**, 313.
Hoshino, T., Matsumoto, U. and Goto, T. (1980). *Phytochemistry* **19**, 663.
Howard, H. W., Kukimura, H. and Whitmore, E. T. (1970). *Potato Res.* **13**, 142.
Hrazdina, G. (1979). *In* "Liquid Chromatographic Analysis of Foods and Beverages" (G. Charalambous, Ed.), pp. 141–159. Academic Press, New York and London.
Hrazdina, G. and Creasy, L. L. (1979). *Phytochemistry* **18**, 581.
Hrazdina, G., Iredale, H. and Mattick, L. R. (1977). *Phytochemistry* **16**, 297.
Imbert, M. P., Seaforth, C. E. and Williams, D. B. (1966). *Proc. Am. Soc. hort. Sci.* **88**, 481.
Ishikura, N. (1975). *Bot. Mag.* (Tokyo) **88**, 41.
Ishikura, N. and Sugahara, K. (1979). *Bot. Mag.* (Tokyo) **92**, 57.
Iwata, R. Y., Tang, C. S. and Kamemoto, H. (1979). *J. Am. Soc. Hort. Sci.* **104**, 464.
Jacques, D., Opie, C. T., Porter, L. J. and Haslam, E. (1977). *J. Chem. Soc.* Perkin I, **14**, 1637.
Jurd, L. (1969). *Phytochemistry* **8**, 2421.
Kamsteeg, J., van Brederode, J. and van Nigtevecht, G. (1978a). *Biochemical Genetics* **16**, 1045.
Kamsteeg, J., van Brederode, J. and van Nigtevecht, G. (1978b). *Biochemical Genetics* **16**, 1059.
Kamsteeg, J., van Brederode, J. and van Nigtevecht, G. (1979a). *Phytochemistry* **18**, 659.
Kamsteeg, J., van Brederode, J. and van Nigtevecht, G. (1979b). *Genetica* (The Hague) **51**, 5.
Keep, E. and Knight, R. L. (1970). *Rept East Malling Res. Stn for 1969*. p. 139.
Kho, K. F. F., Kamsteeg, J. and van Brederode, J. (1978). *Z. Pflanzenphysiol* **88**, 449.
Lea, A. G. H. and Beech, F. W. (1978). *J. Sci. Fd Agric.* **29**, 493.
Luh, B. S. and Mahecha, G. (1972). *Chem. Abstr.* **77**, 60298c.
Maestro, D. and Vazquez, R. A. (1976). via *Chem. Abstracts* **86**, 138186.
Maki, Z. and Inamoto, H. (1972). *Chem. Abstr.* **78**, 82073.
Makus, D. J. and Ballinger, W. E. (1973). *J. Amer. Soc. Hort. Sci.* **98**, 99.
Margna, U. (1977). *Phytochemistry* **16**, 419.

McClure, J. W. (1975). *In* "The Flavonoids" (J. B. Harborne, T. J. Mabry and H. Mabry, Eds), pp. 970–1055. Academic Press, London and New York.
Medrano, M. A., Tomas, M. A. and Frontera, M. A. (1977). *An. Asoc. Quim. Argent* **65**, 59.
Melin, C., Moulet, A-C., Dupin, J-F. and Hartmann, C. (1977). *Phytochemistry* **16**, 75.
Misic, P. D. (1973). *Hort. Res.* **13**, 45.
Nagel, C. W. and Wulf, L. W. (1979). *Am. J. Enol. Vitic.* **30**, 111.
Nozzolillo, C. (1979). *Can. J. Bot.* **57**, 2554.
Oydvin, J. (1974). *Hort. Res.* **14**, 1.
Pifferi, P. G. and Vaccari, A. (1978). *Industria conserv.* **53**, 107.
Pifferi, P. G. and Zamorani, A. (1969). *Ind. Agr.* (Florence) **7**, 51. via *Chem. Abstracts* **71**, 78128.
Prasad, U. S. and Jha, O. P. (1978). *Plant Biochem. J.* **5**, 44.
Proctor, J. T. A. (1970). *Diss. Abstr. Int. B.* **31**, 10.
Proctor, J. T. A. (1974). *Can. J. Plant Sci.* **54**, 499.
Proctor, J. T. A. and Creasy, L. L. (1969). *Phytochemistry* **8**, 2108.
Proctor, J. T. A. and Creasy, L. L. (1971). *J. Amer. Soc. Hort. Sci.* **96**, 523.
Puech, A. A., Rebeiz, C. A., Catlin, P. B. and Crane, J. C. (1975). *J. Food Sci.* **40**, 775.
Puech, A. A., Rebeiz, C. A. and Crane, J. C. (1976). *Plant Physiol.* **57**, 504.
Ribereau-Gayon, P. (1974). *In* "The Chemistry of Winemaking' (A. D. Webb, Ed.). Adv. in Chem. Ser. 137, pp. 50–87. American Chemical Society, Washington D.C.
Robinson, G. M. and Robinson, R. (1932). *Biochem. J.* **26**, 1647.
Ryugo, K. (1966). *Proc. Am. Soc. hort. Sci.* **88**, 160.
Sachse, L. (1973). *Z. Lebensmittel-Unters. u-Forsch.* **153**, 294.
Saito, N. and Hayashi, K. (1965). *Sci. Rept Tokyo Kyoiku Daigaku* **12B**, 39.
Saito, N., Hotta, R. Imai, K. and Hayashi, K. (1965). *Proc. Japan Acad.* **41**, 593.
Saito, N., Osawa, Y. and Hayashi, K. (1972). *Bot. Mag.* (Tokyo) **85**, 105.
Saleh, N. A. M., Poulton, J. E. and Grisebach, H. (1976). *Phytochemistry* **15**, 1865.
Sando, C. E. (1937). *J. biol. Chem.* **117**, 45.
Scheffeldt, P. and Hrazdina, G. (1978). *J. Food Sci.* **43**, 517.
Shrikhande, A. J. and Francis, F. J. (1973). *J. Food Sci.* **38**, 649.
Simpson, T. H. and Garden, L. (1952a). *J. Chem. Soc.* 4638.
Simpson, T. H. and Garden, L. (1952b). *J. Chem. Soc.* 5027.
Statham, C. M. and Crowden, R. K. (1974). *Phytochemistry* **13**, 1835.
Statham, C. M., Crowden, R. K. and Harborne, J. B. (1972). *Phytochemistry* **11**, 1083.
Sun, B. H. and Francis, F. J. (1968). *J. Food Sci.* **32**, 647.
Takeda, K. (1977). *Proc. Japan Acad.* **53**, Ser. B, p. 257.
Takeda, K. and Hayashi, K. (1977). *Proc. Japan, Acad.* **53**, Ser. B, p. 1.
Takeda, K., Narashima, F. and Nonaka, S. (1980). *Phytochemistry* **19**, 2175.
Tan, S. C. (1979). *J. Amer. Soc. Hort. Sci.* **104**, 581.
Tigchelaar, E. C., Janick, J. and Erickson, H. T. (1968). *Genetics* **60**, 475.
Timberlake, C. F. and Bridle, P. (1967). *J. Sci. Fd Agric.* **18**, 473.
Timberlake, C. F. and Bridle, P. (1971). *J. Sci. Fd Agric.* **22**, 509.
Timberlake, C. F. and Bridle, P. (1975). *In* "The Flavonoids" (J. B. Harborne, T. J. Mabry and H. Mabry, Eds), pp. 214–266. Academic Press, London and New York.
Timberlake, C. F. and Bridle, P. (1980). *In* "Developments in Food Colours I" (J. Walford, Ed.), pp. 115–150. Applied Science Publishers Ltd, London.
Torre, L. C. and Barritt, B. H. (1977). *J. Food Sci.* **42**, 488.
Tselas, S. K., Georghiou, K. C. and Thanos, C. A. (1979). *Plant Science Letters* **16**, 81.
Tsukui, A., Kuwano, K., Toshio, M. and Wahachiro, T. (1977). *Chem. Abstr.* **87**, 180667.
van der Westhuizen, J. H., Ferreira, D. and Roux, D. G. (1980). *J. Chem. Soc.* Perkin **1**, 1003.

von Elbe, J. H., Bixby, D. G. and Moore, J. D. (1969). *J. Food Sci.* **34**, 113.
Wann, E. V. and Thompson, A. E. (1965). *Proc. Am. Soc. hort. Sci.* **87**, 270.
Watanabe, S., Sakamura, S. and Obata, Y. (1966). *Agri. Biol. Chem.* **30**, 420.
Wellmann, E., Hrazdina, G. and Grisebach, H. (1976). *Phytochemistry* **15**, 913.
Williams, M. and Hrazdina, G. (1979). *J. Food Sci.* **44**, 66.
Williams, M. Hrazdina, G., Wilkinson, M. M., Sweeny, J. G. and Iacobucci, G. A. (1978). *J. Chromatography* **155**, 389.
Williams, R. R., Church, R. M., Wood, D. E. S. and Flook, V. A. (1979). *J. Hort. Sci.* **54**, 75.
Wrolstad, R. E. and Heatherbell, D. A. (1974). *J. Sci. Fd Agric.* **25**, 1221.
Yoshitama, K. (1977). *Phytochemistry* **16**, 1857.
Yoshitama, K. (1978). *Bot. Mag.* (Tokyo) **91**, 207.
Yoshitama, K. and Hayashi, K. (1974). *Bot. Mag.* (Tokyo) **87**, 33.
Yoshitama, K., Hayashi, K., Abe, K. and Kakisawa, H. (1975). *Bot. Mag.* (Tokyo) **88**, 213.
Zennie, T. M. and McClure, J. W. (1977). *Aquat. Bot.* **3**, 49.

CHAPTER 13

Fruit Biochemistry: Highlights, Perspectives and Prospects

D. GRAHAM

Plant Physiology Unit, CSIRO Division of Food Research, North Ryde, NSW, Australia

In a short summary of the meeting it is impossible to do justice to the many excellent papers, so an attempt will be made here to outline my view of the highlights and set these in the context of current biochemistry with some prospects for the future.

The meeting opened in a fitting manner with Professor Jacob Biale (University California, Los Angeles), a founder of post-war fruit physiology and biochemistry, who brought us up-to-date on the current status of the division of fruits into climacteric and non-climacteric groups. There has been a large increase in the number of species, especially those from the tropics, which has been studied. Although there is a considerable range in the magnitude of the climacteric characteristics (e.g. respiration rise, ethylene production) it struck one as distinctly strange that it is possible to make a clear distinction between the two groups of fruits. The reason for this is presumably to do with the "all or nothing" ethylene phenomenon in terms of its climacteric rise. However, it should be remembered that only about 50 species, mainly the horticulturally important ones, have been examined of the many thousands of fruits available. There is a need, therefore, to continue to expand this area of knowledge since the possibility of finding variant behaviour, perhaps with partial climacteric characteristics, must be high. There is also a particular need for more research on development in non-climacteric fruits which have been relatively neglected so far. Why do they not develop a climacteric? How different is their mechanism of ethylene biosynthesis?

Studies on tropical fruits are in their infancy so it is gratifying to see the emphasis given to post-harvest biology by the United Nations and the developments in the ASEAN group of nations (Indonesia, Malaysia, Philippines, Singapore, and Thailand), where a major post-harvest fruit and vegetable training and research centre is now functioning at the University of the Philippines at Los Banos, together with further post-harvest facilities in the other nations. Cooperation between those with expertise on temperate fruits with their colleagues in the tropics could lead to rapid expansion of this field. The time is perhaps overdue to give the overworked apple a rest.

An outstanding recent achievement has been the elucidation of the pathway of bio-synthesis of ethylene, the ripening hormone, by a team at University of California, Davis. This work was described by Shang Fa Yang. The intermediates in the conversion of the sulphur amino acid, methionine, to the unsaturated hydrocarbon, ethylene, have now been identified; assay methods for them have been developed; some characteristics of the enzymes involved are now known and the control points affected by factors such as nitrogen, temperature, hormones (IAA, cytokinins), CO^{2+}, uncouplers and wounding are established. It will now be possible to study the mechanism of ethylene production and answers should be forthcoming to questions such as:

(i) What triggers ethylene production?
(ii) Is it the first event in ripening of climacteric fruit?
(iii) What is the location of the ethylene producing system?
(iv) Why do non-climacteric fruits lack the ethylene rise found in climacteric fruit.

Progress has also been made on the metabolism of ethylene which was described by Beyer (Du Pont, USA). The product is ethylene glycol with ethylene oxide as an intermediate. Both are inactive in physiological response so the options in the cell seem to be:

C_2H_4 → mode of normal action
C_2H_4 → inactivation by conversion to ethylene glycol

Ethylene glycol is an antifreeze with interesting properties in regard to water structure. An analogy could be drawn with freezing where the compound prevents the ordering of water molecules to form ice crystals. Perhaps there is a similar response in cell membranes, presumably at the plasmalemma, to prevent the normal mode of action of ethylene. It remains to be shown whether the metabolism described is the physiological route in the cell or a diversion from it.

Respiratory metabolism was covered in two lucid papers. The first by ap Rees (Cambridge University, UK) was on low temperature-induced sweetening in a wide range of plant tissues, with particular emphasis on potato tubers. In potatoes, sweetening is a significant commercial problem during storage for the processing industry. Significant progress has been made by ap Rees' laboratory in careful studies including both enzymological and carbon balance experiments showing that a major reason for sweetening (accumulation of sucrose and reducing sugars) is the cold-lability of a regulatory enzyme of glycolysis, phosphofructokinase. This results in diversion of carbon from glycolysis into sugars. Ap Rees was careful to point out that this is not necessarily the only mechanism involved, but the work opens the way for studies on low temperature metabolism. It is interesting that a cold-resistant plant of tropical origin—the potato—has a cold-labile enzyme. Other examples of cold-sensitive enzymes found by other laboratories include RuBP carboxylase, malic dehydrogenase, and PEP-carboxylase.

Lance (Pierre et Marie Curie University, Paris) outlined the situation of cyanide-insensitive respiration by the alternate pathway of electron transport. Although there is some disagreement regarding the physiological usefulness of the CN-insensitive pathway to the plant, it appears to be of widespread distribution. A potential physiological role for this pathway is the re-oxidation of NADH with limited ATP production (from the first phosphorylating site), thereby enabling carbon skeletons to be produced in situations without high ATP production. Recent Japanese studies have shown induction of the CN-insensitive pathway in the cold, which may indicate its potential importance for fruit storage in the cold. The way now seems open to undertake more detailed studies of biochemistry during fruit storage, for example during controlled atmosphere storage. Some work from the East Malling group (Bartley, Hatfield and Knee) along these lines was displayed during a poster session. They showed that development of aroma in stored apples is due to synthesis of simple aliphatic esters, mainly acetates and butyrates of alcohols with up to six carbon atoms. This synthesis requires the production of alcohols, probably from fatty acid metabolism. In low oxygen storage, aroma development is impaired. The intermediary metabolism during fruit storage, and especially during the development of storage disorders, could now provide some insights into how fruit cells regulate their behaviour as they enter senescence.

The second day dealt with softening and fruit texture, a somewhat neglected and difficult field. Knee (East Malling, UK) described the current status of cell wall chemistry with emphasis on the compositon of the primary cell wall and middle lamella in fruits, especially apple and strawberry. The situation seems more complex than in the sycamore cell suspension system. Not surprisingly each fruit poses its own problems in the complex and difficult fractionation and assay methodologies. Nevertheless, Knee was able to report significant progress in characterization of cell wall components during ripening. He suggested that methylation of cell wall components may occur by export of *S*-adenosyl methionine, a highly reactive intermediate in ethylene biosynthesis. How such a reactive compound can move within the cell and to the cell wall remains to be discovered. The methylation is postulated to result in polysaccharide slippage and thus growth and separation of cells which is necessary for fruit development and softening during the ripening process. Further comparative biochemistry is indicated since there seems to be no universal system in the fruit cell wall. We await with interest a study of methylation turnover to substantiate Knee's hypothesis.

The enzymology of fruit softening is now an active research field with several laboratories involved. Hobson (Littlehampton UK) described the enzymology of this problem with emphasis on cellulase, glycosidases, pectin methyl esterases (PME) and polygalactaronases (PG). Again there is wide variation between species in their enzymic complement and in how these change during ripening. PME seems to be universally present but "cellulase" can be very active (e.g. in avocado, banana) or low in activity (e.g. tomato). Although PG cannot be regarded as the only enzyme active in fruit softening, the main difference between quick and slow ripening fruit such as the peach is that the former contain both endo- and exo-PG, while the latter have only exo-PG. The importance of turgor in causing cell separation and formation of intercellular spaces was mentioned. The question was raised as to how Ca^{2+} maintains cell rigidity and how this is lost, with release of Ca^{2+} from the cell wall. This also poses the important problem of how Ca^{2+} is involved in minimizing fruit storage disorders such as bitter pit and low temperature breakdown.

Grierson (Nottingham, UK) described in a masterly presentation how the age of molecular biology has at last arrived in fruit biochemistry. He described the work of his laboratory on the synthesis of the softening enzymes mainly PG and PME, during the ripening of the tomato. Two isoenzymes of PG were described of mol. wt 100,000 daltons (PG1) and 42,000 daltons (PG2), respectively. PG1 appears first and has properties different from PG2 though antibody raised against PG2 cross-reacts with PG1 and also with a variant form from the *Nr* mutant of tomato. The possibility is that PG1 is a dimer of PG2. Synthesis of PG appears to occur *de novo* during ripening. Messenger RNA for PG has been isolated. It is not yet clear whether control for PG synthesis during ripening is at the translational or transcriptional level or a combination of the two. PG appears at the same time as ethylene production which does not accord with the hypothesis of Tigchelaar and McGlasson that formation of PG triggers the ripening process. It will be instructive to see now whether PG has a precursor molecule of higher mol. wt, which now seems characteristic of proteins which have to pass through a membrane barrier to their cellular location, and where PG and the other "ripening" enzymes are synthesized?

It was especially fortunate for the meeting that Professor John Ellis (Warwick, UK) was the recipient of the Tate and Lyle Award, since he delivered a "state of the art" lecture on the molecular biology of chloroplast development which has many parallels to the ripening of fruit. Though the molecular biology of fruit ripening is in its infancy, it was possible to see the fascinating possibilities to which this approach is likely to lead for our understanding of the fruit ripening process. Is it possible that gene cloning of fruit genes (e.g. some of the ripening genes) is just around the corner?

The final day was devoted to phenolics and pigments. These are important components of the fruit, especially for our visual and olfactory senses, and therefore, have significance as quality factors. Mayer (Hebrew University of

Jerusalem, Israel) described the changes in phenolases during fruit ripening. The function of these enzymes is unknown though it seems unlikely that they are involved in an energy-conserving electron transport system. Involvement in resistance to fungal attack, especially during ripening and protection against various predators during initial stages of fruit development were other functions canvassed. As with any other enzyme it is always advisable to differentiate between real changes in enzymic activities and those due to differential extraction or inactivation during the developmental sequence. The properties and mechanism of action of phenolases were discussed by Robb. A number of these enzymes, particularly laccases, ascorbic acid oxidase and tyrosinase, are now well characterized. Rhodes (F.R.I. Norwich) discussed phenol metabolism in fruits and vegetables. Physiological and mechanical stresses can induce the synthesis of various phenolic compounds, with the oxidation of diphenols, production of polymers and lignification ensuing. Cold-induced changes in chlorogenic acid appear particularly significant in potatoes and tomatoes, with the formation (or activation) of enzymes of this pathway. Research into the control of browning reactions is not a popular field but is obviously potentially of economic importance. Knowledge of the mechanisms of cold-induced formation of phenolases would be valuable in studies of chilling injury.

Timberlake (Long Ashton, UK) outlined the chemistry and properties of anthocyanins. The importance of heat damage, causing a loss of colour, and its slow reversal on cooling, emphasized the need for careful handling of fruits, vegetables and cut flowers. Limited nutrition of apple trees enhances the colour in the fruit. Furthering the understanding of anthocyanin biochemistry assumes significance by virtue of the fact that approximately 10,000 tons of anthocyanins are consumed annually, mainly in wine. There does appear to be a relative dearth of information of the physiology of anthocyanins and the regulation of their synthesis in fruits. The other major pigments of fruits and vegetables, apart from chlorophyll, are the carotenoids. Davies (Aberystwyth, UK) gave a comprehensive review of this topic which has not been included in the book. He emphasized the phytochemists' predilection with the chemistry of these fascinating molecules. But he also discussed the importance of their spectral properties which allow their characteristic aesthetic appeal and their important function as photoprotectants in plants. Again we appear to know more of the chemistry of these compounds than their physiology in the context of fruit development and ripening. There is obviously wide potential for plant breeding for colour development and improvement in nutritional quality, since carotenoids are important sources of vitamin A and its precursors.

No meeting of this kind can be comprehensive in coverage. For example, there was little said on the important role of Ca^{2+} in fruits and vegetables, especially in a post-harvest context. Little is presently known on the mode of action of this mineral ion which seems to play, perhaps, a central role in the plant cell.

Controlled atmosphere (CA) storage is a commercial success for storage of fruits without any real understanding of why it works. Although some mention was made of this topic, most fruit biochemistry laboratories have little interest in it. This is unfortunate since CA storage for long periods appears to result in a rather poor quality product, a fact which is becoming apparent to the consumer. This could be a situation where biochemistry could at least define the problem more closely. There are many observations in fruit ripening and fruit storage which remain unexplained, for example, the biochemistry of storage disorders, water loss effects on storage disorders, host-parasite relations in pathology, and the role of Ca^{2+} to mention just a few. A central, and yet unexplained, question is what triggers ripening. As indicated above the key to open this lock may have been forged. It remains to turn that key. One important way in which this may be speeded up is by collaboration. By comparison with other fields of research fruit biochemistry has a relatively small population of workers. Collaboration and especially comunication between the various workers should be encouraged to minimize duplication of work and maximize the returns of our efforts.

Author Index

The numbers in italic refer to those pages where References are listed in full

A

Abe, K., 228, *247*
Abeles, A. L., 91, *104*
Abeles, F. B., 90, 91, 95, 97, *104*, 107, 108, 109, 110, 111, *120*, *121*
Abeles, R. H., 187, *191*
Adams, D. O., 24, *37*, 91, 92, 93, 94, 95, 96, 101, 102, 103, *104*, *105*, *106*
Adato, I., 2, 3, 8, 9, 10, 14, 24, 37, *38*
Aharoni, Y., 6, *37*
Akamine, E. K., 9, 10, 20, 21, 22, *37*
Akerboom, T. P. M., 55, *60*
Akhtor, S., 164, 171, *180*
Albersheim, P., 125, 126, *131*, *132*, 134, 136, 138, 146, *147*, *148*
Allen, P. J., 83, *85*
Amorim, H. U., 162, 165, *179*
Amorim, V. L., 162, 165, *179*
Amrhein, N., 102, *104*
Anderson, D. W., 240, *244*
Antonini, E., 186, *191*
Apelbaum, A., 98, 99, *104*
ap Rees, T., 41, 42, 43, 45, 47, 48, 49, 52, 53, 54, 55, 56, 58, *60*, *61*, 77, *84*, 250
Archer, S. A., 136, 138, *148*
Arron, G. P., 36, *37*
Asada, Y., 194, *218*
Asahi, T., 65, *86*, 193, *219*
Asen, S., 225, 226, 227, 228, 229, *244*
Attridge, T. H., 197, 214, *218*
Avants, J. K., 127, *132*, 141, *148*
Awad, M., 30, 31, *37*, 130, *131*

B

Badenhuizen, N. P., 64, *87*
Bahr, J. T., 67, 72, 75, 77, *84*, *86*
Bailey, R. W., 134, *148*
Bain, J. M., 6, *37*
Baldry, C., 181, *191*
Ballinger, W. E., 242, *244*, *245*
Bangerth, F., 101, *104*
Barcus, D. E., 3, 4, 8, 10, 12, *37*
Barker, J., 55, *60*
Barmore, C. R., 243, *244*
Barnell, F., 163, *179*
Barnell, H. R., 26, *37*, 163, *179*
Barnes, M. F., 139, *147*
Barrett, A. J., 134, 136, *147*, *148*
Barritt, B. H., 237, *244*, *246*
Bartley, I. M., 128, *131*, 133, 139, 140, 141, 142, *147*, 251
Barz, W., 222, *244*
Bauer, W. D., 125, 126, *132*, 134, 136, 138, 146, *147*, *148*
Baur, A. H., 90, 91, *104*
Beckman, C. H., 163, *179*
Beech, F. W., 237, *245*
Beevers, H., 49, *61*, 64, 66, 70, *85*
Behmer, M., 110, *121*
Ben Abdelkader, A., 205, *218*
Ben-Arie, R., 123, 125, 128, 129, *131*
Bendall, D. S., 66, 70, *84*, 181, *191*, 199, 200, *219*
Benjamin, N. D., 164, *179*, 181, *191*
Ben-Shalom, N., 165, 175, 176, *179*, *180*
Benveniste, I., 198, 200, *218*
Berendsen, H. C. J., 186, *192*
Berger, R. K., 146, *147*
Bernlohr, R. W., 57, *61*
Besford, R. T., 130, *131*
Bevenue, A., 47, *61*
Beyer, E. M., 102, *104*, 107, 109, 110, 114, 115, 116, 117, 118, 119, *121*, 227, *244*, 250
Biale, J. B., 1, 2, 3, 4, 5, 6, 7, 8, 9, 10, 12, 13, 14, 16, 20, 22, 24, 32, 33, 34, *37*, *38*, *39*, 81, *85*, 90, *104*, 249
Biederman, W., 163, *179*
Bieleski, R. L., 47, *60*
Billett, E. E., 197, 200, *218*
Bishop, R. C., 237, *244*

Bixby, D. G., 238, *247*
Black, M. K., 81, *87*
Blanpied, G. D., 6, 12, *37*
Blomstrom, D. C., 114, 115, 116, 117, 119, *121*
Bloom, H. L., 129, *132*, 139, *148*
Blumenfeld, A., 12, 16, *38*
Boas, J. F., 186, *192*
Bock, P. E., 57, *60*
Boller, T., 92, 101, 102, *104*
Bonner, J., 130, *131*, 146, *147*, 175, *179*
Bonner, W. D., 64, 66, 67, 68, 70, 72, 75, 77, 79, *84*, *85*, *86*
Bookelman, H., 55, *60*
Bose, A. N., 7, *37*
Bouchilloux, S., 184, 186
Boveris, A., 67, *86*
Bradford, K. J., 99, 100, *104*, *105*
Brady, C. J., 35, *37*
Bridle, P., 222, 225, 236, *246*
Briggs, R. G., 190, *191*
Brooks, D. W., 187, *191*
Brouillard, R., 222, 223, 224, 225, *244*
Bruemmer, J. H., 165, *179*
Bruinsma, J., 125, *132*
Brunori, M., 186, *191*
Buchanan, B. B., 83, *85*
Buche, T., 198, *218*
Bucke, C., 181, *191*
Buckmire, R. M., 243, *244*
Buescher, R. W., 9, 10, *39*, 127, 130, *131*, *132*
Buhler, D. R., 109, 110, *121*
Burg, E. A., 2, 4, 12, 20, 21, 22, 23, *37*, 102, *105*, 108, 109, 111, 119, *121*
Burg, S. P., 2, 4, 12, 13, 20, 21, 22, 23, *37*, 90, 92, 102, *105*, 108, 109, 111, 119, *121*
Burroughs, L. F., 93, *105*
Burton, W. G., 42, 44, 48, *60*
Butt, V., 195, 196, *220*

C

Cameron, A. C., 94, *105*
Camm, E. L., 197, *218*
Cammack, R., 67, *84*, 199, *219*
Campbell, D. J., 99, *105*
Carolus, R. J., 43, *60*
Carpita, N., 146, *147*
Carré, M. H., 141, *147*
Carreno-Diaz, R., 230, *244*
Casoli, U., 230, *244*
Catlin, P. B., 10, 11, *38*, 241, *246*
Chalmers, D. J., 237, *244*, *245*
Chalmers, J. H., 67, *85*
Chan, H. T., 165, *179*, 236, *244*
Chance, B., 68, *84*
Chandler, W. H., 11, *37*
Chaplin, M. F., 164, *179*
Chapman, D., 50, *60*
Chapman, E. A., 50, *61*
Chauveau, M., 65, 67, 70, 71, 72, 77, *84*, *85*
Cherif, A., 205, *218*
Cherry, J. H., 20, *38*
Church, R. M., 237, *247*
Clagett, C. O., 90, *105*
Clegg, C. J., 49, *60*
Clendening, K. A., 12, 13, *37*
Coates, G. E., 109, *121*
Cockerill, A. F., 93, *105*
Cocking, E. C., 124, *131*
Coggan, P., 162, *180*
Coleman, J. E., 183, 184, 186, *191*
Combs, J., 181, *191*
Condorelli, P., 241, *245*
Conn, E. E., 78, 79, *84*, *85*, *86*, 194, 198, 199, *219*
Coombe, B. G., 2, *37*
Coscia, C. J., 205, *219*
Cotte-Martinon, M. G., 199, *218*, *220*
Cousins, H. H., 90, *105*
Craft, C. A., 208, *219*
Crane, J. C., 4, 8, 12, *38*, 89, *105*, 241, *246*
Creasy, L. L., 208, *218*, 233, 237, 243, *245*, *246*
Crowden, R. K., 235, *246*
Cultiera, R., 164, *180*

D

Dall'aglio, G., 230, *244*
Darby, L. A., 150, *160*
Darvill, A. G., 145, 146, *147*
Davies, J. W., 124, *131*
Davis, L., 92, 93, *105*
Dawson, C. R., 187, *191*, *192*
Day, D. A., 36, *37*
De, S., 7, *37*
Deinum, J., 190, *191*
Dekazos, E. D., 238, *244*
Delaporte, B., 222, 223, 224, 225, *244*
Delincee, H., 127, *131*

Delmer, D. P., 146, *147*
Demandre, C., 205, *218*
De Swardt, G. H., 213, *218*
de Troostembergh, J., 72
Dickson, G., 187, *191*
Dilley, D. R., 24, *39*, 99, *104*, 163, *179*
Dixon, M., 188, *191*
Dixon, W. L., 41, 53, 54, 55, 56, 58, *60*
Dizengremel, P., 65, 67, 68, 70, 71, 74, 81, *84*, *85*
Dizik, N. S., 164, *179*
Dolendo, A. L., 29, 30, *37*
Dostal, H. C., 20, 24, *38*
Douce, R., 79, *85*
Drake, S. R., 238, *244*
Du, C. T., 236, 238, 242, 243, *244*, *245*
Dubernet, M., 165, 173, *179*
Dubois, J. E., 222, 224, 225, *244*
Ducet, G., 64, 66, *85*, 199, *218*, *220*
Duchesne, M., 197, 207, *219*
Duckworth, H., 184, 186, *191*
Dull, G. G., 6, 7, 12, 28, 29, 37, *38*, 127, *132*
Dupin, J-F., 238, *246*
Duro, F., 241, *245*
Durst, F., 198, 200, *218*
Duvenage, A. J., 213, *218*
Dyck, R., 37, *38*

E

Eakes, I. L., 6, 23, *37*, *38*
Eason, R., 184, *192*
Ebel, J., 207, *219*
Edwards, D. L., 65, 67, *85*
Edwards, H. H., 83, *85*
Egeter, H., 227, *244*
Eickman, N. C., 186, *191*
El-Beltagy, A. S., 98, *105*
El Hage Chahine, J. M., 224, 225, *244*
Elliot, D. C., 64, *85*
Ellis, R. J., 214, *219*, 252
Elmer, D. H., 90, *105*
el Saifi, A. F., 55, *60*
Englesma, G., 208, *219*
Erickson, H. T., 230, *246*
Eriksson, K. E., 139, *147*
Estabrook, R. W., 206, *219*
Ettlinger, L., 183, 184, 186, *191*
Evans, L. H., 185, 186, *191*

F

Faragher, J. D., 237, *244*, *245*
Fee, J., 190, *191*
Fenton, C. A. L., 94, *105*
Ferreira, D., 232, *246*
Fielding, A. M., 136, 138, *148*
Finazzi Agro, A., 186, *191*
Fink, A., 227, *244*
Fisher, D. F., 90, *105*
Fliege, R., 51, *60*
Fling, M., 182, 183, 184, 186, *191*
Flook, V. A., 237, *247*
Flügge, U-I, 51, *60*
Flurkey, W. H., 163, 172
Forkmann, G., 231, *245*
Forrence, L. E., 109, *121*
Francis, F. J., 230, 236, 238, 242, 243, *244*, *245*, *246*
Franks, F., 41, 58, *60*
Freebairn, H. T., 24, *39*
Frenkel, C., 37, *38*, 65, 74, *85*, *86*, 123, 125, 128, 129, *131*
Friedan, C., 57, *60*
Fritig, B., 197, 207, *219*
Fritsch, H., 231, *245*
Frontera, M. A., 243, *246*
Fuchs, Y., 125, 128, *132*
Fujino, D. W., 102, *105*
Fukui, S., 92, *105*
Fuleki, T., 230, *245*
Fuller, W. A., 54, *60*, 77, *84*

G

Galletta, G. J., 242, *244*
Galliard, T., 81, *85*
Gane, R., 12, *38*, 90, *105*
Gan-Zvi, E., 177, *180*
Garden, L., 225, *246*
Garland, P. B., 67, *85*
Garner, C. D., 190, *191*
Gazit, S., 12, 16, *38*
Geisler, G., 6, 7, 12, *38*
Georghiou, K. C., 222, *246*
Giaquinta, R. T., 116, *121*
Ginsburg, B. Z., 134, *147*
Girardin, J. P. L., 89, *105*
Goeschl, J. D., 2, 8, 17, 18, 20, 22, *38*, *39*, 90, 95, *105*
Golan, A., 163, *179*

Gonzalez, G. A., 163, 164, *180*
Goo, T., 9, 10, 20, 21, 22, *37*
Gortner, W. A., 28, 29, *38*
Goto, T., 226, 227, 229, *245*
Grabham, M., 89, *105*
Graham, D., 249
Grant, G. T., 126, *131*
Grau, N., 230, *244*
Gray, D., 42, *60*
Green, D. G., 43, *60*
Green, M. L. H., 109, *121*
Greene, D. W., 237, *245*
Gregory, R. P. F., 181, *191*
Grierson, D., 128, *131*, 149, 150, 151, 152, 153, 154, 156, 157, 158, *160*, 252
Griffiths, S. K., 65, 74, *85*
Grisebach, H., 195, 197, 199, 200, 207, *218*, *219*, 229, 231, 234, *245*, *246*, *247*
Gross, G. G., 195, 206, *219*
Gross, K. C., 127, 128, 129, *131*, 141, *147*
Gueffroy, D. E., 240, *244*
Gunsalus, I. C., 199, 200, 201, *219*
Gutteridge, S., 183, 184, 187, *191*, *192*
Guy, C. L., 43, 48, *61*
Guzig, H. J., 67, *85*

H

Haard, N. F., 37, *38*
Haas, D. W., 65, 74, *85*
Hackett, D. P., 64, 65, 74, *85*
Haddock, B. A., 67, *85*
Hahlbrock, K., 195, 197, 207, *219*
Hall, M. A., 98, 99, *105*, *106*, 113, 115, 117, 118, *121*, 125, *131*, 145, 146, *147*
Haller, M. H., 12, *38*
Hansen, E., 2, 12, *38*, 109, 110, *121*
Hanson, A. D., 91, *105*, 118, *121*
Harborne, J. B., 222, 230, 234, 235, 237, 238, 242, 243, *245*, *246*
Harding, P. L., 12, *38*
Harel, E., 161, 164, 165, 169, 170, 171, 172, 173, 174, 175, 176, 177, 178, *179*, *180*, 181, 183, 184, 190, *191*, *192*
Harley, J. L., 81, *87*
Hartmann, C., 81, *85*, 238, *246*
Hartmann, H. T., 10, 11, *38*
Hasegawa, S., 128, *131*
Haslam, E., 232, *245*
Hassid, W. Z., 144, *147*
Hathout Bassim, T. A., 233, *245*
Hayashi, K., 227, 228, 230, *245*, *246*, *247*
Heatherbell, D. A., 244, *247*
Heber, U., 44, 45, *60*
Heinemann, S. F., 182, 183, 184, 186, *191*
Heinzmann, U., 235, *245*
Heldt, H. W., 51, *60*
Henderson, L. E., 189, *191*
Henry, M. F., 64, 67, *85*
Herner, R. C., 9, *38*, 92, 101, 102, *104*
Hill, A. C., 193, 196, 198, 199, 200, 207, *219*
Hill, R., 66, 70, *84*
Hinton, D. M., 128, *131*, 141, *148*
Hirth, L., 197, 207, *219*
Hiyama, T., 93, *105*
Ho, D. T., 110, *121*
Hobson, G. E., 2, 32, 33, *38*, 81, *85*, 123, 126, 128, 130, *131*, 139, *147*, 150, 151, *160*, 163, 165, 166, *179*, 252
Hochachka, P. W., 52, *61*
Hoffman, N. E., 95, 96, *105*
Horowitz, N. H., 182, 183, 184, 186, *191*
Hösel, W., 222, *244*
Hoshino, T., 226, 227, 229, *245*
Hotta, R., 230, *246*
Howard, H. W., 230, *245*
Hrazdina, G., 225, 226, 227, 230, 233, 234, 241, *245*, *246*, *247*
Huelin, F. E., 6, *39*
Hughes, J. C., 42, *60*, 134, *148*
Hulme, A. C., 2, 18, *38*, *39*, 81, *85*, 170, *179*, *180*
Huq, S., 66, 67, *85*

I

Iacobucci, G. A., 230, 241, *247*
Ikuma, H., 64, *85*
Imai, K., 230, *246*
Imaseki, H., 91, *105*, 193, *219*
Imbert, M. P., 230, *245*
Inamoto, H., 241, *245*
Ingledew, W. J., 67, *86*
Ingraham, L. L., 188, *192*
Iredale, H., 230, *245*
Isherwood, F. A., 48, 50, 51, 59, *61*, 139, *147*
Ishii, S., 130, *131*
Ishikura, N., 241, 243, *245*
Ivanov, T., 173, *179*
Ivanova, A., 173, *179*

Iwata, R. Y., 236, *245*
Iwata, T., 20, *38*

J

Jackson, M. B., 99, *105*
Jacques, D., 232, *245*
James, W. O., 64, 66, 70, *85*
Janes, H. W., 65, 74, *85*, *86*
Jang, R., 130, *131*, 146, *147*
Janick, J., 230, *246*
Jansen, E. F., 110, *121*, 130, *131*, 146, *147*
Jen, J. J., 163, 165, 172, *179*
Jensen, L. H., 190, *192*
Jerie, P. H., 99, *106*, 113, 115, 117, 118, *121*
Jermyn, M. A., 139, *147*
Jewell, G. G., 124, *131*
Jha, O. P., 243, *246*
Joel, D., 163, 165, *179*
Jolley, R. L., 183, 185, 186, *191*, *192*
Jones, J. D., 170, *179*
Jones, R. L., 146, *148*
Jordan, W. R., 98, *105*
Joshi, P. R., 163, 165, 166, *179*
Joslyn, M. A., 126, *132*, 141, *147*, 162, *179*
Jugmohansingh, K., 164, *180*
Julian, E. A., 240, *244*
Jurd, L., 227, 231, *245*

K

Kader, J. C., 70, 74, *85*
Kahler, K. R., 163, 165, *179*
Kahn, V., 163, 164, 165, 166, 175, 176, *179*, *180*
Kaji, A., 136, *148*
Kakisawa, H., 228, *247*
Kamemoto, H., 236, *245*
Kamsteeg, J., 229, 231, *245*
Kandler, O., 42, *61*
Kannan, K., 189, *191*
Kao-Jao, T. H. C., 236, *244*
Kasai, Z., 110, *121*
Kauss, H., 144, *147*
Keegstra, K., 125, 126, *132*, 134, 136, 138, 146, *147*, *148*
Keep, E., 238, *245*
Kende, H., 91, 92, 93, 101, 102, *104*, *105*, 118, *121*
Kennedy, M. G. H., 51, *61*
Kepner, R. E., 240, *244*
Kertesz, D., 182, 184, 186, *191*
Kho, K. F. F., 229, *245*
Kidd, F., 2, 12, *38*, 90, *105*
Kidron, M., 165, 173, *179*, 184, *191*
Killias, U., 134, *147*
Kirk, Othmer, J., 109, *121*
Kisler, N., 123, 125, 128, 129, *131*
Klein, R. M., 237, *244*
Klingenberg, M., 72, *85*
Knapp, F. W., 164, 165, *179*
Knee, M., 125, 131, *132*, 133, 135, 136, 138, 139, 141, 142, 143, 144, 145, *147*, *148*, 251
Knight, R. L., 238, *245*
Knobloch, K. H., 207, *219*
Koidt, H., 93, *105*
Kono, N., 57, *61*
Konze, J. R., 93, *105*
Kosiyachinda, S., 4, 19, 20, 21, 22, *38*
Koukol, J., 194, *219*
Kraus, B. H., 28, 29, *38*
Kreuzaler, F., 197, 207, *219*
Krishnamurthy, S., 163, *180*
Kröger, A., 72, *85*
Krueger, R. C., 188, *191*
Ku, L. L., 35, *38*
Kubowitz, F., 182, 186, *191*
Kuiper, H. A., 186, *191*
Kukimura, H., 230, *245*
Kulkarni, R. R., 165, *179*
Kunishi, A. T., 90, *105*, 144, *148*
Kushman, L. J., 242, *244*
Kuttner, R., 188, *191*
Kuwano, K., 230, *246*

L

Labavitch, J. M., 139, 145, 146, *148*
Laborda, F., 136, 138, *148*
Lamb, C. J., 198, 200, 202, 204, 207, 215, *219*
Lambowitz, A. M., 67, 72, *85*
Lance, C., 33, *38*, 63, 65, 66, 67, 68, 70, 71, 72, 77, 81, 82, *84*, *85*, *86*, 251
Lanphear, F. O., 43, 44, 48, *61*
Larrabee, J. A., 186, *191*
Laties, G. G., 27, 28, 36, *37*, *39*, 65, 73, 74, 75, 76, *85*, *86*
Lau, O., 95, *105*
Lea, A-G. H., 237, *245*
Leather, G. R., 109, *121*

Lee, M., 127, *132*
Lehle, L., 42, *61*
Lehman, E., 165, 173, 174, *179*
Leopold, A. C., 67, *86*
Lerch, K., 183, 184, 186, 187, 190, *191*
Lerner, A. B., 188, *191*
Lerner, H. R., 164, 171, 172, 174, 175, *179*
Letham, D. S., 134, *148*
Levitt, J., 42, 44, 48, *61*
Lewis, C. E., 125, 128, *132*
Lewis, D. H., 190, *191*
Li, L. P., 175, *179*
Li, P. H., 43, *61*
Lieberman, M., 90, 93, 101, *105*, 144, *148*, 208, *219*
Liljas, A., 189, *191*
Lin, C. H., 78, *86*
Lindskog, S., 189, *191*
Lineberger, R. D., 46, *61*
Linehan, D. J., 134, *148*
Lizada, M. C. C., 94, *105*
Lodh, S. B., 7, *37*
Long, S. P., 181, *191*
Lontie, R., 184, 185, 186, *191*
Lord, W. J., 237, *245*
Lourenco, E. J., 215, *219*
Lozano, J., 163, 164, *180*
Ludwig, G. H., 194, *219*
Luh, B. S., 29, 30, *37*, 165, *180*, 243, *245*
Luhadiya, A. P., 165, *179*
Lürssen, K., 94, *105*
Lutz, J. M., 12, *38*
Lyons, J. M., 4, 8, 20, 22, *38*

M

Macheix, J-J., 163, *180*
MacDonald, I. R., 214, *219*
Macmillan, J. D., 127, *132*
Macrae, A. R., 81, *83*
Maestro, D., 243, *245*
Magar, N. G., 163, *180*
Magyarosy, A. C., 83, *85*
Mahecha, G., 243, *245*
Maki, Z., 241, *245*
Makino, N., 185, 186, *191*
Makus, D. J., 242*245*
Maness, E. P., 242, *244*
Mannella, C. A., 79, *85*
Mapson, L. W., 24, *38*, 90, *105*, 188, 189, *192*
Marbach, I., 163, 165, *179*
March, J. F., 90, *105*
Marei, N., 4, 8, 12, *38*
Margna, U., 197, *219*, 234, *245*
Marshke, C. K., 57, *61*
Martin, F. W., 17, *39*
Mason, H. S., 182, 183, 184, 185, *191*, *192*
Mathews, A. G., 162, *180*
Matsuda, K., 139, *148*
Matsumoto, I., 194, *218*
Matsumoto, U., 226, *245*
Mattick, L. R., 230, *245*
Maxie, E. C., 10, 11, *38*, 89, *105*
Maycock, A. L., 187, *191*
Mayer, A. M., 161, 163, 164, 165, 169, 170, 171, 172, 173, 174, 175, 176, 177, 178, *179*, *180*, 181, 183, 184, 190, *191*, *192*, 195, 252
Mazliak, P., 205, *218*
Mazzoco, F., 164, *180*
McCallion, R. F., 182, *192*
McClendon, J. H., 127, *132*
McClure, J. W., 222, 236, *246*, *247*
McCready, C. C., 81, *87*
McGlasson, W. B., 2, 3, 8, 9, 10, 14, 20, 22, 23, 24, 37, *38*, *39*, 110, *121*, 130, *132*
McMahill, P., 184, 186, *191*
McMichael, B. L., 98, *105*
McMillan, R. T., 243, *244*
McMurchie, E. J., 23, *38*
Medrano, M. A., 243, *246*
Meehan, T. D., 205, *219*
Meeuse, B. J. D., 64, 77, 78, *86*
Melin, C., 238, *246*
Mellenthin, W. M., 20, 22, 24, *39*
Metzler, D. E., 92, 93, *105*
Milde, H., 46, *61*
Miller, E. V., 90, *105*
Miller, J. M., 79, *86*
Miller, W. H., 187, *192*
Millerd, A., 81, 83, *86*
Misic, P. D., 237, *246*
Mizrahi, Y., 20, *38*
Mohr, W. P., 124, 125, *132*
Monro, J. A., 134, *148*
Montezinos, D., 146, *147*
Montgomery, M. W., 163, 164, 166, *179*, *180*, 181, *191*
Moore, A. L., 67, *86*
Moore, J. D., 238, *247*
Moorhouse, R., 81, *85*
Moreau, F., 66, 71, 80, 81, 82, *86*

Moreland, D. E., 74, *86*
Morris, E. R., 126, *131*
Moss, T. H., 186, *191*
Moulet, A-C., 238, *246*
Mueller, W. C., 163, *179*
Muir, R. M., 146, *148*
Murphy, P. J., 205, *219*
Murr, D. P., 101, *105*
Murthy, V. J., 184, *192*
Mussell, N., 130, *132*

N

Nagel, C. W., 239, *246*
Nakagawa, H., 127, *132*
Nakagawa, K., 20, *38*
Nakamura, T., 184, *192*
Nakano, M., 65, *86*
Nakayama, T. O. M., 236, *244*
Narashima, F., 227, *246*
Naumann, K., 94, *105*
Neal, G. E., 139, 144, *148*
Neljubow, D., 89, *105*
Nelson, J. W., 238, *244*
Nether, K., 227, *244*
Newsholme, E. A., 53, *61*
Nicholls, D. G., 78, *86*
Niederpruem, D. J., 65, 74, *85*
Nishimura, M., 49, *61*
Nonaka, S., 227, *246*
Norris, K. H., 225, 226, 228, 229, *244*
Northcote, D. H., 134, 136, *147*, *148*
Nozaki, H., 93, *105*
Nozzolillo, C., 235, *246*
Nukaya, A., 139, *148*
Nyman, P., 189, *191*
Nyns, E. J., 64, 67, 72, *85*

O

Oba, K., 194, *220*
Obata, Y., 230, *247*
O'Connell, P. B. H., 35, *37*
Ogata, K., 20, *38*
Ogura, N., 127, *132*
Ogura, Y., 184, *192*
Olmstead, A. J., 12, 16, 20, 24, *37*
Opie, C. T., 232, *245*
Osawa, Y., 228, *246*
Osborne, D. J., 139, 142, 143, *148*
Osmond, C. B., 81, *86*
Owens, L. D., 144, *148*
Oydvin, J., 238, *246*

P

Padron, M. P., 163, 164, *180*
Palmer, J. K., 163, 164, *180*
Palmer, J. M., 64, 66, 67, 68, 79, *84*, *85*, *86*
Parker, J., 43, *61*
Parpia, H. A. B., 162, 163, *180*
Parrish, D. P., 67, *86*
Passam, H. C., 72, 74, 78, 79, *86*
Patchett, B. J., 139, *147*
Patil, D. L., 163, *180*
Paton, J. Y., 184, *192*
Patwardhan, M. U., 165, *180*
Paull, R. E., 146, *148*
Peacock, B. C., 16, 23, *38*
Pecket, R. C., 233, *245*
Pederson, T. C., 199, 200, 201, *219*
Pellet, N. E., 44, *61*
Penny, D., 134, *148*
Pesis, E., 125, 128, *132*
Peynaud, E., 7, *38*
Pezzoli, C., 144, *148*
Pfeil, W., 57, *61*
Pflander, R., 200, *219*
Pifferi, P. G., 164, *180*, 230, 235, *246*
Pilnik, W., 29, *38*, 127, *132*
Pliny the Elder, 41, *61*
Pollock, C. J., 41, 43, 45, 47, 52, 53, 55, 56, *61*
Pomerantz, S., 184, *192*
Ponting, J., 162, *179*
Poovaiah, B. W., 126, *132*, 139, *148*
Porter, L. J., 232, *245*
Potter, A. L., 47, *61*
Potts, J. R. M., 198, 199, *219*
Poulton, J. E., 234, *246*
Powell, R. D., 98, *105*
Prasad, U. S., 243, *246*
Pratt, H. K., 2, 4, 8, 17, 18, 20, 22, 29, 30, *37*, *38*, *39*, 90, 95, 97, *104*, *105*
Pratt, J. M., 189, *192*
Pressey, R., 42, 46, 47, 51, *61*, 126, 127, 128, *131*, *132*, 141, *148*
Privalov, P. L., 57, *61*
Proctor, J. T. A., 230, 237, 243, *246*
Proebsting, E. L., 238, *244*
Puech, A. A., 241, *246*

Q

Quazi, M. H., 24, *39*
Quesnel, V. C., 164, *180*

R

Radler, F., 6, 7, 12, *38*, 165, 173, *180*
Raison, J. K., 50, *61*
Ramaswamy, S., 165, *180*
Rando, R. R., 92, *105*
Rattanapanone, N., 158, *160*
Ratzlaff, C. D., 43, *60*
Ray, P. M., 139, 145, 146, *148*
Rebeiz, C. A., 241, *246*
Rechtoris, C., 130, *132*
Redgwell, R. J., 47, *60*
Rees, D. A., 126, *131*, 135, 138, *148*
Rege, D. V., 165, *180*
Reid, M. S., 18, 19, 20, *39*, 102, *105*
Reid, P. D., 146, *147*
Reinhammer, B., 182, 190, 191, *191*, *192*
Remmer, H., 206, *219*
Rhodes, M. J. C., 2, 6, 18, 19, 20, *39*, 81, *85*, 90, *105*, 193, 194, 195, 196, 197, 198, 200, 206, 207, 210, 211, 215, 216, *219*, 253
Ribeiro, S., 164, 171, *180*
Riberau-Gayon, P., 7, *38*, 165, *179*, 240, 241, *246*
Rich, P. R., 66, 67, *86*, 198, 199, 200, 202, 204, *219*
Richmond, A., 34, *39*
Riov, J., 146, *148*
Ritchie, G. A., 184, *192*
Rivas, N. J., 165, *180*, 181, *192*
Robb, D. A., 181, 183, 184, 187, 188, 189, *191*, *192*, 253
Roberts, E. A. H., 163, *180*
Roberts, J. B., 164, *180*
Robertson, N. G., 128, *131*, 149, 150, 151, 152, 153, 154, 156, 157, *160*
Robertson, R. N., 74, *86*
Robinson, D. G., 125, 126, *132*
Robinson, G. M., 241, *246*
Robinson, J. E., 24, *38*
Robinson, R., 241, *246*
Roe, B., 165, *179*
Romani, R. J., 6, 7, 12, 35, *38*, *39*, 145, *148*
Rombouts, F. M., 127, *132*
Rose, D. H., 12, *38*
Rosenberg, A. J., 64, 66, *85*
Ross, C., 89, *105*
Rotilio, G., 186, *191*
Roux, D. G., 232, *246*
Russell, D. W., 198, 199, *219*
Rustin, P., 66, 71, 81, 82, *86*
Ruth, J. M., 109, *121*
Rychter, A., 65, 74, *85*, *86*
Ryrie, I. J., 83, 86
Ryugo, K., 238, *246*

S

Saad, F. A., 89, *105*
Sabularse, D., 146, *147*
Sacher, J. A., 2, *39*, 81, *86*
Sachse, L., 230, *246*
Sadovski, A. Y., 163, *179*
Saito, N., 227, 228, 230, *246*
Sakai, A., 43, 45, *61*, 91, *105*
Sakamura, S., 230, *247*
Salaun, J-P., 198, 200, *218*
Saleh, N. A. M., 234, *246*
Salisbury, F. B., 89, *105*
Salminen, S., 26, 27, *39*
Sanderman, H. J., 198, 200, *218*, *219*
Sanderson, G. W., 162, 175, *180*
Sando, C. E., 236, *246*
Santarius, K. A., 44, 45, 46, *60*, *61*
Sargent, J. A., 139, 142, 143, *148*
Sarkanen, K. V., 194, *219*
Saunders, J. A., 78, *86*
Saunders, W. H., 93, *105*
Schafer, E., 197, 207, *219*
Schaller-Hekeler, B., 207, *219*
Scheel, D., 200, *219*
Scheffeldt, P., 225, 226, *246*
Schenkman, J. B., 206, *219*
Schneebeck, D., 102, *104*
Schonbaum, G. R., 67, *86*
Schoot-Uiterkamp, A. J. M., 186, *192*
Schopfer, P., 214, *219*
Schriemer, W. H., 44, *61*
Schroder, J., 197, 207, *219*
Schröder, R., 94, *105*
Schürmann, P., 83, *85*
Schwartz, M., 92, *105*
Schwimmer, S., 47, *61*
Scott, K. J., 81, 83, *86*
Seaforth, C. E., 230, *245*
Seigar, S. G., 199, 200, 201, *219*

Seigler, D. S., 78, *86*
Seitz, U., 235, *245*
Seitz, U., 235, *245*
Sekiguchi, K., 127, *132*
Seybold, S., 42, *61*
Sfakiotakis, E. M., 24, *39*
Sgarbieri, U. C., 163, 164, 166, *180*
Shain, Y., 164, 169, 170, 171, 172, *179*, *180*
Shapiro, K., 92, *105*
Sharon, O., 164, 166, *180*
Shaw, R., 42, 47, 51, *61*
Shimokawa, K., 110, *121*
Shiralker, N. D., 163, 165, 166, *179*
Sho, S., 184, *192*
Shomer, I., 175, *180*
Shrikhande, A. J., 238, *246*
Sieker, L. C., 190, *192*
Simon, E. W., 133, *148*
Simpson, T. H., 225, *246*
Sink, K. C., 9, *38*
Sistrunk, W. A., 127, *131*
Slayman, C. W., 65, 72, *85*, *86*
Smith, B. N., 78, *86*
Smith, C. J., 145, 146, *147*
Smith, E. W., 72, *85*
Smith, H., 197, 200, 207, 214, *218*, *220*
Smith, O., 124, *132*
Smith, P. J. C., 126, *131*
Smock, R. M., 4, 5, *39*
Smolensky, D. C., 128, *131*
Sobotka, F. E., 128, *132*
Soloman, E. I., 186, *191*
Solomos, T., 27, *39*, 64, 65, 68, 73, 74, *86*
Somero, G. N., 52, *61*
Somers, G. F., 127, *132*
Sornsrivichai, P., 26, 27, *39*
Spalding, D. H., 243, *244*
Spencer, E., 182, *192*
Spiro, T. G., 186, *191*
Stafford, H., 195, *220*
Start, C., 53, *61*
Statham, C. M., 235, *246*
Stein, M., 124, 125, *132*, 158, *160*
Stelzig, D. A., 128, *132*, 164, 171, *180*
Stenkamp, R. E., 190, *192*
Steponkus, P. L., 43, 44, 46, 48, *61*
Stewart, R. N., 225, 226, 228, 229, *244*
Stitt, M., 49, *61*
Stockigt, J., 210, *220*
Stocking, C. R., 78, *86*
Stoddart, R. W., 134, *148*
Storey, B. T., 66, 67, 68, *84*, *86*
Stramentinoli, G., 144, *148*
Strand, L. L., 130, *132*
Strandberg, B., 189, *191*
Strothkamp, K. G., 183, *192*
Subramanyam, H., 163, *180*
Sugahara, K., 243, *245*
Sugimoto, Y., 92, *105*
Sulter, A., 199, *219*
Sun, B. H., 236, *246*
Sundin, O., 117, 118, *121*
Swain, T., 163, *180*, 188, 189, *192*
Swanepoel, J. H., 213, *218*
Sweeny, J. G., 230, 241, *247*

T

Tagawa, K., 136, *148*
Tager, J. M., 55, *60*
Takase, S., 227, 229, *245*
Takeda, K., 227, *245*, *246*
Takehana, H., 127, *132*
Talmadge, K. W., 125, 126, *132*, 134, 136, 138, 146, *147*, *148*
Tan, S. C., 237, *246*
Tang, C. S., 236, *245*
Tanner, W., 42, *61*
Tate, J. N., 165, *180*
Thanos, C. A., 222, *246*
Theologis, A., 28, *39*, 65, 74, 75, 76, *86*
Thimann, K. V., 92, *105*
Thom, D., 126, *131*
Thompson, A. E., 236, *247*
Tigchelaar, E. C., 9, 10, 24, *38*, *39*, 127, 130, *131*, *132*, 230, *246*
Timberlake, C. F., 221, 222, 225, 236, *246*, 253
Tindale, G. B., 6, *39*
Tomas, M. A., 243, *246*
Tomlinson, P. F., 74, *86*
Toraya, T., 92, *105*
Torokfalvy, E., 165, 173, *180*
Torre, L. C., 237, *244*, *246*
Toshio, M., 230, *246*
Towers, G. H. N., 197, *218*
Trajovski, V., 165, *180*
Trout, S. A., 6, *39*
Trunova, I., 45, *61*
Tselas, S. K., 222, *246*
Tsukvi, A., 230, *246*
Tucker, G. A., 128, *131*, 149, 150, 151, 152, 153, 154, 156, 157, *160*

Tumanov, I. I., 45, *61*
Turner, D. H., 55, *61*
Turner, J. F., 55, *61*
Turner, J. S., 74, *86*

U

Uota, S., 6, 12, *39*
Uritani, I., 193, 194, *219*, *220*
Uyeda, K., 57, *61*

V

Vaccari, A., 235, *246*
Valdovinos, J. G., 146, *148*
van Brederode, J., 229, 231, *245*
van Buren, J., 163, *180*
van der Deen, H., 186, *192*
van der Meer, R., 55, *60*
van der Westhuizen, J. H., 232, *246*
van Herk, A. W. H., 64, 66, 70, *86*, *87*
van Holde, K. E., 183, *192*
van Huystee, R., 43, *61*
Vanneste, W. H., 183, *192*
van Nigtevecht, G., 229, 231, *245*
van Vliet, W. F., 44, *61*
Varner, J. E., 110, *121*
Vaughan, P. T. F., 184, *192*, 195, 196, *220*
Vazquez, R. A., 243, *245*
Veen, H., 102, *105*
Venkaiah, B., 165, *180*
Vickery, R. S., 125, *132*
Volk, R., 177, *180*
von Elbe, J. H., 238, *247*
Voragen, A. G. J., 29, *38*

W

Wade, K., 109, *121*
Wade, N. L., 2, 3, 8, 9, 10, 14, 24, 37, *38*
Wagreich, H., 188, *191*
Wahachiro, T., 230, *246*
Walker, D. A., 46, 49, *61*
Walker, J. E., 128, 129, 130, *132*, 139, *148*
Walker, J. R. L., 170, *180*, 182, 192
Wallner, S. J., 127, 128, 129, 130, *131*, *132*, 139, 141, *147*, *148*
Wang, C., 20, 22, 24, *39*, 109, 110, *121*
Wang, P. L., 236, 238, 243, *244*, *245*
Wann, E. V., 236, *247*
Wardale, D. A., 90
Warden, J. T., 67, *85*
Watanabe, S., 230, *247*
Webb, A. D., 240, *244*
Webb, E. C., 188, *191*
Wedding, R. T., 81, *87*
Wegmann, K., 227, *244*
Weiser, C. J., 43, *61*
Weklych, R., 198, 199, *219*
Wellmann, E., 197, 207, *219*, 234, *247*
Wenker, D., 102, *104*
Werdan, K., 51, *60*
West, C., 2, 12, *38*, 90, *105*, 205, *219*
Weston, W. J., 47, *61*
Weurman, C., 163, *180*
Whitaker, J. R., 165, *180*, 181, *192*
White, D. B., 44, *61*
Whitmore, E. T., 230, *245*
Whittingham, C. P., 49, *60*
Wilcox, M. S., 208, *219*
Wilkins, M. T., 74, *86*
Wilkinson, M. M., 230, 241, *247*
Williams, D. B., 230, *245*
Williams, M. W., 101, *105*, 227, 230, 241, *247*
Williams, R. R., 237, *247*
Winston, J. R., 90, *105*
Wissenburg, A. K., 127, *132*
Witters, R., 184, 185, 186, *191*
Wong, T. C., 165, *180*
Wood, B. J. B., 188, *192*
Wood, D. E. S., 237, *247*
Woodmansee, C. W., 127, *132*
Woodward, J. R., 141, *148*
Wooltorton, L. S. C., 81, *85*, 90, *105*, 170, *179*, 193, 194, 195, 197, 206, 207, 210, 211, 215, 216, *219*
Worley, J. F., 238, *244*
Wright, B. W., 54, *60*, 77, *84*
Wrolstad, R. E., 244, *247*
Wulf, L. W., 239, *246*

Y

Yahiel, V., 199, *218*, *220*
Yamaki, S., 139, *148*
Yamazaki, I., 182, *192*
Yang, H. Y., 165, *179*
Yang, S. F., 24, *37*, *39*, 89, 90, 91, 92, 93,

94, 95, 96, 97, 98, 99, 100, 101, 102, 103, *104*, *105*, *106*, 250
Yelenosky, G., 43, 48, *61*
Yokoyama, K., 110, *121*
York, G. K., 165, *180*
Yoshida, S., 43, 45, *61*
Yoshitama, K., 228, *247*
Young, R. E., 1, 2, 4, 5, 6, 7, 12, 14, 16, 19, 20, 21, 22, 24, 26, 27, 30, 31, 32, 33, 34, *37*, *38*, *39*, 43, *61*, 81, *85*, 130, *131*
Yu, Y. B., 92, 93, 94, 95, 97, 98, 101, 102, *105*, *106*

Z

Zamorani, A., 164, *180*, 230, *246*
Zannoni, D., 67, *85*
Zauberman, G., 125, 128, *132*
Zenk, M. H., 207, 210, *219*, *220*
Zennie, T. M., 236. *247*
Zeroni, M., 99, *106*
Zito, R., 182, 184, 186, *191*
Züberbühler, A., 183, *192*
Zuurendonk, P. F., 55, *60*

Subject Index

A

Abscission, 146
Actinidia chinensis, 9
S-Adenosylmethionine (SAM), 91–94, 96, 97, 103
  and soluble pectin, 144
Affinity chromatography, 58
Agaricus bisporus, 184
Aldolase, 52, 53
Allium cepa, 236
1-Aminocyclopropane-1-carboxylic acid (ACC), 92–104
  hydroxylase, 93, 94
  synthase, 92, 93, 95, 97, 101, 102, 104
Aminoethoxyvinyl glycine (AVG), 101–104
  and soluble pectin, 144, 145
Aminoxyacetic acid (AOA), 101–103
Amyloplast, 49
Anacardiaceae
  anthocyanins, 243
Anacardium occidentale, 10
Ananas comosus, 10
Annona cherimola, 9
Annona muricana, 9
Anthocyanins
  acylated diglucosides, 230
  acylated glucosides, 228
  acylation, 226–227
  biosynthesis, 229–232
  chemistry, 222–224
  colour augmentation, 225–229
  cranberry, 242
  fig, 241
  highbush blueberry, 242
  HPLC, 239
  Leguminosae, 235
  Liliaceae, 236
  litchi, 243
  mango, 243
  mangosteen, 242
  metal chelation, 227
  Moraceae, 241
  mulberry, 242
  olive, 243
  pomegranate, 243
  Prunus, 238
  Ribes, 238
  Rosaceae, 236
  Rubus, 237
  stabilization, 225–229
  structural transformation, 222–224
  tamarillo, 243
  Umbelliferae, 234–235
  Vaccinium species, 242
  Vitaceae, 238–241
Antimycin, 69
Apium graveolens, 235
Apple
  catechol oxidase, 168–171
  cell wall, 135
  cell wall cohesion, 138
  cell wall degradation, 137
  cell wall hydrolysis, 140
  climacteric pattern, 4
  ethylene biosynthesis, 91
  ethylene and respiration, 18–20
  ethylene synthesis, 96
  β-galactosidase, 140
  pectin metabolism, 143
  polygalacturonate structures, 137
  polyphenol oxidase, 164
  respiratory pattern, 9
  respiratory rate, 12
Apricot
  ethylene and respiration, 20
  polyphenol oxidase, 164
  respiratory pattern, 9
Araceae
  respiratory pathways, 64, 77
Artocarpus altilis, 9
Arum italicum, 64
Arum maculatum, 64
Asimina triloba, 9
Asparagus officinalis, 236

ATP:ADP ratios, 54
Aurisidin, 226
Auxin
  induced ethylene synthesis, 95
Avocado
  cytochromes, 76
  endogenous ethylene, 21, 22
  ethylene and respiration, 19–21
  ethylene synthesis, 95–96
  glycolysis, 27
  nucleic acid metabolism, 34, 35
  phosphate esterification, 34
  pectic enzymes and ripening, 30, 31
  pectins and ripening, 30
  polyphenol oxidase, 164
  protein metabolism, 34, 35
  respiratory pathways, 76
  respiratory pattern, 5, 9
  respiratory rate, 12
Awobanin, 227

B

Banana
  carbohydrate metabolism, 26
  endogenous ethylene, 22, 23
  ethylene effects, 15
  ethylene and respiration, 20
  ethylene synthesis, 95, 96
  glycolysis, 27
  polyphenol oxidase, 164, 166
  respiratory pattern, 9
  respiratory rate, 12
  starch hydrolysis, 27
Berry, 8
Beta vulgaris, 184, 190
Biriba, 8
  respiratory pattern, 9
Blackberry, 237
Blueberry
  respiratory pattern, 10
Boysenberry, 237
Brassica
  phenols, 193
  oleracea, 43, 230, 233
Breadfruit, 8
  climacteric pattern, 4
  respiratory pattern, 9
Bromelia fastuosa, 7
Bromelia pinguin, 7

C

Cacao, 11
  polyphenol oxidase, 164
  respiratory pattern, 10
Caffeic acid, 226
Carica papaya, 9
Cashew apple
  respiratory pattern, 10
Casimiroa edulis, 9
Catechol oxidase
  apple, 168–171
  function in fruits, 178
  grape, 173–175
  olive, 174, 175
  peach, 171, 172
  tissue cultures, 176, 177
Cell expansion, 145
Cell separation, 145, 146
Cell walls
  chemical modification, 138
  composition, 135
  degradation in ripening, 139–142
  enzyme degradation, 136, 137
  galactose changes, 139
  separation, 124
  structure, 123, 126, 134
  synthesis in ripening, 141
  and tissue firmness, 124
Cellulase, 128
  in ripening, 128, 130
  in tomato, 129
Cellulose, 125
  degradation, 139
Cherimoya, 4
  climacteric pattern, 4
  endogenous ethylene, 22
  ethylene and respiration, 19–21
  respiratory pattern, 9
Cherry
  polyphenol oxidase, 164
  respiratory pattern, 6, 10
  respiratory rate, 12
Chinese Goosberry
  ethylene and respiration, 20
  respiratory pattern, 9
Chlorogenic acid
  in potatoes, 209
Cinnamate-4-hydroxylase (CA-4-H), 195, 196, 203, 211, 212
  in swede, 198, 199
Cinnamyl alcohol dehydrogenase, 195

Citrullus lanatus, 9
Citrus fruit
  respiratory pattern, 7
Citrus limonia, 10
Citrus paradisi, 10, 43
Citrus sinensis, 10, 43
Climacteric
  carbohydrate metabolism in, 26
  classification of fruits, 7
  features of, 2–5
  in fruits, 4
  respiratory patterns, 4
  ripening events, 36
Colocasia esculenta, 236
Commelinin, 227
Coniferyl alcohol dehydrogenase, 196
Co-pigmentation, 225–227
Cornus stolonifera, 43
Cucumber
  respiratory pattern, 10
Cucumis melo, 9
Cucumis sativus, 10
Cyanide, 69
  biosynthesis, 78
  metabolism, 79
Cyanide-resistant respiration, *see* Respiration, cyanide-resistant
Cyanidin, 223
  glycosylation, 234
β-Cyanoalanine synthase, 79
Cyanococcus, 242
Cyanogenic glycosides, 78
Cyphomandra betacea, 10, 243
Cynara scolymus, 235
Cytochrome
  mitochondrial, 70
  P420, 203
  P450, 200, 203
    in swedes, 199, 201, 202, 206
Cytokinin
  effect on ethylene synthesis, 95

D

Daucus carota s.sp. *sativa*, 234
Delphinidin, 223
  3,5-diglycoside, 226
  3-glucoside, 226
Dioscorea batatas, 230
Dioscorea tryphida, 230
Diospyros kaki, 9
Drupe, 8

E

Eggplant
  polyphenol oxidase, 165
Enolase, 54
Ericaceae
  anthocyanins of, 242
Ethylene
  and ethylene oxide, 113–115
  autocatalytic production, 23–24
  effects on microsomal components, 201
  effects on phenol metabolism, 197
  effects on respiration, 13, 14, 73
  effects on ripening, 13–14
  endogenous,
    in climacteric, 21–23
    effects on respiration, 16–18
  exogenous effects, 14–16
  inactivation, 117
  mechanism of action, 118–120
  purification methods, 110, 111
  studies of action, 108–109
  studies of metabolism, 109, 110
Ethylene biosynthesis, 89–104
  auxin-induced, 95
  inhibition, 101–102
  in stressed plants, 97–101
  intermediates, 90–94
  pathway, 90–93, 103
  regulation, 95–97
Ethylene metabolism
  effects of enzyme inhibitors, 115
  growth hormone effects, 116
  in pea, 112–116
Ethylene metabolites
  ethylene glycol, 114, 116, 120
  ethylene glycol glucose conjugate, 116, 120
  ethylene oxide, 113, 114, 116, 117, 120
    effect on peas, 120
Ethylene synthesis
  and waterlogging, 99, 100
Eugenia cumini, 10
Eugenia jambos, 10
Eugenia malacensis, 10
Eugenia uniflora, 10

F

Feijoa
  ethylene and respiration, 20
  respiratory pattern, 9
Feijoa sellowiana, 9
Feruloyl CoA reductase, 196
Ficus carica, 9, 241
Fig
  anthocyanins of, 241
  climacteric pattern, 4
  respiratory pattern, 9
  respiratory rate, 12
Foeniculum vulgare, 235
Flavan 3-ols, 226
Flavocommelin, 227
Fragaria, 10
Fructose, 43
Fructose-1,6-biphosphate, 54
Fructose-6-phosphate, 54
Fruit-softening enzymes, 150

G

Galacturonic acid
  in tomato, 129
Garcinia mangostana, 242
Glucose, 43
Glucose 6-phosphate, 54
Glucose phosphate isomerase, 54
Glucose-6-phosphate dehydrogenase, 52, 53
Glyceraldehydephosphate dehydrogenase, 52, 53
β-Glycosidases
  in ripening, 130
Grape
  anthocyanins of, 238–241
  catechol oxidase, 173–175
  polyphenol oxidase, 165
  respiratory pattern, 6, 7, 10
  respiratory rate, 12
  tyrosinase, 184
Grapefruit
  respiratory pattern, 10
Guava, 9
  ethylene and respiration, 20
  respiratory pattern, 9
Guttiferae
  anthocyanins, 242

H

Haemocyanin
  arthropod, 184
  gastropod, 184
Haplopappus gracilis, 229
Hedera helix, 43, 44
Hemicellulose
  degradation, 139
  structure, 125, 137, 138
Homoserine, 91
Hydroxycinnamate CoA ligase, 195, 196, 207
  in swedes, 206, 207
Hydroxycinnamoyl CoA: quinate hydroxycinnamoyl transferase (CQT), 210, 211, 212, 215, 217
  inhibition, 216, 217, 218
Hydroxycinnamoyl CoA reductase, 195

I

Invertase, 47, 51
Ipomoea batatas, 230
Ipomoea tricolour, 228

J

Java plum
  respiratory pattern, 10

L

Laccase
  peach, 171, 172
Leguminosae
  anthocyanins, 235
Lemon
  endogenous ethylene, 22
  ethylene effects, 15
  respiratory pattern, 5, 6, 10
  respiratory rate, 12
Lignin
  biosynthesis, 194, 196
  in swedes, 194
Ligustrum species, 243
Liliaceae
  anthocyanins of, 236
Lime
  endogenous ethylene, 22

Litchi chinensis, 10, 243
Lobelia erinus, 228
Loganberry, 237
Lychee
  endogenous ethylene, 21
  respiratory pattern, 10
Lycopersicon esculentum, 9, 10

M

Malate dehydrogenase, 71, 81
Malate oxidation, 71, 81, 82
Malic enzymes, 71, 81, 83, 84
Malus pumila, 236
Malus sylvestris, 9
Malvidin, 223
  3-*p*-coumaroylglucoside-5-glucoside, 227
  3,5-diglucoside, 225, 226
  3-glucoside, 223, 226
Mangifera indica, 9, 243
Mango
  climacteric pattern, 4
  endogenous ethylene, 21, 22
  ethylene and respiration, 20
  polyphenol oxidase, 165, 166
  respiratory pattern, 9
  respiratory rate, 12
Mammea americana, 9
Mammee apple
  respiratory pattern, 9
Matthiola incana, 231
Melon
  endogenous ethylene, 22
  ethylene and respiration, 17–18, 20
  ethylene synthesis, 96
  respiratory pattern, 9
Methionine, 90, 91, 92, 94, 96, 103
  sulphur cycle, 91
5′-Methylthioadenosine (MTA), 91, 92
5-Methylthioribose, 91, 92
Middle lamella, 124, 125, 134
Mitochondria, 80
  electron pathways in, 67–69
Moraceae
  anthocyanins of, 241
Morus alba, 241
Morus nigra, 241
Mountain apple
  respiratory pattern, 10
Musa sapientum, 9
Mushroom
  tyrosinase, 184, 185, 186, 189, 190
Mulberry
  anthocyanins of, 241
Musk-melon, *see* Melon

N

$NADH_2$—cytochrome *c* reductase, 203
NADH oxidation, 79–80
$NADPH_2$—cytochrome *c* reductase, 203
Non-climacteric pattern
  criteria, 5–7

O

Oleaceae
  anthocyanins, 243
Olea europaea, 10, 243
Olive
  anthocyanins, 243
  catechol oxidase, 174, 175
  composition, 11
  cultivation history, 10
  polyphenol oxidase, 165
  processing, 11
O-Methyl transferase (OMT), 195, 196
Orange
  endogenous ethylene, 22
  respiratory pattern, 5, 6, 10
  respiratory rate, 12
Oxytyrosinase
  relation to tyrosinase, 187
  spectral properties, 185

P

Papaw
  respiratory pattern, 9
Papaya
  endogenous ethylene, 22
  ethylene and respiration, 20
  respiratory pattern, 9
Passiflora edulis, 9
Passion fruit
  respiratory pattern, 9
p-coumarate CoA ligase, 210, 212, 217
  inhibition, 218
p-coumarate hydroxylase, 195

Pea
  ethylene metabolism, 112–116
Peach
  catechol oxidase, 171, 172
  laccase, 171–172
  polyphenol oxidase, 165
  respiratory pattern, 9
  respiratory rate, 12
Pear
  endogenous ethylene, 22
  ethylene and respiration, 20
  polyphenol oxidase, 165
  respiratory pattern, 9
  respiratory rate, 12
Pectic substances
  in ripening, 29–30
  structure, 125
Pectins
  changes in ripening, 30
  solubilization during ripening, 141
  structures, 137
Pectin methyl esterases
  in ripening, 130
Pectinesterase
  properties, 127
Pelargonidin, 223
Penicillium digitatum, 13
Peonidin, 223
Peroxidase, 196, 203
Persea americana, 9
Persimmon
  ethylene and respiration, 20
  respiratory pattern, 9
Petunia hydrida, 229
Petunidin, 223
Phaseolus aureus, 230
Phenols
  in *Brassica*, 193
  metabolism of, 193–218
Phenylalanine ammonia lyase (PAL), 193, 194, 196, 209–212, 217, 218
  activity in swedes, 197
Phenylpropanoids
  metabolism at low temperatures, 208
2-Phosphoglycerate, 54
3-Phosphoglycerate, 54
Phosphoenolpyruvate, 54
Phosphofructokinase, 53–55, 57–59
Phosphoglycerate mutase, 54
Phylloxera, 241
Pineapple
  compositional changes, 28, 29
  endogenous ethylene, 22
  respiratory pattern, 6, 7, 10
  respiratory rate, 12
Pinus strobus, 43
Pisum sativum, 235
Plasmadesmata, 123
Platycodon grandiflorum, 228
Plum
  ethylene and respiration, 17, 20
  respiratory pattern, 9
Polycodium, 242
Polygalacturonase, 126, 127, 137
  correlation with firmness, 127
  in ripening, 128, 130, 141
  in tomato, 129, 151–153
Polygalacturonase isoenzymes
  interconversion, 155
  properties, 152–155
  radioimmunoassay, 157
  regulation of activity, 156
  SDS electrophoresis, 153
  separation, 152–154
Polygalacturonate
  structures, 137, 138
Polyphenol oxidase, 161–178
  distribution in plant, 167
  in fruits, 164, 165
  inhibition, 168
  molecular properties, 168
  systematic distribution, 167
  subcellular distribution, 167
Potato
  p-coumarate CoA ligase, 218
  CQT, 211, 212, 214, 215, 218
  cyanide resistant respiration, 65
  cytochromes, 75
  enzymes of sugar metabolism, 52–59
  glucose metabolism, 56
  glycolysis, 53–59
  glycolytic intermediates, 54
  low temperature effects on phenolic metabolism, 208–218
  mitochondrial oxidation, 67, 68, 74, 75
  PAL, 218
  respiratory pathways, 75
  sugar accumulation, 44
  sugar content, 45
  sweetening, 42–45, 50–61
Propylene
  metabolism, 116, 117
Protein synthesis
  inhibitors, 213, 215, 216

Prunus
anthocyanins, 238
Prunus americana, 9
Prunus avium, 10
Prunus cerasus, 10, 238
Prunus persica, 9
Prunus sargentii, 238
Psidium guajava, 9
Punica granatum, 243
Punicaceae
anthocyanins, 243
Pyrus communis, 9
Pyruvate, 54
Pyruvate kinase, 53, 54, 59

Q

Quercetin, 226

R

Raffinose, 42, 43
Raffinose
synthesis, 47
Raphanus sativus, 230, 234
Raspberry, 237
Respiration, cyanide-insensitive, 64–83
ADP/O ratios, 66, 69
in Avocado, 76
electron carriers, 70
in potato tubers, 73–76
iron–sulphur proteins, 66, 71
Respiration, cyanide sensitive
electron carriers, 70
iron–sulphur proteins, 71
Rhizobitoxine, 144
Rhodanese, 79
Ribes
anthocyanins of, 238
Ripening
carbohydrate metabolism, 26–28
compositional changes, 26–29
degradative reactions, 25
nucleic acid metabolism, 34–35
pectic enzymes, 29
pectic substances, 29
protein metabolism, 34–35
mitochondria, 32–33
phosphate esterification, 33–34
synthetic reactions, 25
m-RNA
synthesis in ripening, 157
translation, 158–160
Robinia pseudo-acacia, 43
Rollinia orthopetala, 9
Rosaceae
anthocyanins of, 236
Rose apple
respiratory pattern, 10
Rotenone, 69
Rubus
anthocyanins of, 237
Rubus crataegifolius, 238
Rubus idaeus, 237, 238
Rubus illecebrosus, 238
Rubus morifolius, 238
Rubus occidentalis, 237, 238
Rubus odoratus, 238
Rutin, 226

S

Salicylhydroxamic acid (SHAM), 69, 72, 75, 82
effect on mitochondrial oxidation, 67
Sapindaceae
anthocyanins, 243
Sapotaceae
anthocyanins, 243
Sapote
respiratory pattern, 9
Sauromatum guttatum, 64
Senecio gruentus D.C., 228
Shikimate dehydrogenase, 196
Shisonin, 227
Silene dioica, 229
Solanaceae
anthocyanins, 243
Solanum intrusum, 230
Solanum melongena, 230
Solanum tuberosum, 43, 190, 230
Soursop, 8
respiratory pattern, 9
Stachyose, 42
synthesis, 47
Starch
balance with sugar, 48
breakdown, 48
Strawberry
cell walls, 135
polysaccharide metabolism, 143

Strawberry – *continued*
  respiratory pattern, 6, 10
  respiratory rate, 12
Sucrose, 42, 43, 51
  accumulation, 42, 47
  conversion to hexose, 47
  synthesis, 46
    site of, 49
Sucrose synthetase, 46, 47, 52, 53
Sucrose phosphate synthetase, 46, 47, 52, 53
Surinam cherry, 11
  respiratory pattern, 10
Swede
  enzymes in homogenates, 203
  microsomal enzymes, 204
    inhibition, 205
  microsomes, 199–202
    effect of cinnamic acid, 206
  p-coumarate CoA ligase, 207
Sweetening, low temperature, 41–60
  frost resistance, 44
  in higher plants, 43
  measurement of, 49–58
  occurrence, 42–44
Swertisin, 225, 226
Symplocarpus foetidus, 64
Synsepalum dulcificum, 243

T

Tamarillo
  respiratory pattern, 10
Theobroma cacao, 10
Tissue cultures
  catechol oxidase, 176, 177
Tomato
  ACC and flooding, 100
  ACC formation, 98
  ACC synthase, 98
  CA4H, 211, 212
  cellulase, 129
  climacteric pattern, 4
  CQT, 211–213, 215
  endogenous ethylene, 22
  ethylene and respiration, 20
  ethylene synthesis, 95, 97, 98
  galacturonic acid, 192
  low temperature effects on phenol metabolism, 208–218
  nor-mutant, 10, 150
  PAL, 211
  p-coumarate CoA ligase, 212
  polygalacturonase, 129, 151–153
  polyphenol oxidase, 163, 166
  respiratory pattern, 9, 10
  respiratory rate, 12
  rin-mutant, 10, 150
  ripening mutants, 150
  RNA and protein synthesis, 158–160
Tradescantia reflexa, 228
Triticum aestivum, 43
Tyrosinase, 181–192
  binuclear copper, 186
  grape, 184
  hamster, 184
  historical, 182–183
  inhibition, 188
  molecular properties, 183–185
  mushroom, 184, 185, 186, 189, 190
  Neurospora, 184, 185, 187, 188, 190
  photo-oxidation, 187
  spectral properties, 185
  spinach beet, 184, 190
  Streptomyces, 184
  Vibrio, 184

U

Ubiquinone, 67
Umbelliferae
  anthocyanins, 233–235

V

Vaccinium angustifolium, 242
Vaccinium corymbosum, 10, 242
Vaccinium macrocarpon, 242
Vaccinium staminium, 242
Vicia faba, 190
  ethylene metabolism, 113, 117
Violanin, 227
Vitaceae
  anthocyanins of, 238–241
Vitis aestivalis, 240
Vitis coriacea, 240
Vitis hybrid, 230
Vitis labrusca, 240, 241
Vitis linecumii, 240
Vitis riparia, 240, 241
Vitis vinifera, 10, 184, 239–241

W

Watermelon
respiratory pattern, 9

Wheat
ACC content, 99
ethylene synthesis, 99